FOUNDRY TECHNOLOGY

FOUNDRY TECHNOLOGY

Dr. Dharmendra Kumar
Former Principal and
Senior Professer of Met. Engg.

Dr. S. K. Jain
Reader in Met. Engg.
Malaviya Regional Engineering College
Jaipur

CBS PUBLISHERS & DISTRIBUTORS PVT. LTD.
New Delhi • Bengaluru • Chennai • Kochi • Mumbai • Pune

ISBN: 81-239-0290-5

First Edition: 1994
Reprint: 1999, 2004, 2007, 2011, 2015

Published by:
Satish Kumar Jain for CBS Publishers & Distributors Pvt. Ltd.,
CBS Plaza, 4819/XI Prahlad Street, 24 Ansari Road,
Daryaganj, New Delhi – 110002, India
delhi@cbspd.com, cbspubs@airtelmail.in • www.cbspd.com
Ph.: 23289259, 23266861, 23266867 • Fax: 011-23243014

Branches:
- ***Bengaluru:*** Seema House, 2975, 17th Cross, K.R. Road,
  Bansankari 2nd Stage, Bengaluru - 560070
  Ph: +91-80-26771678/79 • Fax: +91-80-26771680
  E-mail: cbsbng@gmail.com, bangalore@cbspd.com
- ***Chennai:*** No. 7, Subbaraya Street, Shenoy Nagar, Chennai - 600030
  Ph: +91-44-26681266, 26680620 • Fax: +91-44-42032115
  E-mail: chennai@cbspd.com
- ***Kochi:*** 36/14, Kalluvilakam, Lissie Hospital Road, Kochi - 682018
  Ph: +91-484-4059061-65 • Fax: +91-484-4059065
  E-mail: cochin@cbspd.com
- ***Mumbai:*** 83-C, Dr. E. Moses Road, Worli, Mumbai - 400018
  Ph: +91-9833017933, 022-24902340/41 • E-mail: mumbai@cbspd.com
- ***Pune:*** Bhuruk Prestige, Sr. No. 52/12/2+1+3/2,
  Narhe, Haveli (Near Katraj-Dehu Road Bypass), Pune - 411041
  Ph: +91-20-64704058/59, 32342277 • E-mail: pune@cbspd.com

Printed at :
J.S. Offset Printers, Delhi

Preface

This book is designed to serve as a comprehensive study of foundry technology. This will serve as a bridge between the study of the basic and scientific principles of foundry technology and their applications in the industries. Its further aim is to impart the knowledge of modern techniques of casting production and other developments in the founding of metals. The work is the outcome of long teaching, research, and industrial experience of the authors.

The book is conceived as a text book and work of reference for the use of students and teachers in engineering institutions. This is also intended for higher education in the subject of managerial and technical staff in the casting industry. It is hoped that the book will be highly useful to all the metallurgists and foundrymen. In particular the aim has been to assist engineers and engineering students in realising the importance of casting in design and selection of materials.

The subject has been considered comprehensively to include all aspects relevant in the field of casting to help industry as well a academicians. The book has been arranged in 17 chapters,with the emphasis on need of casting in the first chapter. The second chapters is on patterns, without which casting can not be produced. Three chapters (3-5) have been devoted to the various techniques and materials employed in mould production.

A good coverage has been given on the fundamentals and practical aspects of metal behaviour in three chapters(6-8). These include solidification behaviour, flow of metals, feeding, gating, risering, etc. Different types of furnaces employed in melting various metals and alloys are described in chapter 9. A brief description of cleaning operation for castings is given in chapter-10. Causes and remedies of casting defects have been highlighted in chapter-11.

Since ultimate success of a foundry depends upon proper testing and quality control, a full chapter (12) has been incorporated on this subject. Chapter 13 has been introduced to acquaint foundry engineers and managers to the application of computers in the foundry industry. Three chapters (14-16) have been devoted to describe the specific procedures involved in the production of cast iron, steel and non-ferrous metal/alloy castings respectively. A brief discussion on material handling, pollution control and layout of foundry has been incorporated in chapter- 17.

This book will be very useful for all students of Foundry Technology in different disciplines of engineering educations, particularly for metallurgical, mechanical, production and industrial engineering as also practical foundrymen interested in updating their understanding of this important subject.

DHARMENDRA KUMAR
S. K. JAIN

Contents

1

Introduction to Metal Casting

1.1 CASTING VERSUS OTHER PROCESSES

1.1.1 Various Manufacturing Method

Principal methods for making useful engineering parts and components from metals and alloys may be classified in the following groups:

(i) *Casting* This technique refers to the production of shaped articles by pouring molten metals into moulds having cavity of the desired shape, and allowing the metal to solidify in the mould cavity.

(ii) *Mechanical Working* This technique refers to shaping of metals and alloys in the solid state by plastic deformation above or below the recrystallisation temperature, i.e. hot or cold working. This group includes processes such as forging, rolling and extrusion. Forging produces varied shapes somewhat analogous to castings, whereas other mechanical processes produce standard primary or semi-finished shapes such as bars, plates, sections etc.

(iii) *Joining* This technique refers to the production of structural units by joining of smaller components which are produced by other techniques. The most notable method of joining is welding which is carried out on components cut from standard wrought materials.

(iv) *Machining* This refers to the production of shaped articles by cutting of plain or roughly shaped forms by employing machine tools. Though many articles are wholly shaped by cutting from blanks, machining is usually required as finishing operation on components formed by other processes.

(v) *Powder Metallurgy* This refers to the production of shaped parts by the die pressing and sintering of metal powders.

Consideration of above shaping processes shows that these are in many respects complementary. In many cases, the capacities of different processes overlap and allow an alternative decision, thereby providing the basis for campetition with respect to properties and cost. In general castings have to compete with forgings, weldments and sintered compacts apart from plastics, ceramics and other non-metallic materials. These alternatives pose a challange to the foundry engineer in terms of price as well as functional performance. Castings can also replace other traditional products such as sheet metal constructions by pressure die casting, or of machined components by investment castings. Therefore, full exploitation of casting processes needs thorough study not only of its advantages, but also of difficulties and limitations.

1.1.2 Design Advantages of Casting Process

Casting process satisfies various requirements of stress and strain for objects having varying structural and functional shapes. The main design advantages of casting process are as follows:

(i) *Size or Weight range* Practically there is no weight or size limitation upon castings, the only restriction being the supply of molten metal and means of lifting and handling the casting. The range of products may extend from a casting of about 200 tonnes (cast steel mill housing) down to the precision dental casting or a wire of 0.5 mm diameter. In fact, casting is the only process which can produce massive articles in one single piece.

(ii) *Shape and Intricacy* Casting offers the maximum range of possibilities for producing complex shapes (elaborate contours or intricate detail) which may be either very expensive or almost impossible to manufacture by any other process (machining, forging, etc.) More specialised techniques of casting such as die casting and investment casting can cater more accuracy in shape and size.

(iii) *Weight saving* Casting process results in large savings in weight as the metal can be placed exactly in the required places. Such weight saving leads to economy in production.

(iv) *Production of prototypes* Casting is well suited for the production of models or prototypes since a large number of identical patterns can be easily made and the same pattern can be used to produce more than one casting.

(v) *Composition of material* Using the appropriate foundry techniques, practically all types of engineering alloys can be cast. Earlier, there were some difficulties in casting of reactive and refractory metals, but special techniques have been developed to overcome difficulties in melting and casting of these metals. Further, certain alloys can be formed only by casting, e.g. cast iron around which the foundry industry initially grew. The casting process also offers special advantage for the shaping of hard or tough alloys having poor machinability.

(vi) *Dimentional modifications* Design changes or dimensional modifications can be introduced easily during the development and early production stages of a casting.

(vii) *Dimensional accuracy* Casting process can yield objects to fairly close dimensional tolerances by proper selection of moulding and casting techniques. Depending on the metal to be cast, casting process, shape and size of the casting, the dimensional tolerances as close as ± 0.1 mm can be achieved. The surface finish can also be obtained with in the range of 5 to 50 microns.

1.1.3 Metallurgical Advantages of Casting Process

Many metallurgical characteristics are acquired by castings during solidification which influences the structure and properties. Though properties may be influenced by segregation and microporosity, the basic metallographic features (grain size and distribution of microconstituents) are sensitive to the casting conditions. The structural features can be modified by subsequent heat treatment. The vorious favourable points for castings with respect to metallurgical characteristics are given below:

(i) *Microscopic structure* Wrought metals possess a fibrous structure associated with the direction of working. Castings generally possess more isotropic properties. In some cases specific features of cast structures are very useful e.g. cored structure in bearing bronzes and graphite fiake structure in grey cast iron are responsible for high damping capacity.

(ii) *Grain Size* Mechanical working of wrought metals results in fine grains, but many castings may have grain sizes almost similar to those of mechanically worked metals. In general, nonferrous alloys retain the grain size attained during solidification of the casting. Grain size can be further improved by subsequent heat treatment. In case of high temperature alloys, coarse grains of casting give better creep resistance.

(iii) *Density of Cast Metals/Alloys.* The density of cast metals/alloys is almost identical to that of wrought metals/alloys having the same chemical composition.

1.1.4 Economical Advantages of Casting

In general, casting process is the cheapest method of metal shaping. Casting process is equally suitable for single object or quantity production. In casting process, the pattern takes the form of a positive replica which is usually made of cheap material (wood, plastic, etc.). The cost of pattern compares fovourably with the cost of tools required in other processes. Further, this is more favourable for jobbing and small quantity production, where the versatile methods of moulding can be used to simplify the equipment required. The cost of the more elaborate and durable pattern required for mass production compares still more favourably with that of dies and fixtures used in other processes. The same advantage is also obtained in gravity die casting where simple dies are used. However, pressure die casting needs quite costly equipment but large production compensates the cost.

1.2 HISTORICAL DEVELOPMENT OF FOUNDRY INDUSTRY

Metal founding or casting is one of the oldest processes of manufacturing. The ancient as well as medieval history reveals the manufacture and use of castings from the dawn of civilisation. The simple axeheads, knives, coins, arrows, etc. of copper were cast in open moulds about 5000 years ago as observed from the excavations of Mohenjodaro and Harappa. In the pre-

Christian era, the casting technology was developed to an extent that bronze statues could be produced in two piece · in cored moulds. In early historic times, the casting process was practised by the craftsmen of Greek and Roman civilisations. Since then casting process acquired unique significance. In ancient times, castings of copper and bronze were common, but the use of iron is also evidenced around 2000 B.C., though the use was greatly restricted.

By the end of the medieval period, decorated bronze castings were used in European Church and domestic life, whereas cast iron castings were used in the shapes of cannon shot and grave slabs. One life sized portrait head of cast bronze has been found in *Mesopotamia* antedating about 2250 B.C. The first foundry centre came into existence during the Shang dynasty (1766—1122 B.C.) in China. The religious era started around 500 B.C. and the metals were used for the statues of Gods and Godesses, particularly in bronze. During this time, lost wax process was developed. Subsequently, the use of metals in the form of casting increased in armoury, guns, and war material. The purity of cast metals of that time reveals the ability of precise quality control in refining processes in those days. Perhaps, the greatest break through in the use of metal castings in war material took place at the time when Alexander was intendening to conquer the entire Eurasian continent.[1].

In the sixteenth century (about 1540 A.D.) first time Biringuccio wrote on metal casting giving the details of working conditions and the emotions of foundrymen. At that time moulding boxes and sand were in use. Earlier to this, casting was an art and a craft having all secrets confined to certain families. Amongst the early scientists the work of Reaumur (1683—1757) is of great interest to founders. He worked on cast iron founding, produced malleable iron, and studied the various factors influencing the production of white, grey, and mottled irons. However, the widespread use of cast iron as an engineering material could not be possible till the smelting of iron in coke blast furnace developed by Abraham Darby. This gave the way for the massive use of cast iron in construction during the years following the industrial revolution. The use of cast iron was superseded by the cheap steel available after Bessemer's development of the convertar in 1856.

Despite the interest of a few early scientists, founding was considered to be an art until well into the 20th century. After the industrial revolution, many foundries sprang up, particularly for the manufacture of cast iron castings. The organised casting industry developed around the skills of the moulder and the pattern maker. Though their skill enabled the production of complex shapes, the metallurgical quality and soundness could be possible only after scientific understanding and advent of non-destructive testing in the modern era. It brought a better understanding of the factors influencing quality with improved means of evaluation. The modern techniques of process control superseded the rudementary judgement of the operator and objective measurements of metal temperature, properties of moulding material and other production variables were made. The precise effect of these improvements could be applied to all the cast alloys including cast iron.

A further feature of the modern casting industry has been the shift in the responsibilities of operators and supervisors. In certain foundries, operator was practically a sub-contractor selling his product to the company. The moulder used to receive the pattern and carry out all the operations (sand preparation, moulding, pouring, etc.) himself. Later development increased the responsibilities of shop floor supervisor in determining casting procedures. This type of organisation promoted the best craftsman to guide the operators. Further development of the industry was in bringing the specialist technical staff having the sole duty to establish and record the techniques and procedures of metal casting. The pre-engineering of casting methods is accompanied by metallurgical control based on modern laboratory facilities. The scientific controls are applied at all stages from incoming raw materials to the out going finished product. Hence, primary reliance placed upon craft gave the way to technology and many engineering institutions started metal casting as an independent discipline.

The structure of the modern casting industry is quite complex and directly related to the traditional industry are the jobbing foundries having their capacity for making castings in wide range of sizes and designs. In jobbing foundries, quantity produced is generally small and there is considerable dependence on manual operations even though most of the heavy work is carried out by mechanical aids. The jobbing foundry also faces constantly new problems in the moulding of individual design and in determination of casting procedure to ensure a sound casting at the first attempt.

On the other hand, there are specialised foundries having emphasis either on the mass production of a limited range of articles or on the use of a single special casting technique. Many such foundries are captive to large engineering organisations which use castings in their own finished products. In the mass production foundry emphasis is laid on the close process control to maintain consistency in materials and procedures, e.g. sophisticated pattern equipment eliminates the need for a high degree of skill in moulding. Many companies operate in several fields of foundry in parallel i.e. jobbing and mass production, conventional and highly specialised processes operating in parallel. Similarly, most foundries operate on a limited range of alloys, but several companies may produce castings of many alloys side by side.

Today, metal founding has emerged as an exact science and there are large number of moulding processes, melting equipments, and metals and their alloys. Though the techniques and procedures of production of castings have changed drastically, the basic fundamentals of casting still remain unchanged, Castings possess several specific characteristics due to which their role is well established in modern industries. Today cast metals in various shapes and sizes are required in almost all types of machines and equipments used in transportation, communication, power, agriculture, construction and all other fields.

1.3 OPERATIONS INVOLVED IN PRODUCTION OF CASTING

The various casting processes are sand casting, pressure die casting, gravity die casting, investment casting and centrifugal casting, but the general characteristics can be described conveniently with the production of sand castings by conventional moulding. The Principal steps involved in production of castings from design to finished product are shown in Fig. 1.1. A similar sequence of operations with some modifications can be used to other casting processes. Die casting is an exception in which pattern and expendable mould are replaced by a permanent mould containing a negative impression. In general, the casting process consists of six stages, i.e. (i) Pattern making (ii) moulding and core making, (iii) melting and casting, (iv) fettling, (v) heat treatment and finishing and (vi) inspection and testing.

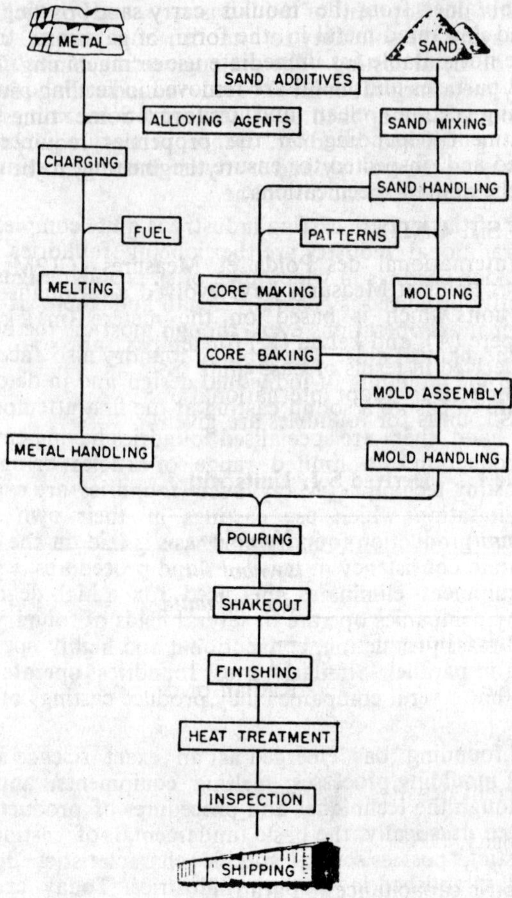

Fig. 1.1 Principal steps involved in production of castings

First a pattern is designed and prepared as per the specifications of casting and depending on the moulding process employed. The material of pattern may be selected depending on factors such as number of castings to be produced, possibility of repeat orders, and the surface finish required. Pattern shop is also responsible to make core boxes required for making cores and other items. Then with the help of pattern the mould is prepared in sand (either by hand or by machines) to form a cavity of the desired shape. Hollow portions of the casting are obtained by the use of cores prepared separately in core boxes. The moulds and cores are dried and baked to impart strength and then assembled for pouring of malten metal. The metal of correct composition melted in a suitable furnace is taken into ladles and poured into the moulds. The casting is allowed to solidify in the mould and then finally taken out of breaking the moulds.

The casting obtained from the moulds carry sand particles adhering to the surfaces and unwanted metal in the form of gates and risers and thus the castings are not suitable for immediate use or machining. Therefore, the unwanted sand particles and metal are removed in fettling section and the surface of casting becomes clean and uniform. Some times, castings are given heat treatment depending on the properties required. Finally the casting is tested and inspected to ensure the casting to be defectless and confirming to the desired specifications.

1.4 USE OF S.I. UNITS

The Comite International des Poids et Measures (CIPM) and Bureau International des Poids et Measures have evolved a rationalised and coherent system of units which is based on the metere (m), kilogram (Kg), second (s), ampere (A), and kelvin (K) for temperature and some others. The units are derived in terms of base units from these basic units. In India it is firmly committed to adopt international units known as S.I. units. The useful derived S.I. units for foundries are given in Table 1.1.

Table 1.1 Derived S.I. Units with Symbol and Definition

S. No.	*Quantity*	*Name and symbol for derived units*	*Units*
1.	Acceleration	—	m S^{-2}
2.	Angle	Radian or degree	rad, degree
3.	Area	—	m^2
4.	Current density	—	A m^{-2} or A mm^{-2}
5.	Density	—	Kg m^{-3}
6.	Electric capacitance	Farad, F	$A\ s\ V^{-1}$ (V = volt)
7.	Electrical charge	Coulomb, C	$A\ s$
8.	Electrical field strength	—	V m^{-1}
9.	Electrical Potential	Volt, V	WA^{-1} (W = watt)
10.	Electrical resistance	Ohm, Ω	$V\ A^{-1}$
11.	Electrical resistivity	—	Ω m

contd.

12.	Energy	Joule, J	N m ($MJ = 10^6$ J) 3.6 $MJ = 1$ KWH
13,	Force	Newton, N	Kg m S^{-2} ($MN = 10^6$ N, k$N = 10^3$ N, $mN = 10^3$ N)
14.	Frequency,	Hertz, Hz	s^{-1}
15.	Frequency, rotational	—	s^{-1}
16.	Impact strength	—	J m^{-2}
17.	Latent heat	—	J Kg^{-1}
18.	Magnetic flux	Weber, Wb	Vs
19.	Magnetic density	Tesla, T	Wb m^{-2}
20.	Power	Watt, W	J s^{-1}
21.	Pressure and stress	Pascal, Pa	Nm^{-2} (Pa = 0.1019716 Kg /m^2)
22.	Sp. heat	—	J Kg^{-1} K^{-1}
23.	Thermal conductivity	—	Wm^{-1} K^{-1}
24.	Velocity	—	m s^{-1}
25.	Viscosity, dynamic	—	Nm s^{-2}
26.	Viscosity kinematic	—	m^2 s^{-1}

2

Technology of Pattern Making

2.1 INTRODUCTION

Patterns are basic tools to form the mould cavity and thereby the casting. Pattern making is often considered as metal trade, but in fact it is a phase of wood working and machining and requires a different type of skill than required in metal casting. However, the pattern maker should have a good grasp of the problems associated in moulding and casting. While making a pattern, due consideration has to be given for the shrinkage of particular metal to be used in the casting and the amount of material to be removed subsequently in machining. The pattern should be made in such a way that it can be removed from the sand mould without causing any disturbance to the mould. Usually, big foundries have own pattern divisions, whereas small foundries are generally dependent on the outside pattern shops to meet their requirements.

2.2 MATERIALS FOR PATTERN MAKING

The main purpose of a pattern is to form a cavity of the desired shape in the moulding. Common materials used in pattern making include wood, metals and alloys, plasters, plastics and rubbers, waxes, etc. The selection of a specific material for pattern making is decided by the following considerations:

(a) Number of castings required and possibility of repeat orders.

(b) Quality of castings required, e.g. intricacy of castings, minimum thickness required, degree of accuracy and finish.

(c) Possibility of design changes in the casting and/or pattern.

(d) Method of casting, moulding, and the equipment to be used.

In order to obtain a good pattern, the materials employed should possess the following characteristics:

(a) The material used should be easily worked, shaped and joined.

(b) To facilitate the handling and working of patterns the density of the material used should be low.

(c) The material should be dimensionally stable and not affected by variation in temperature and humidity.

(d) The pattern should be resuable and repairable.

(e) The material should be capable of taking a good surface finish.

(f) The material should be resitant to wear, abrasion, corrosion, and chemical attack.

(g) The material should be strong, hard and durable.

(h) The material should be easily and cheaply available.

2.2.1 Wood and Wood Products for Pattern making

Wood is widely used for patterns owing to many advantages such as

(i) comparative ease in shaping, working and joining to form complex shape,

(ii) lightness in weight.

(iii) easy availability, and

(iv) relatively cheapness than other materials. However, wood has certain disadvantages such as

(i) ready attack by moisture.

(ii) swelling when wet and shrinkage on drying.

(iii) less wear resistance.

(iv) chances of warpage and

(v) poor strength.

Excessive swelling and shrinkage in wood can be prevented by seasoning of wood before it is used for pattern making. Seasoning minimises the subsequent effect of moisture variation and makes the pattern

(i) dimensionally more stable.

(ii) stronger and lighter.

(iii) resistant to decay, and

(iv) suitable to take preservative coating and polish.

Structure of Wood Basically wood consists of two types of cells i.e. (i) long hollow cells with axes running parallel to the length of tree and (ii) cells radiating from the central axis of the tree. The cells are bonded together by a natural resin called *lignin.* Therefore the behaviour of wood can be considered identical to a bundle of thin walled tubes bonded together. The growth of the tree is directly related to the growth of cells growing rapidly during the spring and early summer. However, the cells laid down during this time are not strong and thick walled as the cells laid down in the later period of the year. Towards the end of summer and early fall the growth of cells ceases. A cross section through a tree shows the rings of annual growth, the dark portion of each ring representing the summer growth.

Seasoning of Wood Processing of wood to prevent excessive shrinkage or swelling is called seasoning of wood. It minimises the effect of subsequent moisture variation by adjusting the moisture content in stable equilibrium with the moisture content of the atmosphere to be used.

When wood is seasoned, most of the shrinkage occurs at right angles to the length of the cell and the amount of shrinkage depends on the thickness of the cells. Because of the cell geometry, the greatest shrinkage takes place across the grain, tangentially to the rings of annual growth and minimum shrinkage occurs longitudinally.

Seasoning of wood can be carried out by natural or artificial methods. In natural method of seasoning, wood is stacked in a suitable manner in open spaces and allowed to air dry for a period of one full cycle of weather conditions. In another method of natural seasoning, the wood is kept immersed in flowing water to wash away a major part of the inherent moisture followed by air-drying for a shorter period.

In artificial seasoning, wood is stacked in drying chambers and hot air is circulated through pipes to air-dry the wood. The air may be heated by steam passing through the pipes. In order to prevent wood from cracks and buckling, the temperature and humidity should be properly controlled. In the initial period of drying, hot air with sufficient humidity in introduced to effect the heating more uniformly through the whole section of wooden pieces avoiding appriciable drying to prevent differential contraction. Afterwords, the humidity of air is reduced gradually to the desired value. The heating chamber is provided with suitable arrangements for air circulation, heating, humidity control, temperature control, and ventilation. Usually a period of one week is sufficient to complete this seasoning process.

In an alternative process of electrical seasoning, large sections of wood are exposed to a high frequency electric field in such a manner as to bring down the moisture content to the desired level within only few minuts. Chemical seasoning has also been employed in some instances in which the wood is kept immersed in a solution of suitable salt and then dried by exposing to air.

Types of woods used for patterns The most widely used woods for pattern work are pine, teak, walnut, mahogany and deodar. Pine wood, though weak, is more popular due to its softness, close grains, extreme lightness, ease in shaping, stability, and ability to take good finish. Pine wood is usually employed when the number of castings to be produced is small (less than 25) and the pettern is under 500 mm in size on any side. Mahogany has uniform grains and can be shaped and joined (with glues) easily with fine natural finish. Mahogany is used for patterns required (i) to withstand rough handling in foundry, (ii) to be used as master patterns, (iii) to produce large number of castings and (iv) for the production on moulding machines. Teak is also classified as hard and strong wood which is unaffected by fungus and is easily available in the country. Deodar is slightly harder than pine and can be shaped easily with good polish.

Compressed wood laminates can also be used for making patterns. These are available as plawoods and as laminated boards which can be glued and pressed together by arranging the grains of alternate laminates in opposite directions. The laminates are stronger, denser, and more stable in dimensions than plain woods.

Another variety is formed by impregnating the wood laminates with resins (to fill all cavities) and then compressing to increase the density and hardness. These overcome the drawbacks of woods and improve various properties such as strength, density, hardness, resistance to wear, and dimensional stability, while preserving the natural advantages of wood. The wood plastic combination (WPC) possesses good properties and can be used to substitute more expensive aluminium patterns. The resins used for impregnation include phenol formaldehyde, methyl methacylate, styrene, isocyanates, polyster, etc.

Some times wooden patterns with metallic coatings are employed where metallic patterns are not economical for production and wooden patterns do not give satisfactory performance. A thin metallic coating (zinc, aluminium, or bismuth) of about 0.2 mm thickness on the wood surface can be obtained by atomization spraying of molten metal by an oxy-acetelyne flame by pneumatic pressure.

2.2.2 Metals and Alloys for Patterns

Metallic patterns are employed for (i) greater durability of the pattern and (ii) repetitive production of large number of castings. Most commonly used metals and alloys for patterns are aluminium, cast iron, steel, bronze and brass. The relative properties of varions metals and alloys are given in Table-2.1.

Table—2.1 Relative Properties of Metals and Alloys Employed for Making Patterns

Property	*Metals/Alloys*				
	Aluminium	*Brass*	*Bronze*	*Cast Iron*	*Steet*
1. Castability	good	good	very good	good	difficult
2. Machinability	very good	very good	excellent	good	good
3. surface finish	very good	very good	excellent	good	good
4. Possibility to modification	good	very good	very good	good	good
5. Density	low	very heavy	very heavy	very heavy	very heavy

contd.

6. Tendency to oxidation	No	No	No	Yes	Yes
7. Cost	medium	high	high	low	low

(i) *Aluminium and its alloys* Aluminium is very popular due to various advantages such as ease of formation and machining, low melting point, low density, good corrosion resistance, and reasonable strength. However, aluminium is soft and patterns get damaged by rough usage. The aluminium alloys recommended for patterns are A-24, A-4 and A-8 (IS: 617-1259). A near equivalent of A-24 is LM 25 WP in B.S.S. which is extensively used due to its easy castability, low shrinkage, and strength, and wear resistance. The properties of aluminium alloys can be improved by solution heat treatment and precipitation hardening. The patterns can also be made by direct machining of rolled forms (duralium is best after giving a stress relieving treatment)

(ii) *Brass and Bronze* Brass and bronze are also widely used in patterns required for high finish and controlled dimensions. These alloys possess high physical strength in addition to resistance to deformation and corrosion and ease of production and joining. However, because of cost and owing to high density, these alloys are not convenient for large size patterns. These alloys can be used for the construction of small patterns fastened to a plate for multiple castings and give large production of castings. Repairs of brass and bronze patterns can also be carried out readily with soft solder.

(iii) *Cast iron* Cast iron may be used for some specific types of patterns. Iron patterns should be protected from rustings and, therefore, should be stored in a dry place. Repairs on cast iron pattern may be made by soldering, brazing or welding. The most suitable cast iron for patterns is grade-24 (IS: 220-1962) having a BHN value of 197-241. However, other cast irons such as heavy duty iron and S.G. iron can also be used.

(iv) *Low melting point alloys* Alloys such as white metal (Pb-Sn), and cerro-alloys (containing and Sn and Bi) have very low shrinkage and the melting points are close to 140 °C. These alloys are quite hard and malleable and are used for moulds used in making epoxy resin or plaster patterns. Due to low melting point, these alloys are free from the chilling effect.

(v) *Electro deposited metals and alloys* The technique of making patterns by electrodeposition has come into vogue quite recently. This method uses an electrically conductive mandrel (master pattern or mould) as an anode. Suitable electrolyte of the required metal salt is used for electrolytic deposition of metal ions. The shape so formed on the mandrel, is removed and employed as mould or pattern. This method is generally used for producing intricate shapes with a very high degree of accuracy. Very fine details can be reproduced and the process can be adopted for large scale production Camparatively high cost of pattern making limits its use to specific applications.

(vi) *Frozen mercury* Frozen mercury has been employed as pattern material in the 'Mercast' process for producing highly intricate and accurate castings. However, the use of frozen mercury is greatly limited due to its very high cost.

(vii) *Production of metallic patterns* Metallic patterns are employed for mass production of castings. The metallic patterns are usually made by casting using a master pattern (made of wood, plastic, plaster or metal). The master pattern should be provided for double allowances for contraction and machining. Aftermachining and required finishing, the patterns are mounted on pattern plates. When the patterns are of the split type, each half may be mounted on separate pattern plate or on either side of the same plate. The pattern plates employed for mounting the metallic patterns can be of wood or metal (cast iron of grade 20 and cast aluminium alloy of grade A-24 are more common). IS: 4604-1968 gives details for pattern plates for machine moulding boxes.

2.2.3 Plastic for Patterns

Plastics are finding applications as a modern pattern material and may be broadly classified as *thermosetting* and *thermoplastic*.

(i) *Thermosetting plastics* These are formed to the required shapes by heat with or without pressure, resulting in a permanent hard product. In the first instance, heat makes the material soft but on further heating or by adding special chemicals, the plastics are hardened by polymarisation and can not be resoftened by heat. The thermosetting plastics include

(i) phenolic resin (phenol formaldehyde or bakelite),
(ii) amino resins (urea formaldehyde and melamin formaldehyde) and
(iii) furnace resins, expoxides and silicones. The phenolic resins and epoxy resin plastics are very suitable for pattern making due to the following advantages:

(a) These are easily castable rendening machining unnecessary.

(b) High strength to weight ratio (5.4 to 4.0 for aluminium alloys and 1.2 for cast iron).

(c) Low cost of making patterns.

(d) Good resistance to wear and abrasion, and

(e) Remain un-affected by moisture and mild heating when in contact with moulding sand.

The plastics are more expensive than many of the other pattern materials. However, the metallic patterns for mass production are being replaced by plastic patterns, as the overall cost of the patterns is lower than that of aluminium patterns due to practically no expenses incurred on casting and machining.

Phenolic plastic is generally used in liquid form to which necessary catalyst and hardening agents are added before use. The mixture is poured into a metal or plaster mould. Epoxy resin is used as a two component material, i.e. liquid binder resin and a liquid hardner. When mixed together, the mixture sets and forms a hard mass in a certain period. These plastics can be used in three forms, viz. (i) surface coat resin, (ii) laminating resin, and (iii) casting resin. In surface coat resin, the mixture (with hardner and catalyst) is used to obtain the surface layer on the pattern which is hardner and stronger than the inner layers. The pattern is formed by first giving a thin cost of release agent and then few layers of the plastic mixture over the surface layer. While using as a laminating resin, the mixture is used to apply few layers of glass fibre or wool laminations (socked in resin) on the back of the release coat. The use of these laminations provides additional strength, toughness, and flexibility to the pattern. The casting resin is used to provide the required backing and to form the complete solid shape. In order to reduce the cost of cast resin, suitable fillers (talc, chalk powder, slate powder, sand, saw-dust, etc.) may be added. For making large patterns, suitable reinforcements and solid fillers of rough shape can also be used to form the internal cavity. All the three types of resin combinations may be used in the making of patterns depending on the service requirements. Various techniques used for making patterns from thermosetting plastics are shown in Fig. 2.1.

(ii) *Thermoplastic type of plastics* Thermoplastic materials do not undergo chemical change while moulding and do not become permanently hard with the application of pressure and heat. These plastics remain soft at elevated temperatures and are hardened only by cooling. They can be remelted repeatedly as paraffins and their plasticity increases with temperature. This group includes a large number of materials such as polyvinyl chloride (PVC), polystrene, cellulosics (acetate, acetate-butyrate, ethyl cellulose), polyethylene, polypropylene, nylon, acrylic and, vinyl resins, etc. For patterns, polysterene foam is most commonly used which becomes soft and finally gassifies on heating. Polystrene is available in different densities in foamed form. This can be easily shaped and fabricated as patterns by gluing. The pattern can be used in conventional form (pattern is withdrawn from the mould cavity), by burning it with a gas torch or breaking it, or as a full mould casting (the pattern is left in the mould cavity) and the molten metal is poured, the heat causing the gassification of the pattern. Polystrene patterns can thus be used for investment castings also. Another thermoplastic material used in patterns is polyurethane foam which is known as flexanes. This material has low thermal conductivity, high strength and almost no shrinkage.

2.2.4 Rubbers for Patterns

All rubbers and rubber like materials are called elastomers. These are usually addition polymerised from unsaturated monomers such as butadiene and cross-linked (vulcanised) with sulphur at about every 500th carbon atom. increased cross-linking gives increased stiffness and reduced elongation. Some specific varieties of rubber, such as silicon rubber are recommended for making intricate type of dies for investment casting. Silicon rubber is also available like epoxy resins in two parts, i.e. a liquid binder and liquid hardener. When the mixture of binder and hardener is poured over a master pattern or into a die, and allowed to cure, a solid shape is produced.

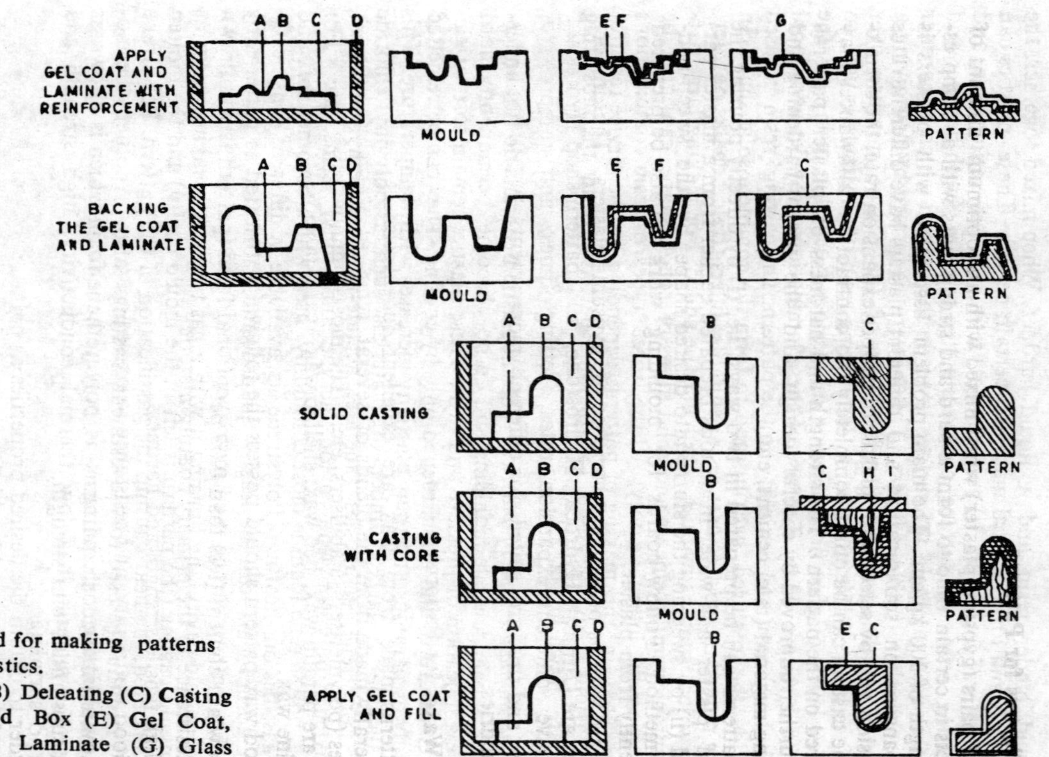

Fig. 2.1 Various techniques used for making patterns from thermosetting plastics.
(A) Mastern Pattern (B) Deleating (C) Casting Compound (D) Mould Box (E) Gel Coat, (F) Glass Fibre Mat Laminate (G) Glass Matreinforad ribs, (H) Wooden Case and (I) Backing plate.

2.2.5 Plasters for Patterns

Plaster of pairs (gypsum plaster) when mixed with an optimum amount of water, sets in certain time to form a hard and strong mass with a compressive strength of 300 kg/cm^2. The main problem associated with plasters is their expansion on solifidication and different plasters have different rates of expansion and by selecting a plaster of proper expansion rate, the shrinkage of the casting can be offset completely (no contraction allowances may be required on the pattern). The strength and hardness of plaster patterns can be further improved by careful design and the use of hardening and reinforcing material (talc, cement, etc.).

The patterns can be prepared in two ways, i.e. (i) by directly pouring the slurry of plaster and water in moulds prepared earlier from a master pattern, and (ii) by sweeping the slurry into desired shape by the sweep-and-strickle method. Follow-boards for moulding work can also be prepared conveniently from plaster.

There are various proprietary varieties of gypsum plasters. The important ones are Ultracal, Hydrocal, Hydrostone, and pattern shop Hydrocal. These have different expansion rates, setting times, and compressive strength and thus are suited for different requirements of pattern shops and foundries.

2.2.6 Waxes for Patterns

The patterns made from wax are excellent for investment casting. These are generally made from the blends of several types of waxes and other additives (polymerisers, stabilisers, etc.). The most commonly used varieties of wax are parafin wax, bee wax, shellac wax, carnauba wax, and microcrystalline wax.

A good wax pattern should possess the following properties:

(i) Low ash content (less than 0.05 pet).
(ii) Resistance to the primary coating material used for investment casting.
(iii) High tensile strength and hardness.
(iv) Good wettability, and resistance to oxidation.
(v) Low shrinkage on cooling.
(vi) A substantial joint strength.

In order to obtain the desired properties as pattern, suitable blending and compounding of waxes with additives is carried out. The determination of actual composition requires sufficient trials and experiments. The wax patterns are generally formed by injecting the liquid or semi liquid wax into a split-die. Injection of solid wax can also be used to avoid shrinkage (caused due to cooling of liquid wax), however, much higher injection pressure will be required to obtain adequate strength. After injection and cooling, the wax pattern is taken out by opening the die. These patterns are generally small in size and a number of patterns can be welded to a common runner or sprue through in-gates. The entire assembly of patterns complete with runner, sprue, ingates, etc. is usually termed as a tree which is used for preparing the investment mould.

2.2.7 Use of Old Castings as Patterns

When a pattern can-not be made owing to urgency, an old casting may be used as pattern. The surfaces of old castings can be glued with thin strips of paper, wood. leather, aluminium etc. to give the required increase in thicknees and other dimensions to provide for contraction, machining and other allowences. sometimes metal spray can be used to provide the metal coating of the required size. The castings employed as patterns should be well cleaned and smoothened to remove undercuts and facilitate their easy withdrawal from the mould. Core prints, if necessary can also be provided on the casting.

2.3 TYPES OF PATTERNS

Wood and metal patterns differ greatly depending on their specific dimensions, method of moulding and particular application. Some patterns are small in size and simple in design, while others are large and very complex, with irregular surfaces, core prints, and loose pieces. The common types of patterns include (i) single or solid pattern, (ii) two pieces or split pattern, (iii) multipiece pattern (having more then two pieces), (iv) loose piece pattern (v) match plate pattern or plate-mounted pattern, (vi) cope and drag pattern, (vii) follow board pattern, (viii) gated pattern, and (ix) special patterns.

2.3.1 Single or Solid Pettern

These are the simplest and easiest to construct and mould and are made in a single piece without joints, partings, or loose pieces. There are the least expensive of all patterns. A typical single pattern is shown in Fig. 2.2. It usually has one broad surface (flat back) serving as the parting surface in the mould. This type of pattern may have very few irregularities (which do not obstruct the withdrawal of pattern), and may or may not have core print. The single pattern is identical in shape to the casting required except the necessary contraction allowance and core prints. The patterns not having flat back, require the use of a follow board (Fig. 2.2(b)) which can support the pattern during moulding and defines the parting surface. The gating system in mould is made by hand tools.

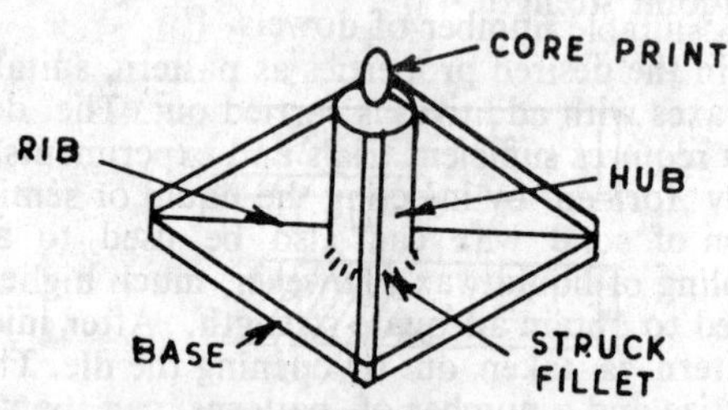

Fig. 2.2 Typical single pattern

2.3.2 Two Piece or Split Patterns

These patterns are made in two parts (Fig. 2.3) which may or may not be identical in size and shape. The pattern is split at a suitable position (determined by the shape of casting) and the surface formed at the line of separation of the two parts is known as 'parting surface', which is also the parting surface of the mould. The two parts of the pattern are held in alignment by two or more dowel pins. The dowels are made loose enough in one piece to facilitate the free lifting of that from the other part without binding. However, the dowels must fit firmly to avoid the movement of two parts of the pattern from side to side. Many two piece patterns (having cylindrical shape) can be turned on lathe. For making mould cavity, one part is moulded in drag and the other one in cope. With split patterns also, the gates are cut by hand tools. A split pattern facilitates easy and accurate moulding and can also be fastened to a match plate for mass production moulding.

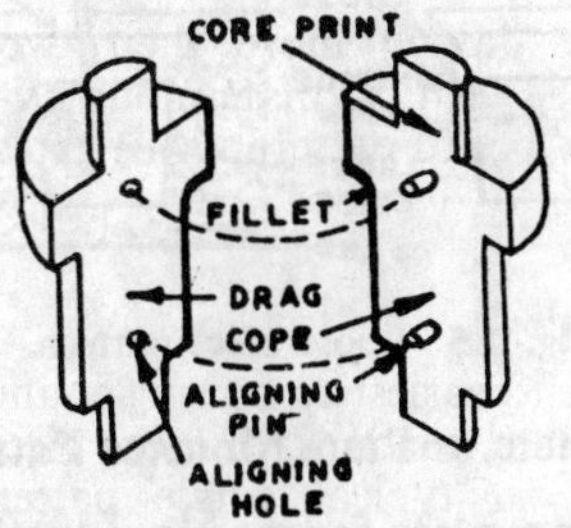

Fig. 2.3 Two piece or split pattern

2.3.3 Multi Piece Patterns

The construction of a pattern for a complex casting may require three or more parts to make the complete pattern (multipiece pattern). These parts may be actual body parts of the pattern, loose pieces (hubs, bosses. etc.) or loose core prints. The multi piece pattern is used when it is not possible to make the casting with single split pattern. The parts of the pattern are held in alignment with a suitable number of dowels. (Fig. 2.4),

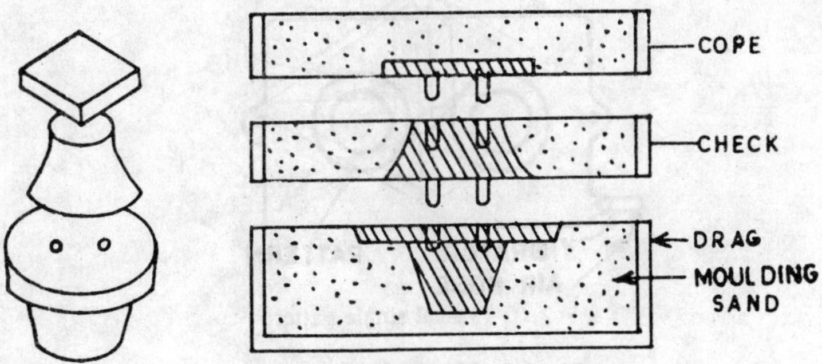

Fig. 2.4 Multi piece pattern

2.3.4 Loose Piece Pattern

Many Castings have a projection on the side which is above or below the parting line, owing to which it may not be possible to show this in the drawing of pattern. This requires the construction of pattern in such a way that this projection is incorporated in the pattern as a *loose piece* with an anchor pin to facilitate the removal of loose pieces during the moulding operation as shown in Fig, 2.5. The pattern is moulded first by placing it on the mould board in the conventional manner and the sand is rammed upto the loose piece. Then the sand is removed from around the piece and anchor pin is withdrawn leaving the loose piece in place. The moulding operation is then continued to completion. On drawing out of the pattern from the mould, the loose piece remains in the mould and is removed with the help of a bent wire, an angle scriber, or a pointed lifter. Construction of some special core boxes also requires the use of loose pieces.

Fig. 2.5 Loose piece pattern

2.3.5 Match Plate Pattern or Plate Mounted Pattern

Patterns are mounted on a plate for mass production of small castings, and are referred as a *match plate* pattern. There should be a perfect matching between the cope (upper half of the mould) and the drag (lower half of the mould) to conform to the parting line. A match plate pattern is illustrated in Fig. 2.6. The plate employed for mounting the patterns may be made of wood, metal or plastics (aluminium is commonly used). Small patterns

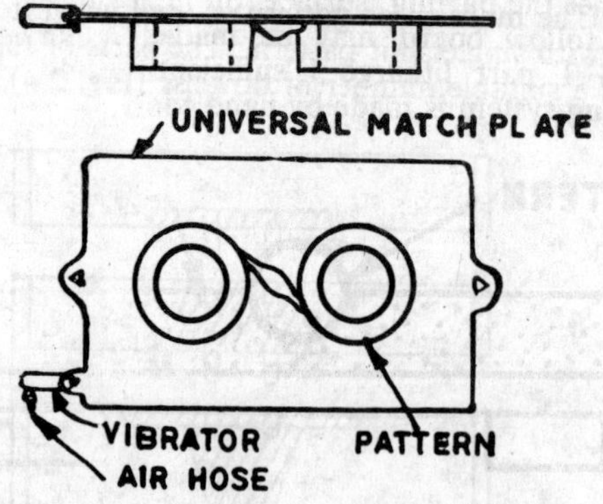

Fig. 2.6 Match plate pattern

having flat backs can be mounted directly on the drag side of the plate. The split patterns (two piece) are mounted with one part on one side and the other part directly opposite on the other side of the plate in such a way that the cavity formed should produce the desired casting. The gating system (sprue bases. runners, and gates) is generally constructed of same material as of pattern, and is always attached to the drag side of the plate in the appropriate position.

The match plate patterns can be prepared by casting as loose patterns which are then machined, finished, and mounted on separately prepared plates. This method, however, is complicated and requires extraordinary skill in mounting the loose patterns with accuracy. The problems of the above method can be overcome by casting the pattern 'integrally' with the pattern plate in sand, plaster or ceramic moulds, as one piece. The casting obtained from plaster mould gives excellent finish and accuracy of dimensions and thus the machining of casting is not required.

The increased cost of the match plate patterns is adequately compensated by the increased production and greater dimensional accuracy of the castings produced. The weight of mould and flask which can be handled by a moulder limits its size.

2.3.6 Cope and Drag Pattern

In this case the cope and drag parts of the pattern are mounted on separate plates and the copes and drags are moulded separately on different moulding machines. This type of pattern greatly facilitates the production of medium and large castings. An accurate alignment of the two mould halves (matching of upper and lower parts) is obtained by means of guides and pins located in suitable positions. This type of pattern is usually required in high speed mechanised or automatic moulding machines and the high cost of these patterns is compensated by high production rate of moulds.

2.3.7 Follow Board Pattern

Loose patterns having an irregular parting line cannot be moulded easily without attaching them to a plate or match known as follow board. This plate serves to support the loose pattern during moulding of the drag part and also establishes the parting surface on removal of the follow board. (Fig. 2.7). The follow board may be made of hard sand (25 parts dry moulding sand + 1 part litharge + sufficient linseed oil or core oil) or plaster. The gating system is made by hand tools.

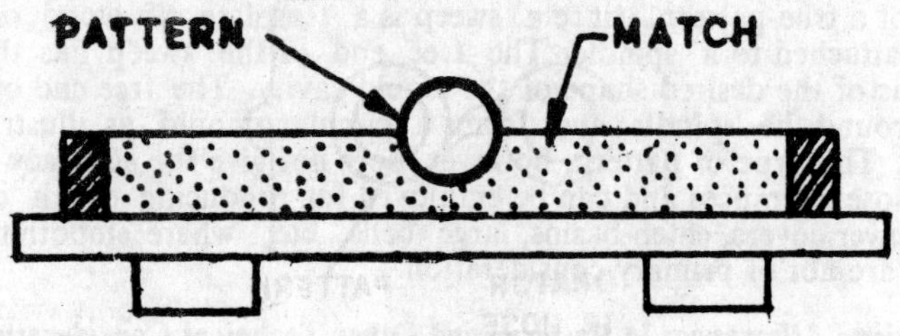

Fig. 2.7 Follow board pattern

2.3.8 Gated Patterns

This is an improved system of ungated loose patterns. The gating system constitutes a part of the pattern in which hand cutting of gates, runners, etc. is eliminated. This system usually consists of a set of patterns (2, 4, 6, etc.) arranged about a common runner and the mould cavities are connected through ingates (Fig. 2.8). This type of pattern permits more rapid moulding of small castings.

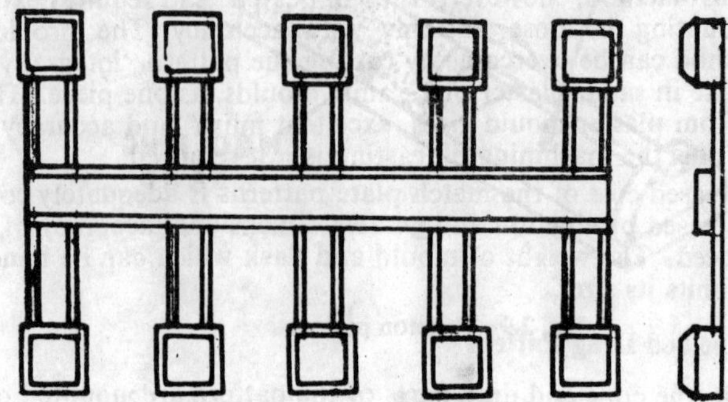

Fig. 2.8 Gated pattern

2.3.9 Skeleton Pattern

This is a ribbed construction of wood forming an outline of the pattern. The required size and shape are formed by ramming sand around the frame work having the general contour and size of the required casting. This type of pattern is suitable when only few castings of large size are to be made with less regid specifications. (Fig. 2.9). To make the surface smooth and of the desired shape, a strike off board known as 'strickle board' (a thin piece of wood or metal used tc scrap off excess sand from skeleton pattern) is used. After making the surface smooth the sand is coated with graphite (plumbago) to ensure a smooth casting surface. Skeleton patterns for round shapes (pies, cylinder, etc) are first prepared in halves and then joined by screws, pins or gluing.

2.3.10 Sweep Pattern

This is not a true pattern, since a sweep is a template of wood or other material attached to a spindle. The free end of the sweep has the same contour as of the desired shape of the mould cavity. The free end of sweep rotates around the spindle and forms a circular mould as illustrated in Fig. 2.10. This type of pattern, however, does not give the accuracy required for close tolerances and can be employed for producing rough castings such as sewer covers, catch basins, large bells, etc, where smoothness and accuracy are not of primary consideration.

2.4 Various Allowances in Patterns and Other Technical Considerations

The pattern is never identical in dimensions with the casting, since many allowances have to be provided for in the pattern for metallurgical and

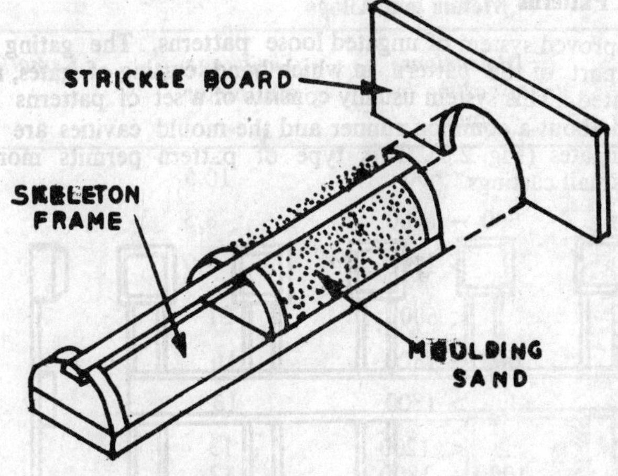

Fig. 2.9 Skeleton pattern

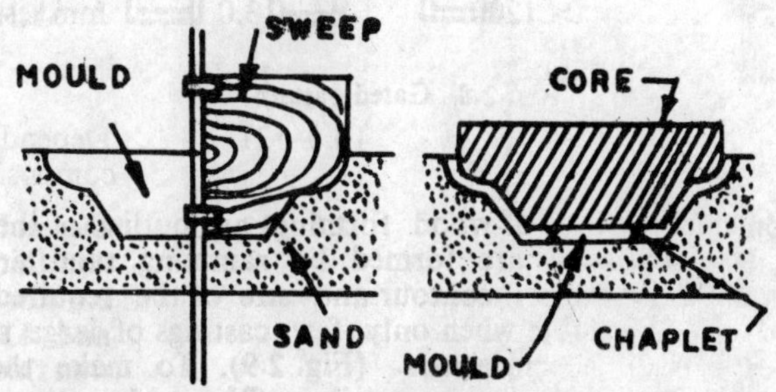

Fig. 2.10 Sweep pattern

mechanical reasons to produce the castings of the required dimensions. The various allowances required on the pattern include shrinkage or contraction allowances, machining allowances, draft or taper allowance, distortion allowance, rapping and shake allowance and dimensional tolerances. Design of patterns with appropriate allowances, reduces the machining cost and avoids rejections.

2.4.1 Shrinkage or Contraction Allowances

It is well known that metals employed in casting expand on heating and contract on cooling. Therefore, the pattern has to be made longer than the casting by an amount equal to shrinkage (known as pattern maker's contraction) caused in cooling. The shrinkage allowance, therefore, is of basic importance in pattern making. The pattern maker uses a special scale known as 'contraction' or 'shrinkage scale' which has the graduations with compensation for shrinkage value. Since, different metals have different shrinkage values, different contraction scales are used for different castings.

Table 2.2 Contraction AlloAances for some Important Metals and Alloys

Metal	*Dimensions of casting in mm*	*Shrinkage mm/meter*	*Remarks*
Cast Iron	< 600	10.5	
	600 – 1200	8.5	
	> 1200	7.0	
Steel	< 600	21	
	600 – 1800	16	
	> 1800	13	
Aluminium	< 1200	13	
	1200 – 1800	12	
	> 1800	10.5	For cored construction it is 1.5 mm less
Magnesium	< 1200	14.5	
	> 1200	13.0	
Brass	—	16	
Bronze	—	10.5—21	Depends on composition
Malleable Iron	—	2.6—11.6	Depends on section thickness e.g. 2.6 for 25 mm and 11.6 for 6 mm thickness respectively.

In actual practice, such scale has two scales of graduations on each side, the total number of scale graduations being four which can be used for most common metals/alloys, i.e. steel, cast iron, brass, and aluminium.

The actual contraction during solidification and cooling depends on several factors such as the composition of metals, moulding method employed, pouring temperature, design and intricacy of the casting, weight and size of casting, etc. A guide line for the contraction of some metals and alloys is given in Table 2.2.

2.4.2 Maching Allowances

Usually the casting is made sufficiently oversize on certain surfaces to permit its machining to obtain the required dimensions and finish. This oversize is referred as machining allowance or finish allowance. The amount of machining allowance to be given depends on the following factors:

(i) The characteristics and composition of metals.

(ii) Method of moulding, i.e. hand moulding or machine moulding, plaster moulding, etc.

(iii) Method of casting, i.e. sand casting, metal mould casting, pressure die casting, etc.

(iv) Size and shape of the casting.

(v) The orientation of casting, e.g. the top surface of the casting in the mould requires greater allowance.

(vi) Application of casting and the degree of acouracy and finish required.

In effect, the machining allowance should be kept to a minimum, since the oversize removed in machining operation is a waste of metal. The machining allowance for machine moulding patterns is usually smaller than for hand moulding as high accuracy of dimensions is obtained in the former case. Table 2.3 represents the typical machining allowances for hand moulding recommended for different metals/alloys.

2.4.3 Draft or Taper Allowance

This allowance is provided on the vertical faces of a pattern to facilitate its removal from the mould without causing any damage to the mould cavity surfaces (Fig. 2.11). The degree of taper depends on (i) length of the vertical size of the patteru (ii) intricacy of the casting, and (iii) method of moulding. The hand drawn patterns usually require a draft of 10-20 mm (1-2 degrees) per meter on the exterior surface, whereas the draft for the interior surface may be 40-60 mm per meter (4-6 degrees). The machine drawn patterns requirea draft of only about 10 mm per meter (1 degree). In some cases the patterns upto 200 mm deep may be drawn by machine withont providing any draft if the pattern is very smooth and clean, and the drawing equipment is properly aligned.

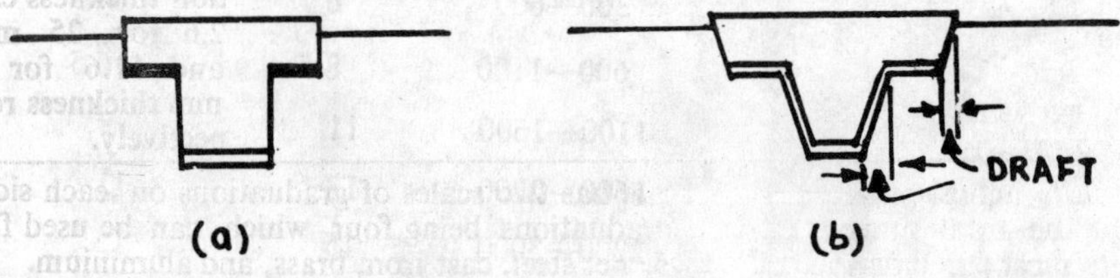

Fig. 2.11 Draft Allowance (a) Pattern having no draft (b) Pattern having draft

2.4.4 Dimension Tolerances

The variation permitted on a certain casting size is termed as dimension or size tolerance. This is equal to the difference between the minimum and maximum limits permitted for any given dimension. The typical maximum tolerance values for heavy castings are given in Table 2.4 and are approximate. As a general rule, size tolerance should be atleast half the shrinkage allowance. However, with some casting processes, these values may be much less.

2.4.5 Rapping and Shake or Drawing Allowance

'Rapping' and 'drawing' operations are required in all moulding work before a pattern is withdrawn successfully. The rapping may increase or

Table—2.3 Typical Machining Allowances for Commercial Metals and Alloys

Casting alloys	*Pattern size* (mm)	*Bore* (mm)	*Surface* (mm)
Grey Cast Iron	<200	3.5	3.5
	200—400	4.5	4
	400—700	6	5
	700—1100	7	6
	1100—1600	9	7
	1600—2200	10	8
	2200—3000	12	9
Malleable Iron	<150	1.5	1.5
	150—220	2.5	1.5
	220—300	2·5	2.5
	300—600	4	3
	600—750	4.5	4.5
Cast steel	<300	4.5	3
	300—600	6	4.5
	600—1100	8.5	6.5
	1100—1600	11	9
	1600—2200	13	11
	2200—3000	15	13
Common non-ferrous metal castings	<400	2	1.5
	400—700	3.0	2.0
	700—1100	3.5	2.5
	1100—1600	4.0	3.0

decrease the dimensions of casting if it is over done. Internal sections of mould may be decreased whereas external sections are increased in size with excessive rapping (Fig. 2.11). To compensate this increase in size, the pattern should initially be made slightly smaller than the desired size. However, this allowance is usually arbitrary, since the extent of rapping varies from one moulder to another. In general, no rapping allowance is considered by the pattern maker. The moulder, thus raps the pattern just enough to facilitate its removal from the mould without any distortion of the mould.

Table 2.4 Typical Size Tolerance for Casting Weighing 400 kgs or more

Metal	*Tolerance* mm
Grey Cast Iron	1.5
Malleable Iron	2.5
Cast steel	4
Alluminium alloys	2
Magnesium alloys	4.5
Brass	2.5
Bronze	3

2.4.6 Distortion or Warpage Allowance

Some castings tend to warp during cooling depending on their shape, and type of metal. For examgle, a U-form casting will tend to contract at the closed end causing the vertical legs out of parallel (slightly incilned). This can be prevented by making the legs of U-pattern converge slightly. The distortion allowance is usually considered for castings tending to warp and having irregular shape. This warpage is of particular disadvantage for castings requiring precision machining. Allowance can be made on pattern for an extra metal on the surfaces which are likely to be affected. The distortion in casting may occur due to (i) internal stresses in casting resulting due to unequal cooling rates of different sections of the casting, (ii) restricted conraction from cores and (ii) unequal heat transfer rates.

Usually. the allowance for warpage are not made on the pattern except in some specific cases where the castings tend to distort. The distortion is usually prevented by the following measures:

(i) Designing a casting to avoid abrupt changes in sections, areas of heat and stress concentration, and intersecting ribs, etc.
(ii) Selection of metal having minimum contractions.
(iii) Controlling pouring temperature.
(iv) Using appropriate moulding procedure, use of chills for more uniform cooling rates and avoiding the use of rigid cores
(v) Giving adequate machining allowance to cover the distortion effect.
(vi) Straightening the distorted castings (if ductile) by pressing either in cold or heated (400-500°) condition. For small and medium sized castings, cold pressing is sufficient whereas large castiags may require heating.
(vii) Providing appropriate allowance known as camber or distortion or warpage allowance to the pattern.

It is extremely difficult to work out the amount of distortion, as it depends on several parameters and the usual approach is of trial and error.

The camber graphs for grey cast iron machine tool castings have been published and the amount of camber for a given length, wall thickness, and depth of casting can be computed. For other metals and alloys, no published data on camber is available.

2.5 Core Prints

Core print is an added projection on a pattern forming a seat in the mould in which dry sand core is placed during pouring. The core print should be of proper size and shape to form a seat large enough to provide a stable support and grip between the core and the mould. The core placed should not have a tendency to shift or rise due to the buoyant effect of the molten metal produced by its displacement by the core. For example, the buoyancy force for molten steel can be calculated as follows:

Weight of molten steel is 7800 kg/m^3

Weight of core is 1600 kg/m^3

Then, buoyancy force (or core buoyancy) – 7800 – 1600 = 6200 kg/m^3

Then the ratio of buoyancy force to weight of core (for steel),

$$K = \frac{6200}{1600} = 3.88 \text{ gm/cm}^3$$

This ratio can be similarly determined for cast iron (3.5 gm/cm^3), Copper (4.5 gm/cm^3), brass (4.25 gm/cm^3), aluminium (0.66 gm/cm^3), etc. For metals having with high value of K, there is greater tendency for the core to float, and thus more secure method is required to hold the cores. Therefore, the core prints are longer for heavy metals (iron, steel, brass, bronze, etc.) than required for light metals (such as aluminium). Sometimes, even *chaplets* (metal pieces placed between mould and core surfaces) may be needed for casting heavy metals for reason of buoyancy.

Once the core buoyancy is known, the core print size can be calculated as follows:

Unsupported load = core buoyancy – core print holding force

Where, core print holding force = compressive strength of moulding sand × core-print surface area.

If the compressive strength of sand is considered as 0.6 kg/cm^2.

Then core-print holding force = 0.6 × core print surface area,

In order to ensure stability of core in the mould, the value of unsupported load should be zero or negative.

Therefore, 0 = core buoyancy – 0.6 core-print surface area.

and thus minimum core-print surface area

$$= \frac{\text{core buoyancy}}{0.6} = 1.66 \times \text{core buoyancy}$$

There are several types of core prints which are characterised mainly by their position in the mould, and likewise cores are also classified. The common types of core prints are (i) horizontal core prints, (ii) vertical core prints, (iii) balanced core prints, (iv) hanging or cover core prints, and (v) wing core prints.

2.5.1 Horizontal Core Prints

This is the most common type core print which forms a seat in the mould in which the core is laid horizontally, and is usually found on a split or two piece pattern. (Fig. 2.12(a)).

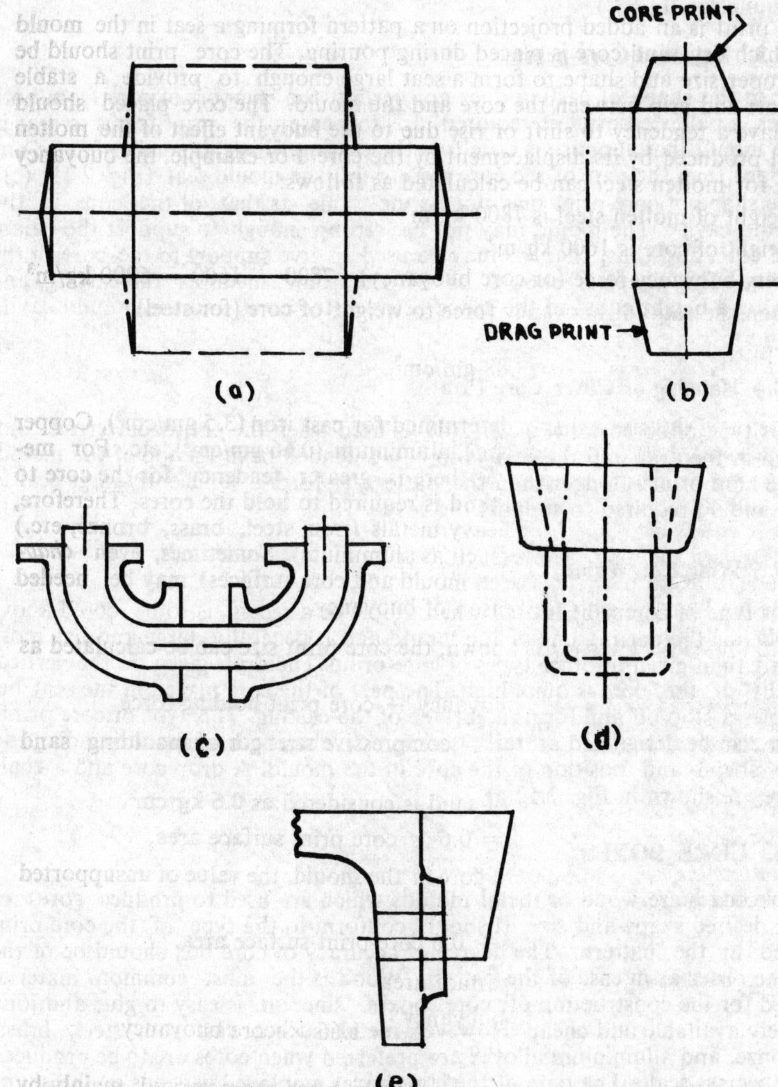

Fig. 2.12 Common types of Core prints: (a) Horizontal Core Print (b) Vertical Core Print, (C) Balanced Core Print, (d) Hanging or Cover Core Print, (e) Wing Core Print.

2.5.2 Vertical Coe-print

This is another commonly used type core print and forms a seat in the mould that stands vcrtically (Fig. 2.12 (b)). This is also referred as a cope and drag print. A considerable draft (10-15°) is given on the core-print, particularly on the cope end to facilitate the withdrawal of the pattern and the replacement of the cope on the drag. The draft given on the drag end is quite small (2-3°).

2.5.3 Balanced Core-print

When the shape of casting does not permit the support of core from both sides, a balanced print is required. In this case, the print forms a seat in the mould that supports a core for the casting which is smaller in size and weight than the part of the core resting in the mould seat (Fig. 2.12 (c)). The shape of core-print may not be the same as that of the core in the mould cavity. The mould may not be strong enough to support the balanced core, and thus, chaplets may be used to give support to the core in the cavity. The chaplets should be of the same metal as the casting, and having a height equal to the space between the core and the mould cavity wall.

2.5.4 Hanging or Cover Core Print

This type of coreprint (Fig. 2.12(d)) is used when the entire pattern is rammed in the drag and the core is to be suspended from the top of the mould. The core print is longer than the pattern to form a sufficiently large seat in the mould required to support the core.

2.5.5 Wing Core-Print

This type of coreprint is used when a hole or a recess is to be cored above or below the parting line of the mould and is generally large and has more draft than given in other types of core-prints.The draft given on the vertical walls of the core is quite high. The part of the core placed in the seat becomes a stop off and forms a surface of the casting. This type of core prints can also be designated as 'tail', 'saddle', or 'chair' coreprints, depending on the shape and position of the core in the mould. A drop core and a chair core are shown in Fig. 2.12(e).

2.6 CORE BOXES

Core boxes are wood or metal moulds which are used to produce cores of the desired shape and size. It should conform to the type of the core print used in the pattern. The degree of accuracy of core box should be of the same order as in case of the pattern. Wood is the most common material used for the construction of core boxes, since it is easy to glue and join, freely available and cheap. However, metals such as cast iron, steel, brass, bronze, and alluminium alloyes are preferred when cores are to be produced on a mass scale. The type of the core boxes employed depends on the shape of the core. Some commonly used core boxes are (i) half core box, (ii) dump or slab core box, (iii) split core box, (iv) strickle type core box, (v) right and left handed core box, and (vi) loose piece core box.

2.6.1 Half Core Box

The half core box is most commonly used and is employed for cores having a shape that can be prepared in identical halves. (Fig. 2.13(a)). The two identical halves of core are rammed in the half box, and after drying and baking, the core halves are pasted together to form a complete core.

2.6.2 Dump or Slab Core Box

This is similar to half core box in appearance. However, this is used to form a complete core requiring no pasting (Fig. 2.13(b)). If the core required is of slab shape, the core box is referred as slab core box.

2.6.3 Split Core Box

When the core box is constructed in two or more parts and a complete core can be obtained by a single ramming, the box is referred as split core box. In order to remove the core after ramming, the box is separated in two or more parts, as shown in Fig. 2.13(c). In some special cases, one end of the split core box may be closed. The split core box is usually held together with the help of a clamp during the ramming of the core from one end. The correct alignment of the parts is aided by the use of guide (dowel pins) in one part and corresponding holes in the other. Sometimes, the parting surface is made on an irregular line to eliminate the use of guide pins for correct alignment of parts.

2.6.4 Strickle Core Box

This type of core box is used to make cores having an irregular surface, which may not be possible by other methods. In a strickle box, the desired irregular shape of the core is produced by striking off the core sand from the top of the core box with a piece of wood (strickle board having the contour corresponding to the core shape (Fig. 2.13(d)). The use of strickle board, in general simplifies the construction of a core box.

2.6.5 Right Hand and Left Hand Corebox

When the core is to be made in two parts which are not identical, two different coreboxes of the half core type are employed for each part of the core. A corebox used for such purpose is called right hand and left hand corebox. The core halves formed in these boxes are pasted together to form a complete core. An example of this type is the core hole in the pipe elbow (Fig. 2.13(e)).

2.6.6 Loose Piece Corebox

When two parts of the core are not identical, the cores can also be made from a single core box with the help of loose pieces of wood (called stop-offs) as shown in Fig. 2.13(f). The recess in the corebox is made in the form of a cross. In one of the two recesses, stop-offs are used to close the end of the box that is not in use at the time of ramming each of the alternate halves of the core. The first half is rammed by placing the stop-off in the left hand recess and then stop-off is shifted to the opposite side of the core box when ramming the second half. This type of corebox may be used to make the core required for pipe 'T'.

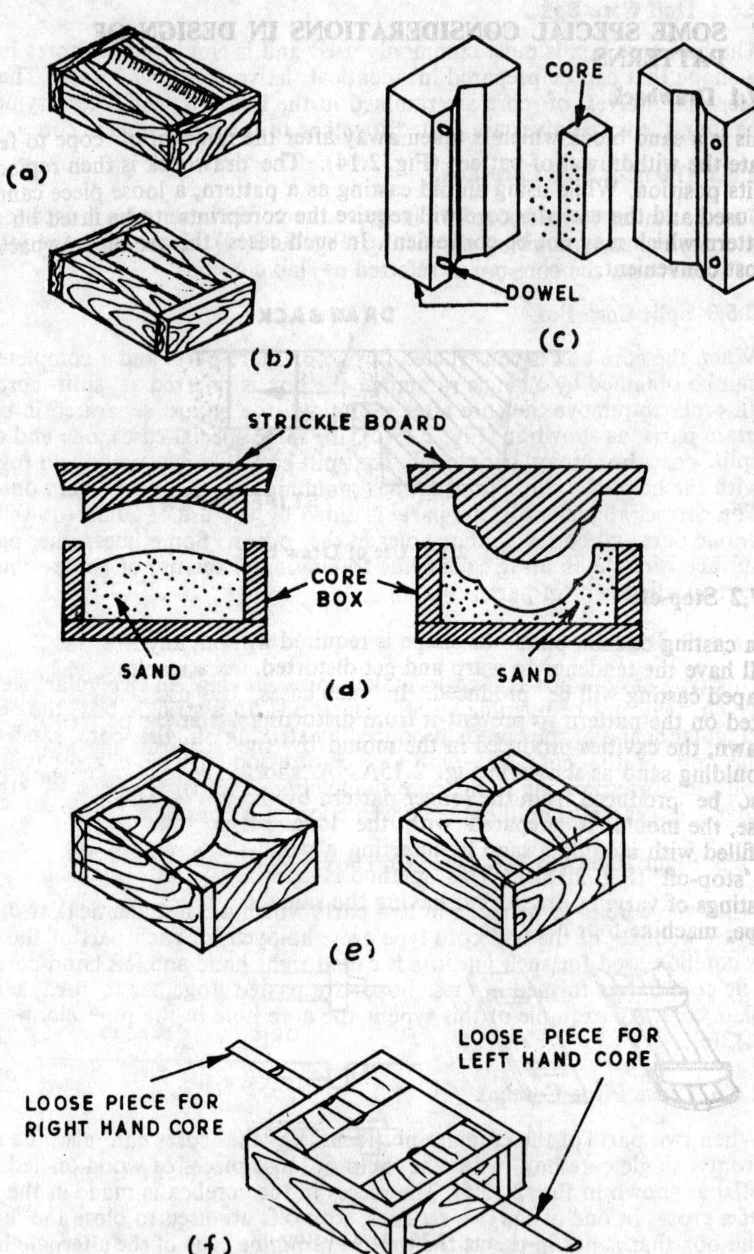

Fig. 2.13 Types of core boxes. (a) Half corebox (b) Dump or slab core box, (c) Split core box, (d) Strikle core box, (e) Right hand and left hand core box and (g) Loose piece core box

2.7 SOME SPECIAL CONSIDERATIONS IN DESIGN OF PATTERNS

2.7.1 Drawback

This is a sand block which is taken away after the removal of cope to fecilitate the withdrawal of pattern (Fig. 2.14). The drawback is then replaced in its position. While using an old casting as a pattern, a loose piece cannot be used and the use of a core will require the coreprints to be fitted on the pattern which may not be convenient. In such cases the use of drawback is most convenient.

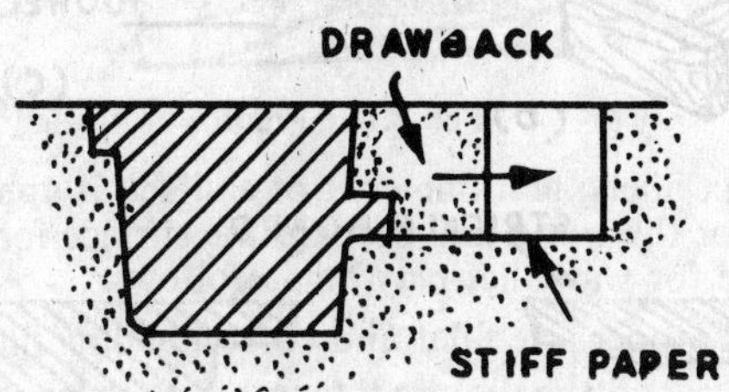

Fig. 2.14 Use of Draw back

2.7.2 Stop-off

If a casting of thin, plate like shape is required without any ribs, the pattern will have the tendency to warp and get distorted, as a result of which a deshaped casting will be produced. In such cases, the use of pieces may be fitted on the pattern to prevent it from distorting. After the pattern is withdrawn, the cavities produced in the mould by the stop-offs are field with moulding sand as shown in Fig. 2.15A. A shorter length of casting can also be produced from the longer pattern by the use of stop-offs. In this case, the mould is prepared with the long pattern and the extra lngth is filled with moulding sand by inserting a wooden mending piece, known as 'stop-off' (Fig. 2.15B). This method is specifically suitable to produce castings of varying lengths but having the same cross-section, for example, pipe, machine-tool beds, etc.

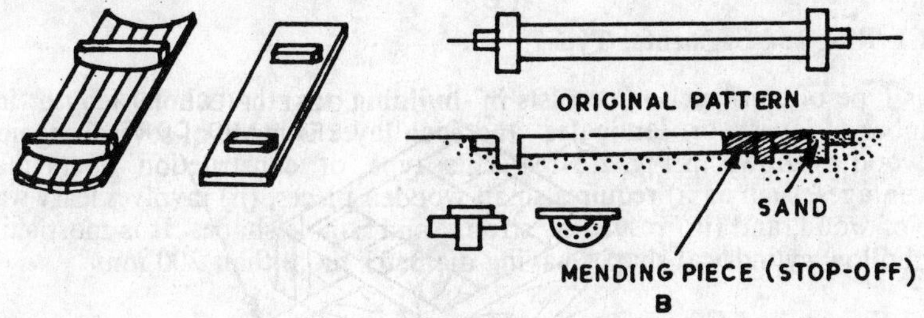

Fig. 2.15 Use of stop-off

2.7.3 Fillet

This is the rounding of an internal corner resulting from two intersecting surfaces as shown in Fig. 2.16. The shape of this connecting surface is made cancave. A smooth fillet is made at the sharp corners to provide (i) easy flow of metal into the mould, (ii) uniform cooling without stress concentration, (iii) more sound casting, and (iv) easy withdrawal of pattern from the mould. The fillets can be made of wood, wax, leather, plastic, or a combination of sawdust and glue.

Fig. 2.16 Fillet

For making wax fillets, first one coat of shillack is applied to the pattern to seal the grain of the wood and prepare the surface for the wax fillet. The general procedure for wax fillet may be as follows:

(i) Selection of proper size fillet and corresponding fillet iron.
(ii) The fillet is cut to the required length for its location on the pattern.
(iii) Fillet iron is heated by a suitable source (alcohol torch) to a temperature sufficient to soften but not to melt the wax fillet.
(iv) The fillet is placed in the corner and then it is rubbed with the heated iron applying firm pressure until it is smooth and even.
(v) The excess wax is removed with a knife.

Leather fillets are used when durability is a major factor. The leather fillet is held in place with glue (without a coating of shellock on the pattern).

2.8 CONSTRUCTION OF PATTERNS AND CORE BOXES

Various types of construction emplyed for patterns and core boxes include (i) ring or segmental construction, (ii) tongue and grove construction, (iii) disc construction, (iv) header and stave construction, (v) box construction, and (vi) combination of wood and metal. These are discussed below.

2.8.1 Ring and Segmental Type

This type of construction consists in building up the whole length in a number of layers or laminates and each layer is made of several segments of wood as shown in Fig. 2.17(a). This type of construction offers many advantages, such as (i) requires small wooden pieces, (ii) involves least wastage of wood, and (iii) results in strong and stable shapes. It is most suited for hollow cylindrical shapes having diameter more than 200 mm.

2.8.2 Tongue-and Groove Type

In this case a groove is cut to accommodate the tongue of the board as shown in Fig. 2.17(b). This type of construction is suitable for making the patterns of a circular pulley or fly wheel where the hub and therim are joined by arms.

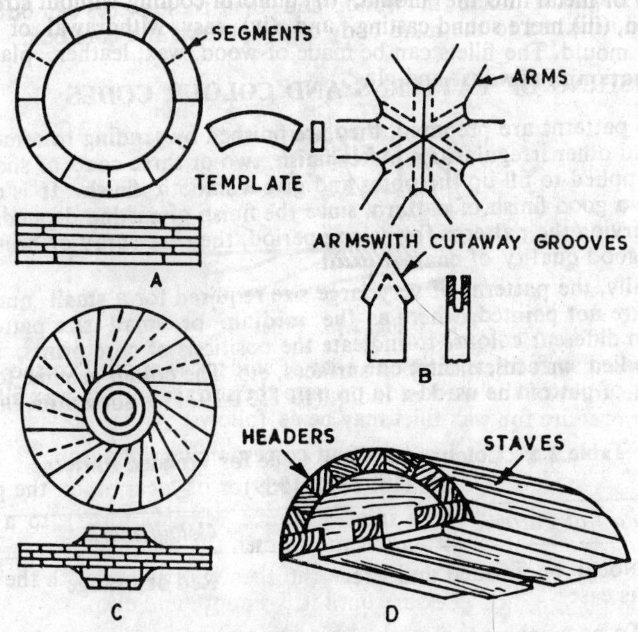

Fig. 2.17 Various types of construction of patterns and core boxes: (A) Ring and segmental type (B) Tangue and grove type (C) Disc type, (D) Header and stave type.

2.8.3 Disc Type Construction

In this case also, the laminates are used to build up the desired thickness. Each laminate is formed by joining small pieces of wood laid down in a non-radial manner. (Fig. 2.17(c)). This type of construction provides the structural strength and the pieces remain firm even during machining.

2.8.4 Header and Stave Type Construction

In this case, the ends of hollow pattern are closed by pieces called header on which the length wise sections called staves are fixed (Fig. 2.17(d)). The headers and staves are glued and shaped finally to the required dimensions. It is used to make large shapes which are too expensive to be made completely solid.

2.8.5 Box Type Construction

This type of construction is useful when the inside of the pattern can be made hollow and the pattern is of large size. The weight of the pattern and the quantity of wood required is reduced.

2.8.6 Wood and Metal Composite Construction

In this case, metal plates or strips are provided in the areas of extensive wear during moulding. The projecting pieces such as ribs, vanes, webs, etc.

are quite fragile, if made in wood, and thus these pieces can be made of metal and attached to the main body of the pattern and core box.

2.9 FINISHING OF PATTERNS AND COLOUR CODES

After the patterns are prepared, they are finished by sanding to remove tool marks and other irregularities. After finish, two or three coats of shellac are usually applied to fill up the pores and give a smooth finish. It is essential to obtain a good finish of pattarn, since the finish of casting depends on it. For preserving the patterns for a long period, they are spray or brush painted with good quality of enamel paint.

Generally, the patterns of very large size required for a small number of castings are not painted, where as the medium or small size patterns are painted in different colours to indicate the pòsitions of core prints, stop-off pieces, chilled surfaces, machined surfaces, etc. The colour scheme for different parts of pattern as used in India, (IS 1531-1971) is shown in Table-2.5.

Table 2.5 Colour and Mark Code for Wooden Patterns (IS: 1513-1959)

S.No.	*Part of Pattern*	*Colour/mark*
1.	Not to be finished and left as cast	Red or orange
2.	To be machined	Yellow
3.	Unmachined openings	Black on core prints
4.	Machined openings	Yellow strips on black on core prints
5.	Pattern joint (split patterns)	
	A—Cored section	Black
	B—Metal section	Clear varnish
6.	Touch core:	
	Cored shape	Black
	Legend	Touch
7.	Seats of and for loose pieces and look coreprints	Green
8.	Stop off	Diagonal black strip with clear varnish
9.	Chilled surfaces:	
	Outline in legend	Black
		Chill

2.10 TOOLS FOR MAKING WOOD PATTERNS

The pattern making is mainly wood working, since most of the patterns used are made in wood. Therefore, most of the tools required are the same as used by the cabinet maker with the addition of some special tools used only in pattern making. The various tools generally employed are measuring and marking tools, sawing tools, planing tools, boring tools, clamping tools, corebox plane, pattern maker's saw, gauges, pattern fillet irons, and other miscellaneous tools.

2.10.1 Measuring and Marking Instruments

In marking patterns and core boxes, the various instruments used for measurements and marking include shrinkage rule, rule, trysquare, bevel, marking gauge, dividers, outside callipers, inside callipers, vernier callipers etc. These are shown in Fig. 2.18.

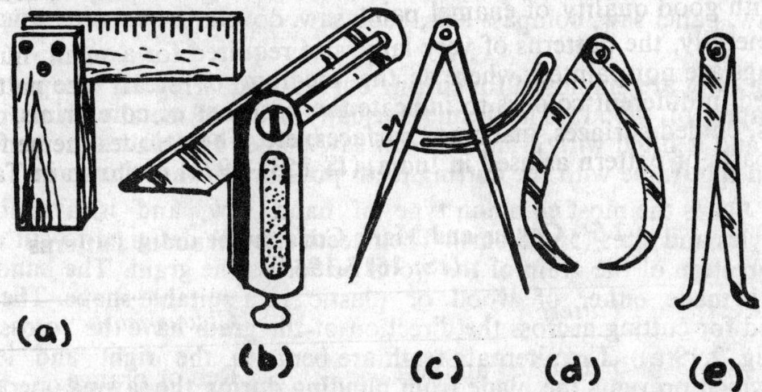

Fig. 2.18 Measuring and marking instruments (a) Square, (b) Bevel, (c) Divider, (d) Outside calliper, (e) Inside calliper

The shrinkage (contraction) rule is used to measure the dimensions of work pieces of patterns and core boxes. This rule is made with suitably over sized graduations to take care of allowance for the contraction of castings. A separate rule is available for each metal, viz., cast iron, steel, brass and aluminium. A single rule may also be provided with four graduations, two on each side, contraction rules are also available with specified contraction rate for example, 0.6 mm per meter, 1.0 mm per meter, 1.5 mm per meter, 2.0 mm per meter etc. The shrinkage rules are usually made of stainless steel.

The try-square is used to check right angles and make lines normal to each other on bars and planks. It consists of an arm (support), and a thin rule fastened to it at right angle. The bevel is a metal or wooden rule employed to check various angles and mark the wood pieces. It consists of a prop and rule pivoted on a bolt.

The *marking gauge* is used for marking and drawing parallel lines on plank. A steel pointed marking pin is provided to scribe a line at the desired distances (by setting of the adjustable head). Marking gauges are usually made of wood with some metal parts at the points showing possibility of wear. However, all metal marking gauges are also available. A double bar gauge consists of two independent bars with a marking pin or roller on each. The roller is particularly useful in marking across the grains.

The *dividers* are used for marking circles, arcs of circles and for transferring measurements from one place to another. A divider is identified by its largest opening, e.g. 10 cm divider can be used to scribe a circle of 20 cms diameter. *Trammels* or beam compass is used to layout the distances and scribe the arcs and circles beyond the capacity of ordinary dividers. Steel trammels with fine adjustments and beams ranging from 20 cm to 50 cm and extensions upto 180 cms are particularly useful for large pattern work.

The *outside calliper* is used to measure outside dimensions of circular shapes and the thickness of pieces. The inside callipers are employed to measure the hole diameters, recesses, and the distances between the individual parts of a pattern.

2.10.2 Sawing Tools

The various sawing tools employed in pattern making include coping saw, tenon saw, panel saw, compass saw, back saw, dovetail saw, pattern maker's saw, etc.

Coping saws are employed for cutting curves and irregular shapes from thin shapes of wood. The frame is usually made of metal with a wooden handle, and is fitted with a fine, narrow saw blade. The blades are generally placed in the frame with the cutting teeth pointing toward the handle.

Panel saw is the most common type of hand saw, and is available in many styles and sizes (56-65 cms). Their teeth may be designed to cut either in the direction of the grain of the wood or across the grain. The handle of the saw is made either of wood or plastic in a suitable shape. The saws employed for cutting accross the direction of the grain have the "cross-cut" teeth (Fig. 2.19(a)). The alternate teeth are bent to the right and left, a system which prevents the blade from blinding during the sawing operation. The saws employed to cut in the same direction as the grain of the wood, have 'rip teeth' (Fig. 2.19(b)) which are chisel-like and cut faster than cross-cut teeth.

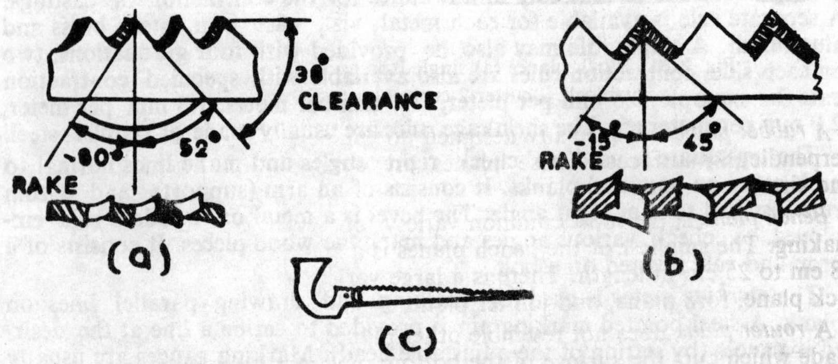

Fig. 2.19 Various sawing tools (a) cross cut, (b) rip teeth (c) compass saw

Compass saw (Fig. 2.19(c)) is employed for making cutouts on the inside of a piece of wood. This pointed saw is long and narrow. Both straight and curved cuts can be made on large work.

Back saw refers to the saw having a unusually thin blade re-inforced with a rigid steel back. The blade used is hard and tough having a row of fine cross-cut teeth filed along the cutting edge.

The *dovetail saw* is extremely thin (0.5 cm) with very fine teeth (6 points per cm). This is also reinforced by a rigid steel back. The dovetail saw is employed to make the finest possible cuts in wood, especially when a dovetail joint is made.

Pattern maker's saw is made of a thin steel blade (0.7 mm thick, 200 mm wide, and 300 mm long) and is designed for the accurate work required in pattern making.

Miter saw is an oversized backsaw which is used in conjunction with the miter box (an arrangement for holding work piece and guiding the saw accurately for square or angular cuts.

2.10.3 Planing Tools

There is a great variety of planing tools used in pattern and core making. The important ones include block plane, rabbet plane, bench plane, router plane, circular plane, corebox plane, spokshave, etc.

The *block plane* is a small, light weight planning tool with an adjustable blade (Fig. 2.20). This is made to be worked with one hand. The cutting blade rests on its seat at an angle of about 20° to the bed of the plane. These planes are particularly suited for small work of accuracy.

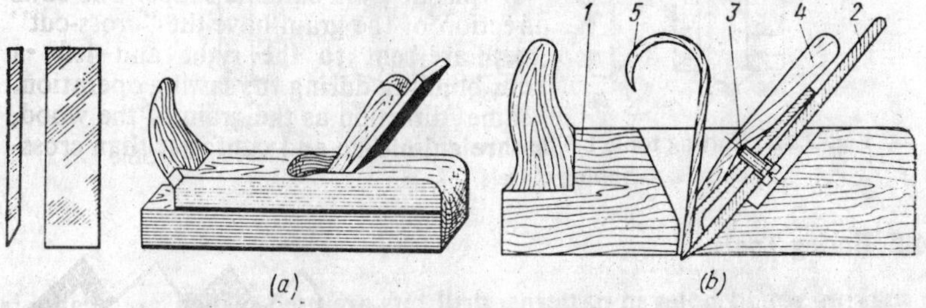

Fig. 2.20 Block planes (a) single iron and (b) double iron 1-block, 2-cutter 2-curve plane, 4-wedge, 5-shaving

A rabbet plane is specially designed to plane into corners. or against perpendicular surfaces. These planes are available in a variety of shapes and sizes.

Bench plane is the most common variety of plane employed in pattern making. The smallest of the bench planes is a smooth plane measuring from 18 cm to 25 cm in length. There is a large variety of bench planes, such as jack plane, fore plane, and jointer plane.

A *router plane* does not resemble other planes. It consists of a chisellike blade which cuts the surface of the bottom of grooves or other depressions and planes the recesses parallel to the surface of the work.

A *circular plane* has a flexible steel bottom adjustable for convex or concave surfaces of a wide range of curvature. These planes are particularly useful in smoothing circular surfaces of patterns.

A *core-box plane* is used to plane a semicircular groove in a pattern. Corebox planes have two beds at right angles to each other, which guide the cutter and make the semicircular cut.

A *spokeshave* is a small plane with a very short bed. Its name is derived from its first use made for shaping of spokes of wagon wheels. These are available in different styles for making inside or outside curves. The spokeshave is very useful in forming patterns having irregular shapes.

2.10.4 Cutting Tools

Variety of chisels are used for cutting work in pattern making. The important chisels include side chisel, round nose chisel, profile chisel, ripping chisel, etc. as shown in Fig. 2.21.

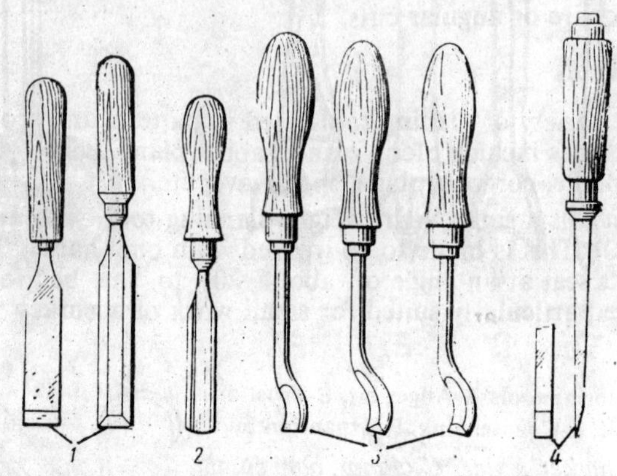

Fig. 2.21 Cutting tools 1-Side chisels, 2-round nose chisels, 3-profile chisels, 4-ripping chisels

2.10.5 Boring Tools

For making round holes in patterns, drill bits are used which are available in a variety of shapes and sizes. The drill bits are held in *brace* which is a crank-shaped tool. The various types of drills are auger bit, twist drill, counter sink, centre bit, expansion bit, and multi-spur machine bit as shown in Fig. 2.22. *Auger bits* are used to cut clean holes and their twist helps to remove shavings from deep holes. A *twist drill* is used for drilling small holes in wood or metal and are available in many sizes. A *counter sink* is a tool employed to shape the edge of a hole to receive the flat head of a screw and has an included angle of 82°.

Multi-spur machine bit is especially useful for large holes through a piece of wood. It also causes less split of wood when boring near the edge or end of the work and is available in sizes from 12.5 mm to 100 mm.

A *bit gauge* permits the boring of holes to a predetermined depth. It may first be adjusted for the desired depth and then several holes can be bored to the some depth. A doweling jig is of great importance in pattern making. Its use is indispensable when two or more members of a pattern are to be doweled by having holes in adjacent pieces in a perfect line.

2.10.6 Clamping Tools

One of the simplest clamping tool is *hand screw* (15 cm to 50 cm) which is used for innumerable purposes. Other equipments employed for clamping

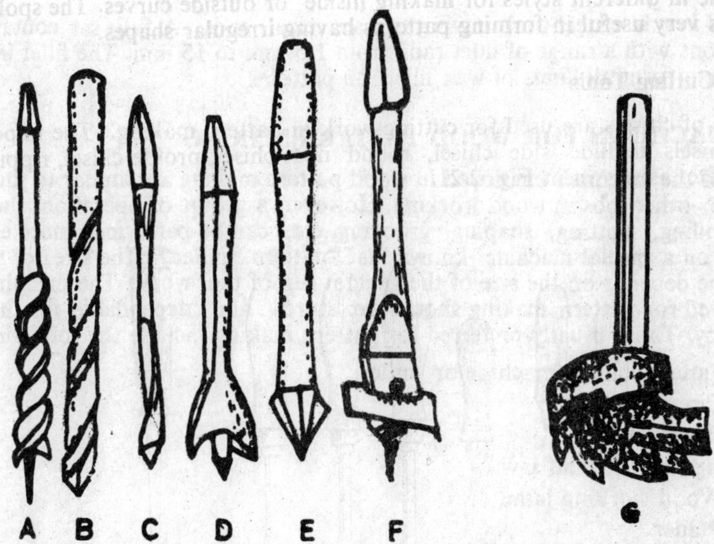

Fig. 2.22 Boring tools A-Auger bit, B-Twist drill, C-Twist bit, D-Centre bit, E-Counter sink, F-Expansion bit, G-Multispur machine bit

include *pattern maker's vice. C-clamp, bar clamp*, and *pinch dog*. C-clamp shaped like the letter-C' is handy and finds many uses in the pattern shop. An adjustable bar clamp is very useful for gluing wide pieces of stock. Pinch dog is a shaped clamp which is used to hold the pieces of stock together while working or gluing. The clamping action of pinch dogs is due to their wedge shaped prongs.

2.10.7 Miscellaneous Tools

There is a large variety of tools required in pattern making for various specific jobs. The tools commonly used are claw hammer, mallet, gauge, cabinet file, rasp, screw driver, nail set, brad awl, brad pusher, fillet iron, cornering tool, etc. The *claw hammer's* head has two main parts, i.e. the face for driving nails, and the claw for withdrawing the nails. The claw hammers are available in various sizes ranging from 200 gm to 600 gm. *Mallet* is handy tool made of hard wood in various sizes, and is used for striking chisels and gauges. The *gouge* is a circular chisel which may be sharpened in two ways, i.e. either on the inside or on the outside of the curved blade. Gouges are also available in the same sizes as chisels. *Cabinet file* is a special file used for wood, the half-round cabinet file being very popular. The files are identified by their length (without handle). *Rasp* is a special coarse file used to remove large amounts of stock. A *brad awl* is a handy tool having a sharp pointed tip and a hardwood handle. This is used for layout and starting holes for brads, screws, etc. A brad pusher holds small brads in its barrel with a magnetic action. When the wooden knob is pushed toward the work, a small brad will be driven into place.

The cornering tool is employed to round the sharp edges, since patterns are usually required with uniform rounded corners.

Fillet iron is a steel tool having two spherical ends. A full set contains four irons with a range of fillet radii from 1.5 mm to 15 mm. The fillet iron is used to form all kinds of wax fillets on patterns.

2.11 MACHINES FOR WOOD PATTERN MAKING

Most of the machines employed in wood pattern making are similar to those used for other jobs in wood working. However, a group of operations (boring, milling, slotting, shaping, grooving, etc.) can be performed more efficiently on a special machine known as "pattern miller". The size of the machine depends on the size of the general run of the work. The machines employed for pattern making should be sturdy and dependable for their accuracy. Those usually preferred for pattern making include the following:

(i) Pattern milling machine or miller
(ii) Circular saw
(iii) Band saw
(iv) Jig saw or scroll saw
(v) Wood working lathe
(vi) Planer
(vii) Jointer
(viii) Shaper
(ix) Disc and bobbin sander
(x) Abrasive belt machine
(xi) Drill press
(xii) Wood trimmer
(xiii) To grinding machine

2.11.1 Pattern Milling Machine

This is illustrated in Fig. 2.23. It consists of a large base on which a broad column is supported. In front of the base, a table support is provided which slides on nails and can be maintained in any desired position. The work piece is mounted on the table and it can be moved in horizontal as well as vertical directions. The column carries an overhanging arm which can be raised or lowered as desired. The outer end of the arm carries a spindle head and a cutter spindle. Spindle speeds may be provided varying from 850-4200 rpm. This machine is equipped with a large number of tools required in working of a pattern. Since this machine is provided with a wide range of tools, flexibility of spindle, table movements, etc., it is capable of performing a large number of operations. The important operations to be carried out by the milling machine include boring, drilling, facing, slotting, milling, grooving, shaping, fillet cutting. worm cutting, radius forming, angular cuts, and forming of other shapes.

2.11.2 Circular Saw

This is a very important single machine in the pattern shop. It consists of mainly (i) a cast iron table or flat which gives the support to the work piece when it is cut. (ii) a circular cutting blade supported on bearings and

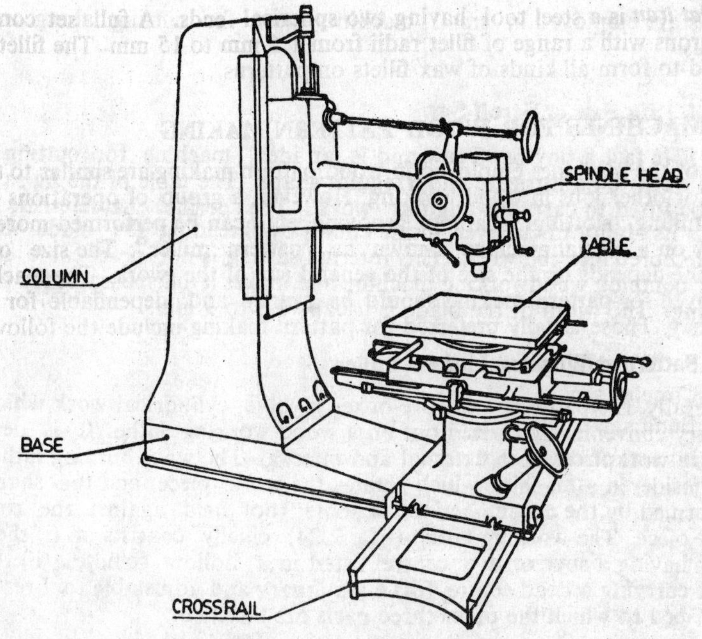

Fig. 2.23 Pattern milling machine

rotating at high speed, (iii) a driving arrangement for the saw blade (an electric motor with a set of pulleys mounted on shafts), (iv) a cut-off guide required during cross cutting to steer the piece towards the saw blade, and (v) a ripping fence acting as a guide during sawing along the grains of wood. Provision is made for tilting the table (to 45°) to provide the angular cuts. The size of thc circular saw is indentified by the diameter of the cutting blade. The most commonly used circular saw is a 300 mm saw which is employed for small and medium sized work. The cutting speeds may vary from 1000 to 3000 meters per minute depending on the hardness of the wood.

2.11.3 Band Saw

It is designed to cut wood by means of an endless metal saw band travelling over the rims of two or more rotating pulleys. The number of operations performed by band saw are much less than the operations performed by a circular saw. However, it is preferred for making irregular or curved cuts. The main parts are similar to those of a circular saw. The band saw is available in two models, i.e. horizontal and vertical. In case of horizonal saw, the two wheels are arranged alongside each other and the table is underneath, whereas in the vertical type, the wheels are arranged one above the other and the band passes through the table mounted in a central position between the two wheels. The size of a band saw is identified by the distance from the saw blade to the frame, which is almost equal to the diameter of the wheels. The maximum width of the saw depends on the size of the machine and usually varies from 6 mm to 50 mm. Narrow bands are employed in small machines used for cutting along a small radius.

2.11.4 Jig Saw or Scroll Saw

This is in fact a tiny band saw and is an ideal machine for cutting small sized work to an intricate and irregular shape. The table of the jig saw can also be tilted to carryout angular work. The speeific characteristic of this saw is its ability to cut even inside curves by threading the blade through a hole bored into the work. This internal sawing is not practicable in any other common woodworking machine and makes it indispensable for pattern making. The width of the blade is usually 1.5 to 9 mm.

2.11.5 Wood Working Lathe

Generally, the patterns and core-boxes involve cylindrical work which can be very conveniently carried out on a wood working lathe. It is designed to turn workpieces both external and internal. The wood turning lathe may be considered a machine which rotates the work piece and the shaping is performed by the cutting action of some tool held against the rotating work piece. The working lathe (Fig. 2.24) chiefly consists of (i) the head stock having a spur or live centre fitted in a hollow spindle, (ii) the tail stock carrying a dead centre, (iii) a stationary and adjustable tool rest, and (iv) a bed to which the other three parts are fastened.

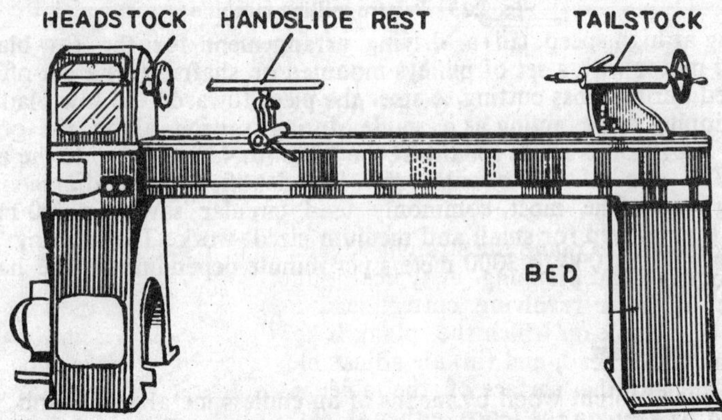

Fig. 2.24 Wood working lathe

The wood working lathes are generally supplied with a number of additional accessories which increase their usefulness and adaptability. Some of the important and typical accessories include different types of centres, (drive centre, cup centre and screw centre), a face plate, a 3-jaw self-centring chuck. 4-jaw independent chuck, a tool holder, and a set of wood turning tools.

2.11.6 Planer

The planer is employed for planing bars and planks. The main part of a planer is a cutter drum (head) located on the table which is driven by an electric motor. The thickness of shavings cut of the plank can be adjusted by shifting the table by secrews (Fig. 2.29 (b)). The plank is fed by means of feed rolls along a table against a revolving cutter head. A variety of planers are available. The panel planer (Fig. 2.25) is employed to smooth out the surface of a plank to make it even in thickness. The panel planer has a vertically movable table to adjust it for cutting off the plank to the desired thickness. The modern planers automatically surface the wood to desired thickness.

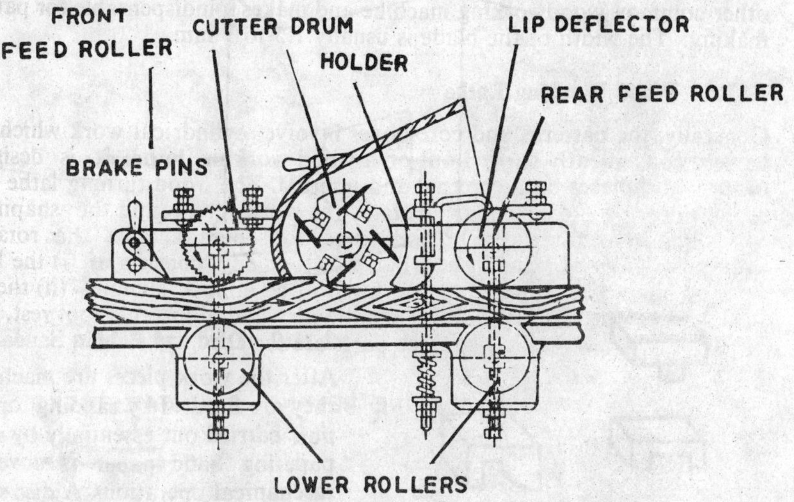

Fig. 2.25 Panel planer

2.11.7 Jointer

This is designed to plane the straight edges and surfaces of planks avoiding the labour in hand planning. A typical jointer is shown in Fig. 2.26 which consists of (i) a revolving cutter head fitted with three or more cutter knives. (ii) a table on which the plank is kept pressed and fed against the revolving cutter head, and (iii) an adjustable fence guiding the board at a desired angle to the surface of the table. Angular and bevel cuts can be obtained by keeping the fence either at 90° to the table, or inclined at any other angle. The depth of the cut can be regulated by lowering or raising the cutter head by means of a hand wheel. The purpose of wood jointer is similar to planer.

2.1.18 Shaper

This machine also consists of (i) cutter head fitted with knives and rotating about a vertical axis and (ii) a horizental table similar to that of a jointer. The plank is hand fed along the table against the cutter which is guided by an adjustable fence. The shape of the cut on the surface of the wood is the same as the profile of the cutter teeth. The cutter can be designed to produce a variety of shapes as shown in Fig. 2.27.

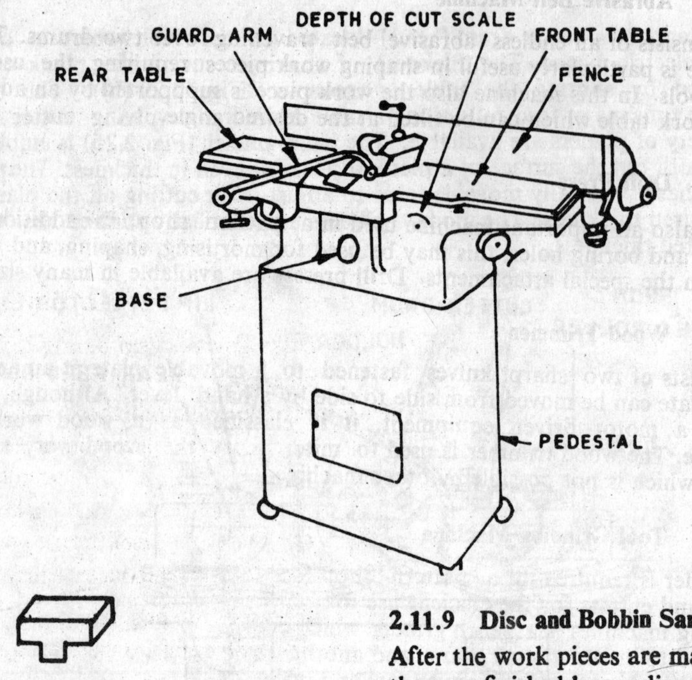

Fig. 2.27 Cutters to produce variety of shapes

2.11.9 Disc and Bobbin Sander

After the work pieces are machined, they are finished by sanding operation carried out essentially by sand papering. Sand paper is moved by mechanical operations. A disc sander consists of a large disc (500 mm diameter) rotating at high speed (3000 rpm). A sand paper of suitable grade is glued and fixed on the outer face of the disc. The work piece to be sanded is placed and pressed on an adjustable table provided against the disc. A bobbin sander consists of a cylindrical bobbin (75 mm diameter and 200 mm long) having sand paper around it. The bobbin also moves up and down through a short stroke of about 50 mm along with its rotation at high speed. A table to support the work piece is also provided around the bobbin. Bobbin sander is used to sand curved surfaces which cannot be sanded on the disc sander. Both the functions can also be combined in a single unit known as disc-cum-bobbin sander.

2.11.10 Abrasive Belt Machine

This consists of an endless abrasive belt travelling over two drums. This machine is particularly useful in shaping work pieces requiring the use of hand tools. In this machine also the work piece is suppported by an adjustable work table which can be tilted at the desired angle.

2.11.11 Drill Press

This is also an important machine used in a pattern shop. In addition to drilling and boring holes, this may be used for mortising, shaping, and nutting with the special attachments. Drill presses are available in many sizes.

2.11.12 Wood Trimmer

It consists of two sharp knives fastened to a movable plate at an angle. (The plate can be moved from side to side by a hand lever. Although, this is not a motor driven equipment, it is classified as a wood working machine. The wood trimmer is used to miter (cut) the wood very accurately, which is not possible by other machines.

2.11.13 Tool Grinding Machine

A grinder is required in a pattern shop for sharpening and shaping of tools, and cutters for the efficient use of various machines. The simplest grinding machines is a bench grinder which consists of two grinding wheels fastened to an *arbor* (one coarse and another fine). Other essential machines include (i) sharpener for circular saw and band saw blade, (ii) a band saw butt welder and (iii) planer knife grinder.

2.12 Instruments for Metal Pattern Making

The tools and instruments required for making metal patterns are different than those required for wood patterns. The metal patternshop should have many types of hand tools for measuring. marking, fitting, finishing and fabrication. The measuring tools include standard rule (300 mm and 600 mm), contraction rule, vernier bevel protractor, fixed and adjustable squares. micrometers, inside micrometer set, vernier, sine bar, radius gauge, screw pitch gauge, wire gauge, etc.

Marking tools iuclude surface plates, marking table, angle plate, vee blocks, callipers, dividers, trammels, marking gauge, scriber, punch, screw jacks, level, punches for figures and letter, etc.

Cutting and finishing tools are chisels, flat and diamond point, punches, electric hand drill, pneumatic drill, electric hand grinder, pneumatic grinder, drill bits, grinding wheels, flat files, rotatry files, rotary cutters for pneumatic operation, dies and taps, hack-saw, counter sinks, counter bores, etc.

2.13 Metal Pattern Making Machines

Working of metal patterns is different from that of wooden patterns. First the patterns are cast in a foundry in their approximate dimensions (over size) and shape. Then the cast pieces are machined and hand finishing by a number of operations. The metal pattern making shop should have facilities for machining and hand finishing. Various important machines for metal pattern making include universal milling machine, vertical milling machine, shaping machine, column type drilling machine, radial drilling machine, hydraulic backsaw machine, metal band saw, surface grinding machine, tool and cutter grinder, and many other special machines.

3

Moulding and Core Materials, Properties and Testing

3.1 FUNCTIONAL REQUIREMENTS (PROPERTIES) OF MOULDING MIXTURES (SANDS)

1. *Green strength* The strength of moulding mixture after mixing of water, is referred as green strength. This is important for making and handling of the mould. The green strength and plasticity of moulding materials should be adequate so that there is no distortion or collapse of the mould when the pattern is withdrawn. The required green strength will depend on (i) degree of support from box bars, lifters and core irons, (ii) shape of the pattern and (iii) dimenssions of the pattern. Less green strength is required for a shallow core supported on a corner than in case of deep mould projection.

2. *Flowability* This is the measure of the ability of moulding material to be compacted to a uniform density. The degree of flowability required largely depends upon the method of compaction which may vary from hand ramming with tools to ramming on moulding machines and high velocity delivery on sand slingers and core blowers. High flowability is required for non selective ramming action of the moulding machines.

3. *Dry strength* This is a significant property in dry sand practice as well as in case of green sand practice. In the latter case, in two instances, i.e. (i) on standing before casting and (ii) when casting is poured, the sand adjacent to hot metal quickly loses its water. Thus when a casting is poured, the mould must withstand erosive forces and metallostatic pressure of molten metal.

4. *Hot strength* In the production of castings, a considerable layer of moulding sand rises to a temperature at which normal mechanical properties are not the criterion responsible for dimensional stability and resistance to contraction. If the moulding sand does not possess adequate hot strength, the metallostatic pressure and flow of molten metal may cause mould enlargement, erosion, cracks or breakage.

5. *Collapsibility* Collapsibility refers to the readiness with which the moulding material will yield after solidification of casting and breakdown in knockout and cleaning operations. If the heated sand becomes hard and rocklike, removal of sand from the casting will be difficult and tear or crack may be caused in the casting owing to its failure to poor response to collapsibility.

7. *Fineness* To prevent metal penetration and produce smooth casting surfaces, the mould surface should be smooth for which fine sand should be used. Since permeability as well as fineness are the functions of grain size and distribution and place conflicting demands, a compromise is generally necessary. All the measures used to achieve fineness (such as use of fine grained sands, continuous grading of sand, or incorporation of filter materials) reduce permeability. In an alternative approach to achieve fineness without affecting permeability a highly permeable moulding sand is used in which mould coatings are employed to obtain smooth mould surfaces.

8. *Thermal stability* The sand surface in contact with hot metal undergoes rapid expansion. If the sand is not stable dimensionally under such conditions, the mould surface may crack, buckle or flake off.

9. *Refractioness* This is the ability of sand to withstand high temperatures without fusion or other major physical changes. Refractioness is of greatest importance in the production of castings in high melting point alloys, particularly steels and ferrous alloys. In case of low melting point metals and alloys (e.g. aluminium and its alloys), refractoriness may be less important.

10. *Bench life* This refers to the ability to retain moulding properties on standing or storage.

11. *Durability* This refers to the capacity to withstand repeated cycles of heating and cooling in integrated sand systems. This will decide the extent of reusability of moulding sand.

12. *Thermal Properties* Moulding sand should also possess relatively good conductivity to allow dissipation of heat from the cooling casting.

The functions of moulding material can not readily be defined in terms of simple phisical properties. In general, bulk properties of complex aggregates are of greater importance. Some of these properties can be measured directly by simple tests upon sand compacts. Other properties can be assessed with the help of specially developed empirical tests in the foundry. Bulk properties and the special tests provide the basis for the control and development of moulding material qualities.

3.2 TYPE OF MOULDING SANDS

The combination of properties required in the moulding sand depends on (i) weight of the casting, (ii) composition of casting alloy, (iii), moulding practice, (iv) quantity of castings, and (v) quality requirements of casting. Combination of required properties is obtained by blending refractory sands with bonding materials, water and special additives. Broadly moulding can be classified into two categories depending upon the base sand employed, i.e. natural moulding sands and synthetic sands.

32.1 Natural Moulding Sands

These sands occur naturully and consist of sand grains associated with the clay required for bonding the sand grains. The addition of water alone develops good moulding properties. Relatively high and variable clay contents reduces refractoriness and permeability of the moulding sand.

Natural moulding sands are prepared by crushing and milling of soft rocks which breaks down the clay aggregates, and clay particles get uniformly distributed over the sand grains. Since the clay content is usually inexcess of the minimum required for desired properties, crude silica sand relatively free from clay may be mixed with natural sands.

Due to ease of availability and low cost of natural moulding sands (widely available in diff. parts of the country), are commonly used for casting of both ferrous and nonferrous metals/alloys. According to IS: 3343-1965, the natural moulding sands are classified into following three grades depending on clay content and fusion temperature:

	Pct. of clay.	Sintering Temperature °C
Grade A	5—10	1350—1450
Grade B	10—15	1200—1350
Grade C	15—20	1100—1200

These grades are further sub-divided into 11 sub-grades depending on the distribution of sand grains. The grain shape of natural moulding sands should be sub angular to round. In case of natural moulding sands an accurate adjustment of moisture is difficult. Natural moulding sands are suitable for hand moulding, particularly for light castings.

3.2.2 Synthetic Moulding Sands

These are based mainly upon silica sands containing little or no binder and the strength properties are developed by separate additions. In this case, the amount of binder required is much less than that present in natural moulding sands. Synthetic moulding sands provide greater freedom in the control of moulding sand properties. The sands are blended with clay (3-5 pct) and other additives required to impart specific properties. Synthetic sands are more refractory, require less moisture (3-5 pct.) and are widely used in steel foundry. Synthetic sands are also employed for casting of nonferrous metals where close control of properties is required.

3.3 CONSTITUENTS OF MOULDING SANDS

The effects and requirements of various constituents of molding sends are discussed below:

3.3.1 Refractory Sand

This principally consists of 50-95 pct. refractory sand which is generally silica sand. Some times special sands (zircon sand, olivine sand, chromite sand, etc.) may be used fully or partly to meet the specific requirements. The critical characteristics of refractory sand are (i) chemical composition and (ii) mechanical grading and grain shape.

(i) Chemical Composition of Refractory Sand

Most moulding sands are based upon the mineral quartz which is abundant in nature and refractory to temperatures upto 1700 °C. However, phase transformations are involved leading to volume changes at lower temperatures. Generally, silica sand (+99.8 pct.) is considered the most refractory and thermally stable. Presence of iron oxide, felspars, mica, alkali oxides etc. is undesirable as these impurities lower the fusion point in sands.

According to IS 1987-1974, high silica sands used in foundries can be classified in following three grades depending on the silica content.

Grade	*Constituents in pct.*				
	Silica	*Alumina*	*Iron oxide*	*Ca and Mg oxides*	*Alkalies.*
		max	max	max	max
A	98	1.0	1.0	1.0	0.5
B	95—98	1.5	1.0	1.0	0.5
C	90—95	2.0	1.5	2.0	1.5

For the production of special steel and other high quality castings, silica sands may not give required characteristics. In such cases special sands are employed. These include zircon sand, olivine sand, chamotte, chromite and chrome—magnesite sands.

Zircon sand ($ZrSiO_4$) possesses certain specific characteristics distinguishing it from other sands. These are (i) low thermal expansion (1/6th that of silica sand) responsible for producing castings free from scabs, (ii) chemical inertness to the action of molten metals, particularly important for alloy steels, (iii) high heat conductivity (almost twice that of silica sand) resulting in quick solidification and fine grained structure of cast metal, (iv) higher density which prevents metal penetration, and (v) high fusion temperature which prevents sand burns and permits the casting of high melting point metals. Zircon sand needs less binder and water (2-3 pct each). Zircon sand find restricted use as facing sand for steel and alig steel castings because greater quantity is require due to high density. In India, zire on sand is found in beach sands of Kerala (Quion).

Olivlne sand is orthosilicate of iron and magnisium, (Mg Fe) O, SiO_2. Only small deposits are found in India. The specifications are covered by IS 7297-1974.

Chamotte is calcined high grade fire clay which (Calcination at about 1100 °C) is crushed to obtain the required grain size. The specifications of chamotte are covered by IS 7295-1974 (Chemical composition is Al_2O_3 4O pct. min., SiO_2 30 pct min. Fe_2O_3 4 pct max., CaO and MgO 2pct max.

and loss on ignition 0.5 pct. max.). Chamotte is much cheaper than zircon and olivine. Specific characteristics of chamotte (fusion point 1780 °C, water absorption 7 pct. max., low thermal expansion and negligible affinity to liquid steel) make it valuable for heavy steel castings. A typical chamotte sand for steel castings consists of 70 pct. chamotte, 20 pct. fire clay, 1 pct graphite, and 8-10 pct, water.

Chromite sand is crushed chrome ore containing various proportions of of different minerals ($FeO \cdot Cr_2O_3$, $MgO\ Cr_2O_3$, $MgO\ Al_2O_3$ serpentine etc.). In general, the composition of chromite sand is 45-55 pct. Cr_2O_3, 14-20 pct. iron oxides and rest Al_2O_3, MgO, SiO_2, etc. Chromite sands possess high refractoriness high density, good conductivity, and nonwettability to molten metal and require very low amounts of binder (3 pct.) and temper water (2.5 pct.). These sands are particularly useful for castings of austenitic manganese steel, and in cases when chilling tendency is to be increased for controlled solidification. Chromite sands are also employed as facing sands in mould for steel castings. The specifications of chromite sands are covered by IS 6788-1973. In India chromite ore is found in Orissa and Karnataka.

Carbon sand is produced by grinding a hard and low sulphur coke to the desired grain size. Due to high refractoriness and many other desirable properties such as nonwettability of grains carbon by molten metal, dense packing, freedom from mould wall movement, good chilling power, low thermal expansion, etc., carbon sand is used as a special facing additive for complex castings. These can be bonded with clay, bentonite, cereals, pitch, etc. and are generally mixed with silica sand upto 25 pct. to prevent metal penetrations.

(ii) Grain Size and Mechanical Grading

Permeability and surface fineness of mould are most affected by grain size and mechanical grading of base sand. Coarse and graded sands give high permeability, whereas fine grained sands give low permeability and surface finish. In continuously graded sands, finer particles are disposal in the voids between the coarse particles and these decrease the permeability. Coarse and continuously graded sands result in high flowability and maximum refractoriness of mould for a given chemical composition. The grain size of base sand also influences the strength of mould. In clay bonded sands, there is an inverse relationship between compression strength and grain size.[4] Average fineness of the sand grains establishes the fineness of moulding sand. Grain sizes of foundry sands usually fall within the range of 1 mm to to 0.1 mm. The material of less than 0.1 mm size is known as clay.

The shape of sand grains may be rounded, angular, subangular or compound depending on the geological origin (Fig. 3.1). The compound grains are conglomerate of cemented particles and are unsuitable for foundry processes. The shape characteristics of sand grains can be obserbed under the microscope or measured as the coefficient of angularity. Grain shape is responsible for the flowability and strength of the moulding sand. Angular sands give steeper bulk density gradient showing a lower flowability in moulding. Some times, rounded grains are superior to angular grains, as they give narrow disparity in strength.

The grain size and shape together determine the *specific surface* of grains (Total surface area per unit mass) which gives a rough guide to the amount of binder necessary to coat the grains.

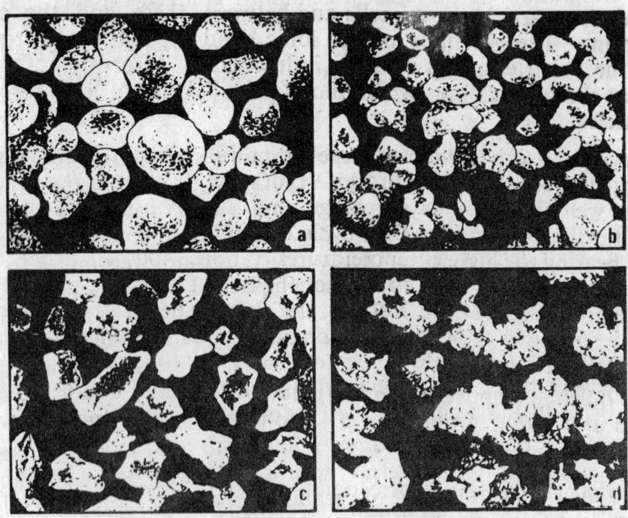

Fig. 3.1 Shape of sand grains (a) Ronded, (b) Subangular, (c) Angular and (d) Compounded

3.3.2 BONDING MATERIALS

Bonding material produces cohesion between the sand grains in the green as well as dried condition. Since the fusion point of bonding materials is usually low, the necessary strength should be obtained with the minimum possible addition of binder.

Clays and a number of other materials such as organic oils and resins, cereals, silicates, etc. possess bonding properties. The bonding agents may be used singly or in combination. Clay bonded sands can be recirculated in closed systems where the bond is regenerated by the addition of water, whereas, in case of using other binders, the action is irreversible and the moulding sand is descarded after its use.

The primary object of sand preparation is to achieve a thin and uniform coating of binder (in liquid form or very fine state) around each grain. When moulding sand is compacted, the binder coatings form lens shaped masses at the contact points of the grains and the strength of bond is dependent on the number and areas of these contacts. As the binder layers become continuous, with progressive thickness of binder layer, proportionately less advantage is achieved. Fig. 3.2 illustrates the general graph of strength with binder content. Further addition of minimum binder with adequate strength is desirable as permeability falls with increasing binder content. Important binders are described below:

(i) Clays—Clays form the second most important constituent for green as well as dry sand monlding particles. These are hydrous aluminium silicates obtained as weathered products of silicate rocks and when tempered with water, produce a plastic or semiplastic mass. Clays have the general structure form of plate or flakes in the approximate particle size range 0.01-1 μm in breadth. Net attractive forces are generated between charged hydrated clay particles or miscells and between these and the surfaces of sand grains. The strength of the ionic bond depends on the total surface area of the particles and is strongly influenced by adsorption of exchangable cations at the free surfaces which modify the balance of local forces between the particles.

On drying, adsorbed water is expelled and shrinkage of lattice is produced which srtengthens the bond further. Hydration of clays is reversible to temperatures well above the drying range and hence bond can be regenerated by addition of water after each cycle of mould production. However, progressive heating to higher temperatures (above 400 °C) expels chemically combined water and results in permanent loss of bonding property. At still higher temperatures (obove 700 °C), the clays undergo drastic mineralogical changes resulting in the crystallisation of alumina and cristobalite and finally the formation of mullite,

Based on the composition and structure, the clays used as binder can be classified into three groups, i.e. (i) kaolinite, (ii) montmorillonite, and (iii) illite.

(i) *Kaolinite* Kaolinite corresponds to the general formula Al_2O_3 $2SiO_2$ $2H_2O$ and is formed by weathering of feldspar or other aluminous minerals, Kaolinite is the principal constituent of china and ball clays and fire clay. Relatively high alumina content of these clays make them reasonably refractory (softening point 1600-1700 °C). Irreversible dehydraction occurs in the temperature range 400-650 °C. Bonding properties of these clays are not high as compared to other clays and hence higher binder content is required, i.e. 10-20 pct. Therefore, the use of kaolinite in foundry is very limited

(ii) *Montmorillonite* This can be represented by the basic formula $Al_2O_3 \cdot 4SiO_2 \cdot H_2O$, but a proportion of Al^{+++} ions are replaced by Mg^{++} ions in isomorphic substitution. Whenever Mg^{++} replaces Al^{+++} by substitution, there is capacity for adsorption of exchangeable cations such as Na^+ and Ca^{++} to which the properties of clay are particularly sensitive. The typical chemical formula can be written in the following form:

$$M^+_{0.33}\ (Al_{1.67}\ Mg_{0.33})\ Si_4O_{10}\ (OH)_2 \quad (M = \text{Na or Ca})$$

Montmorillonite is the principal mineral constituent of bentonites. These clays possess a high capacity for water absorption and exceptionally favourable bonding properties (3-5 pct clay is sufficient to derive strength properties). Use of bentonites offers the following advantages:

(a) Bentonite can be circulated in closed systems and the bond is regenerated by the addition of water.

(b) Patterns can be stripped easily and thus mould can be made quickly.

(c) Bentonite resist erosion of moulds.

(d) Volumetric contraction of bentonites helps in compensating the expansion of silica grains.

(e) Bentonites retain their capacity for water absorption upto 500-700 °C and are thermally more stable than other clays and thus have longer life in closed systems.

Depending on the kind of substitution metal present, bentonites are of two types i.e. sodium bentonite known as Western bentonite and calcium

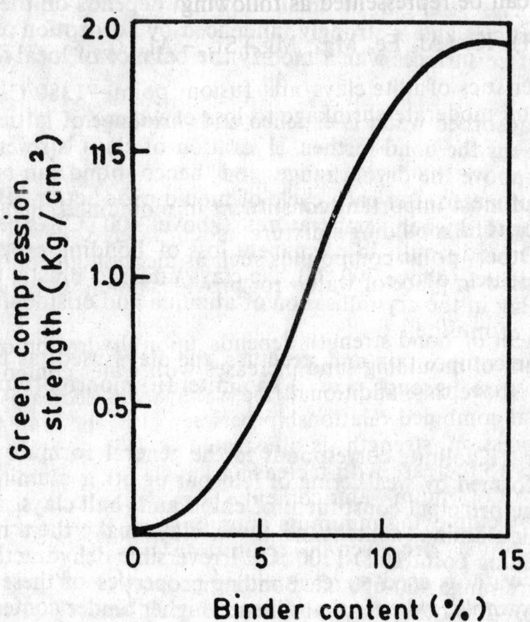

Fig. 3.2 Graph showing strength vs. binder content

bentonite known as Southern bentonite. The notable characteristics of sodium bentonites are high swelling capacity, high liquid limit, low plasticity, low green strength, and high level of dry and hot strength. The characteristics of calcium bentonites include low swelling, low dry and hot strength, low liquid limit, high plasticity, and high green strength.

Bentonites of both types are available in India. Sodium bentonites are found in Bhavnagar and Kutch (Gujarat) and Jodhpur (Rajasthan) whereas calcium bentonites are found in Rajmahal (Bihar), Bhavnagar and Kutch (Gujarat), Madras, and Kashmir. The most common bentonite used in foundries is obtained from Rajasthan. This bentonite is pale yellowish in colour and possesses relatively poor base exchange capacity. The typical composition is SiO_2—52.93, Al_2O_3—20.87, Fe_2O_3—8.80, TiO_2—2.24, MgO—3.8, CaO—0.75, alkalies—1.48, and loss on ignition—13.87. Other properties of this hentonite are: fusion point—1180 °C, pH 9.2 to 7.7 after 24 hours, flowability—78, sp. gr.—2.43 and swelling index 5.56. IS: 3021—1965 describes the specifications of two types of bentonites categorised as tope A and type B. The specifications include pH value, moisture content allowed, percentage of Ca-ions, gel index, Na_2O/CaO value, liquid limit, green compression strength, and swelling index.

(iii) *Illite clays* These are produced by the weathering of micas and are principal source of bond in the natural moulding sands. Though the swelling of these clays is poor, they give reasonable strength properties. Irreversible dehydration occurs in the temperature range 500—550°C. The chemical formula of illite can be repressented as following

$$(OH)_4\ Ky\ (Al_4\ Fe_4\ Mg_4,\ Mg_6)\ Si_{8y}-Al_y)\ 0_{20}.$$

Other characteristics of illite clays are fusion point—1380°C, moderate base exchange, and moderate shrinkage to loss of water.

3.3.3 Water

Water is the third most important constituent in moulding sand as it bonds the sand and clay. The moulding mixture is considered to be sand-clay water system. Other polar compounds such as alcohol, glycol, acetic acid etc. can also be used in place of water to provide bonding between sand and clay.

The development of bond strength depends upon hydration of clay and thus green strength of moulding sand increases with water content upto an optimum value. Above this, additional free water again decreases the green strength. A typical combined relationship between clay and water contents and green compression strength is illustrated in Fig. 3.3. However, dry strength continues to increase with increasing moisture content. Therefore, by determination of optimum water content, the required strength properties can be achieved by employing minimum amount of clay. Such relationships for various types of clay are given in literatures.

In some cases, water is used to react with specific reagent and result in bonding. For example, when ethyl silicate is used, water gives hydrolised bonding by the formation of amorphous silica gel and ethyl alcohol. Similarly, in CO_2 process, CO_2 used forms a weak solution of carbonic acid which decomposes the sodium silicate to form silica gel bond.

3.3.4 Additives.

Besides the above-menticned three ingredients (refractory sand, clay, and water), other materials known as additives are added to moulding and core sands to improve some specific properties (gas permeability, refractoriness, yieldability, collapsibility etc.). The important additives used in foundry practice are discussed below.

(i) *Cereals* Cereal binders are mainly starch and dextrine made from corn. Starch is made by separating the starchy portion from corn by dry or wet milling followed by cooking to gelatinate the starch and finally grinding to required size. Dextrines are produced from corn starch by acidifying with hydrochloric acid and heating in large steam jacketed roasters. This treatment makes the starch soluble in cold water. Sometime dextrose (crystalline corn sugar) is also used. The cereal binders develop a gelationus bond with water and are normally employed along with other binders, e g. in clay bonded moulding sands and in oil and resin bonded core sands. In moulding sands the cereals contribute to dry strength, green strength, and collapsibility. An additional advantage is increased resistance of green sand moulds to friability on air drying. Addition of cereals results in strong moisture retention which delays air drying of sand mixtures and improves bench life. Use of cereal also eliminates scabbing and other expansion defects.

In core sands, cereal binder gives improved green strength and collapsibility with ease of knockout. However, they increase gas evolution and

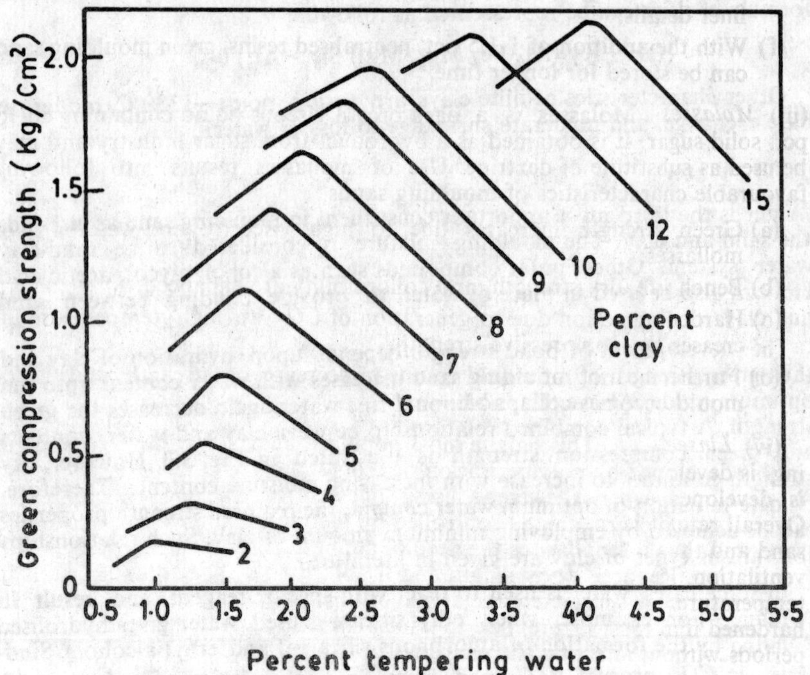

Fig. 3.3 Typical relationship between clay and water contents and green compension strength

decrease the resistance of dried cores to the absorption of moisture.

Dextrine is more effective than starch and it gives the same strength with less moisture. Specifications of dextrine are covered by IS: 4269—1967.

Sulphite lye (by product of pulp industry) may also be used to improve dry as well as hot strength and collapsibility of moulds.

(ii) *Resins* Resins are high melting point gums produced by chemical alterations of rosins (low melting point gums obtainable from pine trees) or produced synthetically. Synthetic resins are of two types, i.e. (a) *furans* which are urea formaldehyde resins modified with furfuryl alcohol in varying proportions, and (b) phenolics which ara phenol formaldehyde modified with urea in varying proportions. In general, furans are more stable even in warm weather, where as phenolics are cheaper and are resistant to heat and moiture. A large number of resins made in different ways are available suited for particular jobs. Resins are not soluble in water and also do not swell in water. When resins are mixed in the moulding sands, following improvements are obtained.

(a) good surface finish of casting.

(b) improved appreciably the dry strength properties.

(c) improved frictional properties.

(d) decreased toughness and brittleness of green moulding san.l.

(e) increase flowability of sand and improved dispersion of fines (bentonites, etc.) throughout the sand mix and thus moulds are produced with finer details.

(f) With the addition of 1-1.5 pct. neutralised resins, green moulding sand can be stored for longer time.

(iii) *Molasses* Molasses is a dark brown viscous liquid containing 60-70 pct. solid sugar. It is obtained as a byproduct from sugar industry and may be used as substitute of dextrine. Use of mollasses results into following favourable characteristics of moulding sands.

(a) Green strength increases due to high viscosity and wettability of mollasses.

(b) Bench life, dry strength, and collopsibility are enhanced.

(c) Hardening action due to generation of CO_2 at high temperature increases hot compressive strength.

(d) Further heating of mould results in decrease of strength and hence mould becomes collapsible.

(iv) *Linseed oil* This is the most popular binder for cores. The hardening is developed by polymerisation or cross-linking. The necessary strength is developed when the cores using lineseed oil are heated to 200-240°C. Overall rate of hardening is controlled by the rate of oxygen supply through sand and thus it emphasises the need of permeability, core venting, corestove ventilation. Reduced dry strength and friability result due to excess baking temperature. A specific characteristic of linseed oil is the resistance of its hardened film to moisture, due to which cores can be stored for prolonged periods without affecting the properties.

(v) *Sodium Silicate* This is the most important inorganic binder which is mainly used with carbon dioxide hardening for facing sands, block moulding, and cores. SiO_2—Na_2O—H_2O system forms a series of substances having varying viscosities depending upon SiO_2—Na_2O ratio (most common molecular ratios are 1.5 and 3). When CO_2 is passed through the sodium silicate bonded sand, carbonic acid is produced in aqueous solution. This results in a rise in ratio of SiO_2—Na_2O and formation of colloidal silica gel giving the hardening effect due to formation of bond between sand grains. The reaction may be represented as follows.

$$Na_2\,SiO_3 + CO_2 \rightarrow Na_2\,CO_3 + SiO_2$$

Though the reaction is simple, the chemistry of the process is complex. In the early stages, the reaction proceeds at a fast rate, but the compression strength of a given sand attains a maximum at some critical amount of gas passed. If gassing is continued beyond this point, the strength of the bond decreases. Further, a given amount of gas causes higher strength when passed at low pressure and for a longer time than under the reverse conditions. Hardening effect also continues on standing after completion of gassing. Thus high strength can be most economically achieved by undergassing and ageing.

The inorganic bond has no inherent collapsibility and bence retained strength of binder products may be quite high. Therefore, to achieve the breakdown of the sand mixture, separate additions such as coal dust, cereals, pitch, wood flour, clay or glucose are used. It has been shown that upto 12.5 pct CO_2 sand can enter a system without affecting much the mechanical properties of reconditioned sand. However, when moulding practice is based fully on sodium silicate bonded sand, reclamation of sand is not feasible.

(vi) *Special additives to moulding and core sands* Other substances are invariably added to impart some specific properties such as improved resistance to specific defects, surface finish, collapsibility, etc. There is a large number of such additives and the important ones are given below.

S. No	*Sucbstance*	*Purpose of addition*
1.	Pitch or coal tar available as by product of coal carbonisation.	To improve hot strength of moulding sands and surfacefinish of castings.
2.	Asphatt available as by product of petroleum	-do-
3.	Ethylene glycol	To enhance bench life and resistance tc drying out.
4.	Iron oxide	To improve hot strength.
5.	Silica flour	To improve hot strength, surface finish and resistance to metal penetration.
6.	Sulphur	To inhibit metal-mould reaction.
7.	Boric Acid	-do-
8.	Ammonium bifluoride	-do-
9.	Coal dust	To improve surface finish and resistance to metal penetration.
10.	Saw dust and wood flour	To improve collapsibility and resistance to expansion defects
11.	Fuel oil	Used in small amounts (0.01 to 0.10 pct) to improve mouldability.
12.	Graphite	Used in small amounts 0.2-2 pct) to improve mouldability of the sand and surface finish of the castings.

3.4 TESTING AND CONTROL OF MOULDING SANDS

Control over the properties of moulding and core sands is of great importance for the quality of catings. In order to exercise proper control there is need for systematic evaluation of moulding sand qualities under foundry conditions which require a variety of tests to be carried out. Some of the tests are concerned with basic chemical and physical properties but most of the tests are designed to measure bulk properties which are sensitive to small variations in mixing conditions and specimen preparation. Hence rigid standardisation is required at all stages. In order to have better reproducibility, two or more determinations should be carried out and systematic techniques for collection and storage of samples. are used to minimise error. The various tests performed are discussed below

3.4.1 Frequency of Sampling

Bulk tests should be carreied out on standard specimens compacted on samples taken from production batches. The frequency of sampling should not be less than two times an hour in case of high-volume production foundry (using sand more than 100 m^3/h). In foundry consuming sand less than 100 m^3/h the samples should be taken at hourly intervals. At the foundry producing small castings and consuming sand less than 50 m^3/h the frequency of sampling should not be more than twice in an hour. The samples of sand are taken from each batch after half the batch has gone off the mixer, whereas backing sand is tested two or three times, the samples taken at the site of moulning.

Core sands are usually sampled two times per shifs. The samples are taken when half the batch has been discharged from the mixer. In case of continuous type mixers, the samples are taken at an interval of 2 hours Moulding and core sands which harden in the moulding box, should be checked two times every shift. The amount of sample drawn is not less than 1 kg which should be kept in sealed containers.

3.4.2 Standard specimen for Bulk Testing

Most bulk properties are tested on specimens compacted to a bulk density similar to that obtained in a propery rammed mould. The tests utilise the A.F.S. standard cylindrical specimen having the dimensions 5 cm dia and

Fig. 3.4 Standard rammer for production of specimens

5. cm height. The standard specimen is prepared by subjecting a weighed quanty of sand to a selected number of blows (3 times) from a standard rammer (Fig. 3.4), having a weight fall of 6.35 Kg (14 pounds) from a height of 5 cm. The weight of sand is adjusted to produce a specimen having close tolerance (diameter $d = 50 \pm 0.8$ mm). The rammed specimen is expelled from the tube on a stripping post. The standard specimen so produced may be used for various tests such as permeability, green and dry strength, hardness, etc. Behaviour during ramming may be used as critrion of flowability.

3.4.3 Strength Tests

Strength is determined under compression, tensile, and transverse loading of green, dry, and baked sands. Compression test in the green strength range can be carried out on a simple, manually operated, spring loaded machine as shown in Fig. 3.5. However, most strength tests are performed on universal sand strength testing machine as shown in Fig. 3.6. The machine is usually provided with the provision for changing the strength range. But the use of alternative types of specimen holder, any type of the strength test can be performed. Hydraulic machines of higher capacity may be used for very high dry compression strengths. For green strength tests newly made specimens are used, whereas for dry strengths specimens baked at 105-110°C are used. In case of core of sand testing, the specimes are hardened in a standard manner as required for a particul?r bonding material.

Fig. 3.5 Simple sand testing machine

Fig. 3.6 Universal sand testing machine

For *compression tests,* the standard specimen (5.08 cm dia × 5.08 cm height) axially loaded through flat faced holders as shown in Fig. 3.7 (a). For shear test, compressive loading system is modified to provide off set loading of the standard specimen (Fig. 3.7b). In general, the results of shear test are closely related to those of compression tests. *For tensils tests,* a special waisted specimen (Fig. 3.7c) is loaded in tension through a pair of grips. For *transverse test,* a plain rectanqular specimen is employed which is supported on knife edges at the ends and loaded centrally to fracture (Fig. 3.7d). Tensile and transverse tests are usually performed on high strength core sands for which conditions are more relevant to the stresses encountered in cores during handling and casting. The transverse stregth is found by the following formula

$$M = 3\ \frac{WI}{2bd^2}$$

where M = modulus of rupture, W = total load at which failure occurs, I = distance between supports, b = width, and d = thickness of test piece.

3.4.4 Flowability of Moulding Sand

Generally, moulding aggregates do not posses a high standard of flowability due to frictional and other forces which assist in bonding. Therefore, flowability tends to bear an inverse relation to strength properties. This is nonstandard testing as several tests have been developed for flowabilty test which are based on (i) compaction of standard cylindrical test specimen, (ii) extrusion of sand through an aperture in a tube under standerd load (iii) measurement of hardness differentials on the sloping base of a modified cylindrical tube mould and (iv) incidence of *supervoids* in the surfaces of compacted specimens.

The simplest and most promising method is the dietert method which uses the standard rammer with a fixture attachment. This method is based on the movement of moulding sand between the fourth and fifth blows of the rammer head in the preparation of a standard test specimen. This movement is assigned an arbitrary scale of values, zero travel corresponds to maximum flowability, i.e. compaction of highly flowable material is fully accompalished by the earlier blows.

Data on flow properties can also be derived from studies of the influence of ramming upon bulk desity distribution. To assess flowability, longitudinal

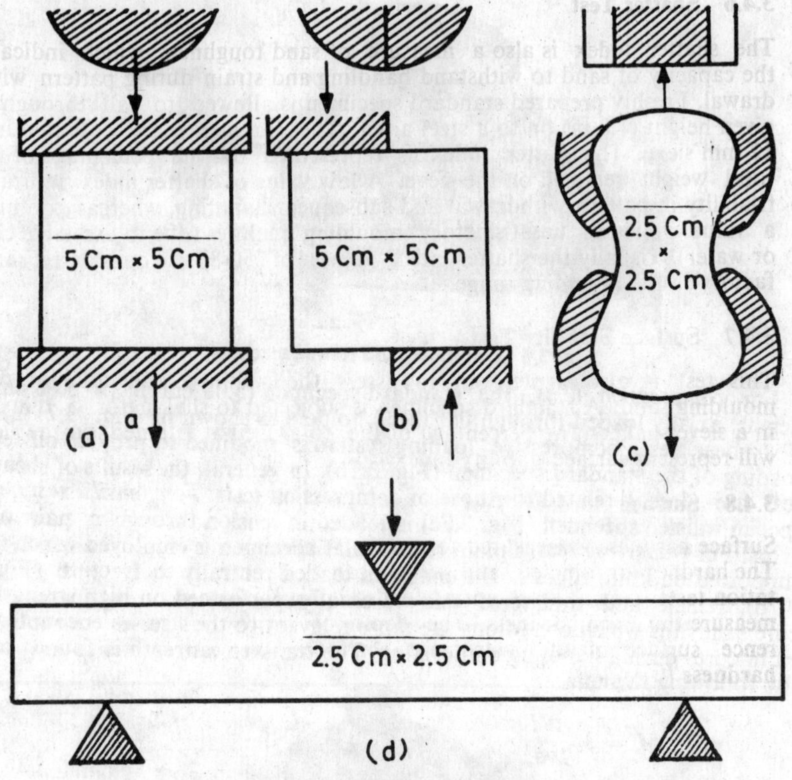

Fig. 3.7 Loading of standard specimen (a) for compression test (b) for shear test (c) for tensile test (d) for transverse test

bulk density gradient is determined by progressive dismantling of the sectional core box (in which sand is compacted by steady pressure) and weighing the successive segments of compacted sand. The slope of the bulk density gradient is inversely related to flowability. This is influenced by a number of sand characteristics such as grain shape, grading, etc.

3.4.5 Deformation and Toughness

The deformation or plasticity is measured from the decrease in length of tho sand specimen during the green compression test just before the failure of the specimen. The property can be related to capacity of moulds to withstand hydrostatic pressure and to the contraction of casting. The product of green compression strength and deformation is used as an index of toughness.

Sand toughness number $= d \times S$

where $d =$ deformation in mm

$S =$ green compressive strength in kg/mm

3.4.6 Shatter Test

The shatter index is also a measure of sand toughness which indicates the capacity of sand to withstand handling and strain during pattern withdrawal. Freshly prepared standard specimen is allowed to fall through a given height (1.8 m) on to a steel anvil. The fragments are collected in a 12 mm sieve. The shatter index is represented by the percentage of the total weight retained on the sieve. A low value of shatter index indicates friability in pattern withdrawal and subsequent handling, whereas too nigh a value indicates unsatisfactory moulding qualities (due to excessive clay or water). Usually the shatter index values of 50-85 are considered satisfactory in the moulding range.

3.4.7 Surface Stabtlity Test

This test is also carried out to assess the friability and toughness of moulding sand. A standard specimen is subjected to shakingfor 5 minutes in a sieve shaker. The percentage of the total weight retained on the sieve will represent surface stability index.

3.4.8 Surface Hardness Test

Surface hardness is measured to check the ramming density of moulds. The hardness of rammed sand surface can be directly measured by an indentation tester (Fig. 3.8). A spring loaded spherical indentor may be used to measure the green hardness. The depth of penetration from the flat reference surface of the instrument corresponds to an empirical scale of hardness (0 – 100).

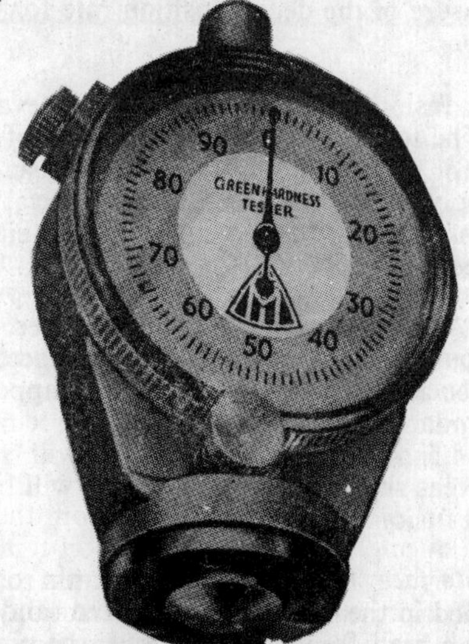

Fig. 3.8 Indentation tester for sand hardness

3.4.9 High Temperature Properties

High temperature tests are relevant to conditions under which a mould or core is used in the foundry. The various high temperature properties to be tested include hot strength, hot deformation, collapsibility, expansion and refractoriness. These are described below:

(i) *Hot strength* For hot strength testing, a cylindrical test specimen measuring 2.86 cm diameter × 5.08 cm long $\left(1\frac{1}{8}\text{ in dia.} \times 2\text{ in}\right)$ is used. For hollow specimen, this has in addition a bore of 1.27 cm. The test measures the compressive load required to fracture the specimen after certain soaking period (usually 20 min) at the testing temperature. Sintering of clay bond shows a marked strength peak which reaches at a certain temperature above which the strength again decreases due to the formation of liquid between the sand grains. The use of dielectric heating eliminates the effect of prolonged soaking. For this purpose, a special furnace called dilato meter is used in which the specien is heated to a test temperature uniformly and rapidly. For the measurement of retained strength, application of the load can be delayed until the specimen has cooled.

(ii) *Hot Deformation* During the measurement of hot strength, longitudinal deformation of specimen is measured. The total deformation to failure is taken as the criterion. *The hot deformation rate* (deformation/load) is also employed as an index of the behaviour of mould at some fixed load (usually below half that producing failure).

(iii) *Collapsibility* The hot strength test is unsuitable for core sands using organic binder as the decomposition of binder causes rapid collapse on loading. The time to failure at a particular load and temperature can be determined as a measure of the decomposition rate for an organically bonded sand.

(iv) *Expansion* Expansion behaviour of sand is relevant to certain types of casting defects. The total longitudinal expansion of specimen (of same dimensions as used for hot strength test) can be measured at the test temperature without applying any significant load. Special apparatus is available for expansion test in which the specimen may be either free standing or confined within fused silica tube.

(v) *Refracteriness* A cone fusion test may be used as a guide to high temperature behaviour of bonding clays. The refractoriness of a compacted sand mass may be assessed by observing the appearance and dimensional change of standard specimen (5.08 cm dia × 5.08 cm height) fired for 2 hours at 1550 °C. A refractory specimen will retain its shape with slight expansion, wereas shrinkage and distortion will be caused due to excessive sintering and fusion.

(vi) *Mould-metal interface reaction* A metal pin of 1.25 cm dia × 1.9 cm length is embedded in the centre of a standard sand specimen. The sample is then heated in a high frequency unit and the pin is allowed to melt at controlled rate. After cooling, the mould metal interface can be observed for the reaction.

(vii) *Metal penetration test* This test has been standarised by AFS committee on the principle that metallostatic pressure is the main reason of metal penetration. The test can be carried out by varying sprue heights of a standard mould assembly and the relative resistance to metal penetration may be estimated.

3.4.10 Permeability Test

The permeability number of sand may be defined as the volume of air in mililiters per minute passing through a specimen of 1 cm length and 1 cm^2 cross sectional area under a pressure difference of 1 cm water gauge. Green permeability is determined by measuring the rate of flow of air through a compacted A.F.S. standard cylindrical specimen (5 cm dia × 5 cm height) retained in its ramming tube. The permeability meter shown in Fig. 3.9 consists of a graduated bell of 2000 cm^3 air capacity over water. A tube from the air enclosure communicates directly with the specimen tube wich is placed over a mercury seal. The air is allowed to escape through the specimen with the descent of the bell and the time required for 2000 cm^3 of air to pass is determined. Permeability number, P, can be given by the following formula

$$P = \frac{Vh}{atp}$$

where V = volume of air, cm^3 = 2000 cm^3

h = height of specimen, cm = 5 cm

a = cross sectional area, cm^2 = $\frac{\pi(5)^2}{4}$

= 20.268 cm^2

t = time, min

p = pressure difference, cm water

= 10 cm in standard apparatus.

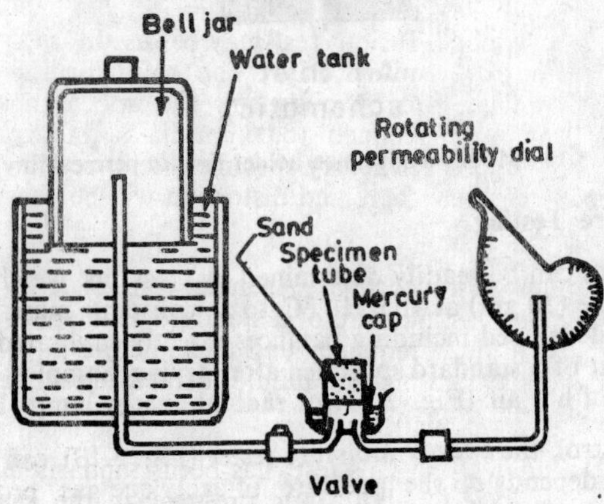

Fig. 3.9 Permeability meter

Therefore, using the standard apparatus and technique the permeability number may be derived directly from the formula

$$P = 3007.2/t$$

where t is expressed in seconds.

For routine control of sand permeability, a rapid measurement may be made, the principle of which is shown in Fig. 3.9 (b). For this, air is made to flow under a constant pressure of 10 cm of water through an orifice into the open end of the specimen tube and out through the sand specimen. The pressure in the specimen tube (orifice discharge pressure) can be related to the permeability (pressure will be maximum i.e. 10 cm of water for completely inpermeable specimen, and pressure will be zero for completely permeable specimen). Hence, the orifice discharge pressure (measurable by manometer) may be caliberated in terms of permeability of the sand specimen. Permeability number can be directly read with the help of a calibrated sector scale which can be rotated until it intersect the meniscus of the water column in manometer.

Dry permeability can be determined by using a modified specimen holder of the type as shown in Fig. 3.10. In this a mercury seal is used between the dry specimen and the walls of the holder.

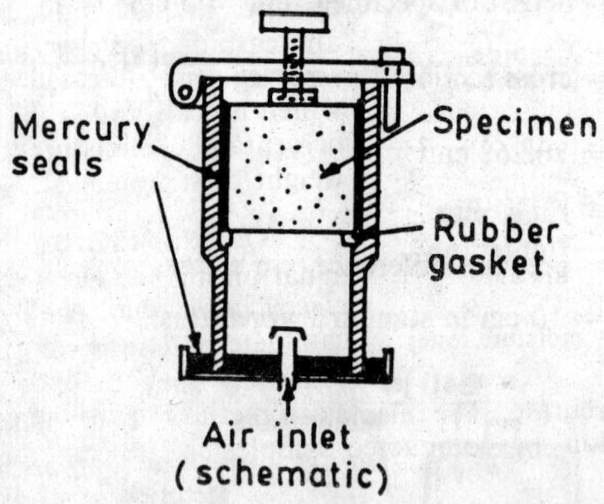

Fig. 3.10 Modified specimen holder for dry permeability

3.4.11 Moisture Testing

Moisture content can be readily determined by loss in weight on drying a weighed sample (50 gm) at 105-110 °C to a constant mass. Many rapid aids have been developed including balances directly calibrated to read the moisture content of a standard specimen after drying. Samples can be dried by the passage of hot air (Fig. 3.11) or radiant energy lamp (Fig. 3.12).

For shop control, the *speedy* moisture tester (Fig. 3.13) can be used. The working of this depends on the pressure of acetylene gas produced by the following reaction of calcium carbide with the moisture present in the sand.

$$CaC_2 + H_2O = Ca(OH)_2 + C_2H_2$$

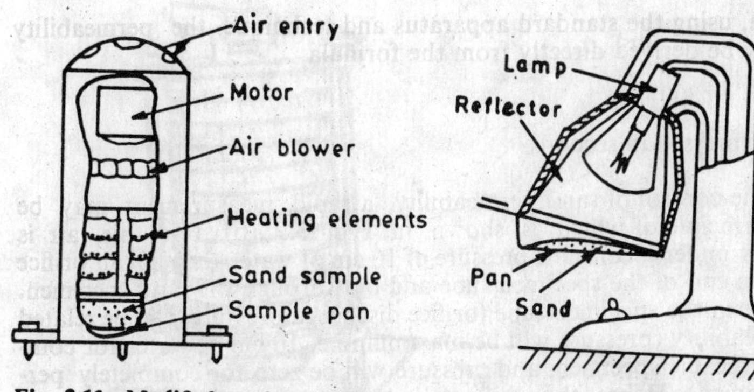

Fig. 3.11 Modified specimen holder for dry permeability

Fig. 3.12 Radiant energy lamp for moisture testing

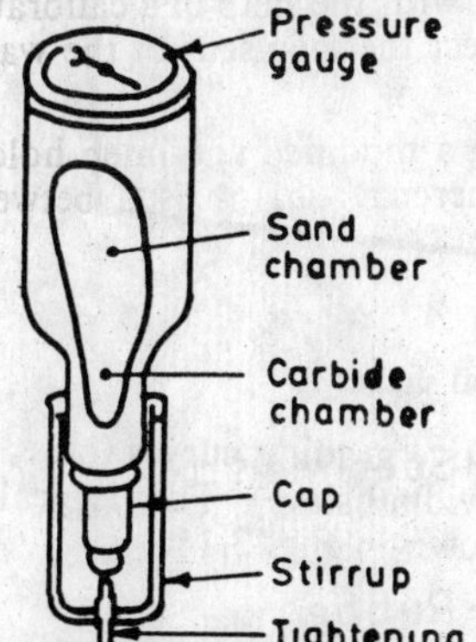

Fig. 3.13 Speedy moisture tester

A weighed quantity of sand sample is taken in the cap of a metal flask fitted with a pressure gauge. An excess of the calcium carbide is placed in the flask body. The flask is closed, clamped and shaken to mix the sample and carbide. The resulting pressure of acetylene is indicated on the gauge which is callibrated directly in the percentage of moisture for a standard weight of sample.

Other methods of moisture determination include *electronic probe* and *electrical method*. The former method can determine the moisture content of loosely heaped sand, although, the assessment is not accurate. The electrical method using measurements of microwave absorption in compacted samples can be employed in mechanised system sands.

3.4.12 Mechanical Grading (Grain Size) Test

Grading of sand is based on the use of standared sieves to separate selected fractions for weighing. The first step is the separation of clay and moisture from the sand under test. Clay is separated by boiling the sample in dilute ammonia followed by cooling and siphoning off the resulting suspension in repeated rinses. The sand residue is dried, weighed, transferred to a uest of sieves and mechanically shaken (Fig. 3.14) for a definite length of time. The nest of sieves consists of standard sieves (280 mm dia) having one above the other with varying but known number of meshes. The coarsest sieve (usually 2.5 mm) is placed on the top and finest (0.1 mm) at the bottom. The sand retained on each sieve and the bottom pan is weighed and the percentage of each fraction in the total sample determined. The percentage of these fractions can be represented by direct plots of frequency distribution

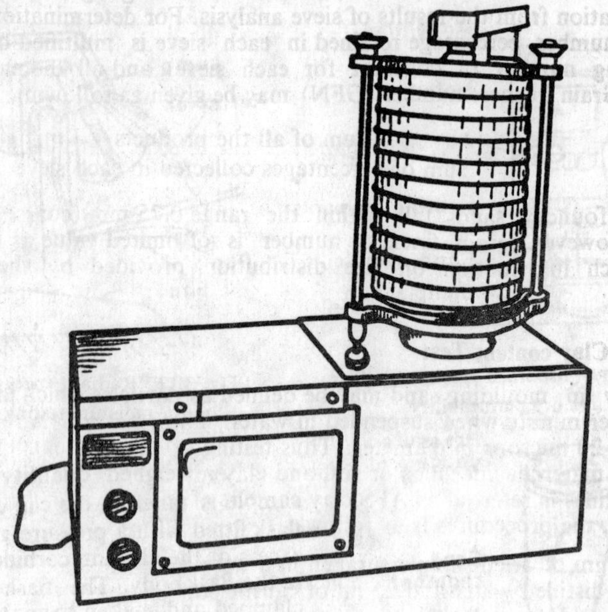

Fig. 3.14 Mechanical shaker

(retained sand on each sieve) or by cumulative grading curves (based on the total weight of material coaser than individual sieve). The typical curves for both the types of representation are shown in Fig. 3.15.

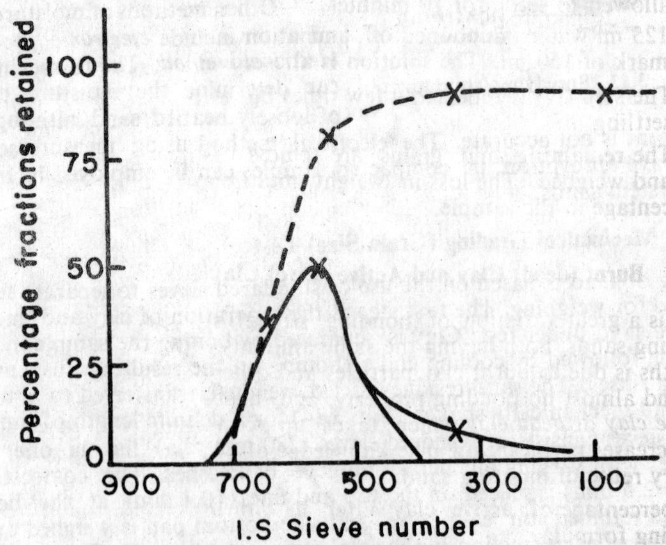

Fig. 3.15 Representation of direct and cumulative grading plots

The grain fineness number is a measure of average grain size derived by calculation from the results of sieve analysis. For determination of grain fineness number, percentage retained in each sieve is multified by its own multiflying number (a constant for each sieve), and all the products are added. Grain fineness number (GFN) may be given as following

$$\text{GFN} = \frac{\text{sum of all the products}}{\text{sum of percentages collected in each sieve}}$$

Most foundry sands fall within the range 0.75 mm (coarse)-0.15 mm (fine). However, grain fineness number is of limited value as it does not give much information on size distribution provided by the complete analysis.

3.4.13 Clay content Test

The clay in moulding sand may be defined as particles which fail to settle 2.5 cm per minute when suspended in water. These particles are generally less than 20 microns in diameter. Thus testing of clay content includes all very fine material, fine silica or silt, and clay minerals. The total percentage of these fines is referred as AFS clay content of the sand. For clay determination, the procedures is as follows:

(i) 50 gm of dried sample is taken in a wash bottle or jar and to this 475 ml distilled water and 25 ml of caustic soda solution (25 gm per litre) are added.

(ii) The wash bottle is securely covered and sealed and then is agitated for 5 min. in mechanical stirrer or shaker so that solution is shaken vigorously.

(iii) After shaking, it is diluted with water to a height of 150 ml (mark of bottle) taking care of particles adhering to sides. The solution is allowed to settle for 10 minutes.

(iv) 125 ml water is siphoned off, and water is again added to the original mark of 150 ml. The solution is allowed to settle for 10 minutes.

(v) The step (iv) is repeated a few times but allowing only 5 minutes for settling.

(vi) The remaining sand grains are removed from the wash bottle, dried and weighed. The loss in weight multiplied by 2 gives the clay percentage in the sample.

3.4.14 Burnt (dead) Clay and Active (Live) Clay Test

There is a great variation of bonding strengths between fresh and used moulding sands, both having the same amounts of clay. This variation in strengths is due to burnt clay particles and other fines with little active surface and almost no bonding property and hence burnt clay is referred as *inactive clay* or *dead clay* which takes up extra water without being useful but decreases permeability. The knowledge of dead clay is essential for the foundry reuse of molding sand.

The percentage of active clay after its *n*th use can be found out by the following formula.

$$\text{Active clay (per cent)} = \frac{A\,(100 - D)^n}{100} + B \times \frac{1 - \left(\dfrac{100 - D}{100}\right)}{1 - \left(\dfrac{100 - d}{100}\right)}$$

where A = initial bentonite percentage,

D = sand lost or removed in per cent after each cycle,

d = sand burnt in per cent during each cycle, and

n = number of cycles of use.

Active clay can also be estimated by nickel sulphate test and methylene blue test. The former method is based on the absorption of nickel ions by the active clay in a solution of nickel sulphate. The excess nickel is estimated by the addition of dimethyl glyoxime and measuring the intensity of colour by spectrophotometer.

In a modified method of methylene blue test, tetra sodium pyrophosphate solution is used in place of sulphuric acid. The procedure may be summarised as below.

5 gm sample of sand to be tested is taken in a 250 ml stainless steel beaker to which 50 ml of 2 per cent tetra sodium pyro phosphate solution is added. The mixture is mixed thoroughly and then a standard methylene blue solution is added to the beaker through a burette till the end point is reached (a blue green halo surrounding the blue circle is indicated by end point, otherwise only blue circle exists). The active clay can be estimated with the help of calibrated charts prepared from samples of known active clay contents.

The difference of total clay (Section 3.4.12) and active clay will give the amount of dead clay. In order to find out the water requirements of the moulding sand, contents of both dead clay and active clay should be determined.

3.4.15 Other Tests on Moulding Sands

Many new tests are developed and devised to evaluate behaviour of sand under actual casting conditions. These include tests for resistance to erosion and gas evolution from cores. In addition, sand testing is frequently used for comparison of base sands and bonding materials. Such tests are particularly useful in the development of new mixtures or as a check on incoming materials. The various tests of importance include determination of pH value of sand and gelling index, swelling capacity and liquid limit of clay. The details of these tests are given in literature.

3.5 MOULDING MIXTURES

Moulding and coremaking practice is generally based upon mixtures of the refractory sand, bonding materials, water and special additives (discussed under section 3.3.4). A wide range of sand mixtures are employed in the different sections of the foundry. In practice, the main division is between natural and synthetic sands. The general trend is towards increasing use of synthetic sands, particularly in mechanised foundries, due to their favourable characteristics.

3.5.1 Sand Mixtures for Green and Dry Moulds

In steel foundries, synthetic sands are most widely employed. Relatively coarse grained sands are bonded with bentonite and cereals to impart the required properties. Details of typical green facing sand and synthetic dry sand mixtures used in steel foundry are given below:

	Green sand mixture	Dry sand mixture
Composition	Medium silica sand 5 per cent bentonite 0.75 per cent starch	50 per cent coarse silica sand, 50 per cent medium silica sand. 6 per cent ball clay 3 percent bentonite, 0.5 per cent dextrin.
Moisture content	3.0 per cent	6.0 per cent
Permeability	160—180	100—110
Green compression strength	0.5 kg/cm^2 5.0 kN/m^2)	(about 0.6 kg/cm^2 (About 60 kN/m^2)

In the case of dry sands, relatively less permeability serves the purpose, and hence use of mixed gradings and fillers is more common. In order to develop maximum strength, dry sands are worked at higher moisture contents.

In cast iron and non-ferrous founding, the casting temperatures are lower and the rate of gas evolution is greatly reduced, and thus finer sands can be used with relatively low permeability (25—90 being common). The green strengths usually range 0.35—0.7 kg/cm^2 (about 35—70 kN/m^2). The distinctive feature of sands in the cast iron foundry is the addition of about 2 per cent coal dust to give good surface finish. The coal used should contain high volatile matter to give a reducing atmosphere at the metal—mould interface. Another feature of cast iron foundry practice is the reclamation of a high proportion of floor sand.

In case of aluminium and copper alloys both refractoriness and permeability are of secondary importance and hence naturally bonded sands are still in common use. However, for magnesium alloys synthetic sands containing reaction inhibitors are the main moulding sands due to strong tendency of metal—mould reaction.

3.5.2 Sand Mixtures for Cores

Irrespective of alloy being cast, core sands are usually designed for high permeability by the use of uniformly graded sands. Other notable characteristics of core sands should be low green strength and high dry strength. However, now a days, distinction between moulding and core sands is diminishing and many of the materials originally used as core sands are finding wide application in moulding, particularly in the event of high standard of accuracy. A typical core sand mixture for general purpose with its properties is given below.

Composition	100 per cent medium silica sand
	1.5 per cent starch
	1.5 per cent core oil
	2.5 per cent moisture
Permeability	200

Green compression strength 0.15 kg/cm^2 (about 15 kN/m^2)

Dry compression strength 75 kg/cm^2 (about 7 MN/m^2)

3.5.3 Other Moulding Materials

Apart from silica sand based moulding mixtures, there are some other materials employed for moulding in special circumstances. The materials required for very heavy steel castings should possess exceptionally high refractoriness, volume stability and mechanical strength and for this purpose a material traditionally employed was a bonded aggregate mode from crushed used refractory products with addition of fire clay and graphite. A later development is the use of calcined aluminous clay (known as chamotte) in the crushed and graded form and bonded with raw clay. Moulds made from these materials are finished with thick slurries of fine refractory material and finally dried at high temperatures. In the past, these materials were extensively used in the production of heavy steel castings, but now a days, these are replaced almost entirely by dry sands and other types of refractory sands (zircon sand, chromite sand, etc.).

Zircon sand can be bonded with bentonite for use as a facing material, or with core binders to use it as ramming mixtures for confined pockets of cores in positions which are susceptible to metal penetration. Crushed olivine sand is particularly suitable for resistance to slag reaction in the casting of high manganese steels. Chromite sand has also been widely used as a steel foundry facing material, particularly in Russia.

In addition to high refractoriness and low thermal expansion, non-siliceous sands offer the advantage of moderate chilling effect which is valuable for the production of sound surface regions in long freezing range nonferrous alloys.

3.5.4 Cement Bonded Sands

A distinctive moulding practice known as Randupson process uses cement as a binder. A silica sand based moulding mixture with about 11 per cent and 6 per cent water exhibits quite different characteristics compared to those of normal moulding mixture. This mixture has low green strength but develops high dry strength on setting. After hydration of the cement, the excess water is allowed to evaporate and the final air dried mould is strong, permeable, and highly, refractory. Setting and drying take several days. Reclamation of the sand is limited as the setting reaction is irreversible. After removal of fines by crushing and screening, the sand can be used as a weaker backing mixture with the addition of new cement. Hand as well as machine moulding can be used and the process has found limited application in ferrous as well as non-ferrous foundry practice.

3.5.5 Loam Moulding Mixtures

The constituents of loam moulding mixtures vary widely and usually include fine sands, finely ground refractories, clays, graphite and fibrous reinforcements. These materials are mixed with upto 20 per cent water to form a thick slurry of highly plastic consistency. The slurry is coated to an underlying structure of bricks, and steel. The finished moulds are thoroughly dried before pouring the metal.

3.6 MOULD SURFACE COATINGS

Mould surface coatings are applied to sand mould to provide smooth and inpervious surface which prevent metal penetration into intergarnular voids in the mould face. The materials used are known as mould wash, mould coating, spray, blacking, etc. The use of surface coatings offer benefits such as better surface finish, case of cleaning the casting, and less surface defects.

A coating may be applied either as a dry solid or as a wash. In the latter case, refractory particles are suspended in a liquid medium (suspension and bonding may be assisted by the addition of other substances). The common materials used for surface coatings and rammable facings include silica, alumina, magnesia, chromite, zircon, talc, graphite, plumbago, blacking, sillimanitc, chamotte, molochite, etc.

Silica flour and other refractories are used for the high melting point alloys, whereas talc and alumino-silicate are mainly used for non-ferrous alloy of low melting points. Carbonaceous coatings are preferred in iron foundry due to their reducing effect at the metal mould interface, and inhibition to wetting of the mould material and metal penetration.

Mould washes are usually applied by spraying, swabbing, brushing, or painting of wet materials and dusting of dry materials. The specific gravity and viscosity of the suspensious should be controlled to a consistency suitable to the mode of application. In general, a layer of 0.5 mm thick is desirable to prevent metal penetration, but much thicker coatings may be required for very heavy castings. Thicker layers can be obtained by using successive applications of coating (chammotte paints and blackings are usually applied in this way).

The liquid medium is evaporated to consolidate the point coating. Hot air, gas torch or infra-red heating can be employed for drying of water based paints. The paints can also be applied to green moulds before stoving. In case of green sand practice, coatings are applied by dusting powders on to the mould surface. Alternatively, volatile suspending liquids such as isopropyl alcohol based coatings can be applied to green as well as dry moulds. The surfaces can be dried by ignition as the isopropyl alcohol is inflammable.

The coating applied should dry without cracking or peeling to give hard, tough layer, otherwise local failure of mould coating may result in metal penetration into the underlying sand. The use of oils, sulphite lye or molasses in moulding sand having excessive amounts of bentonite or cereals alone, avoids the drying cracks. Cracking tendency diminishes with increasing dry strength of the underlying sand.

3.7 PREPARATION OF MOULDING AND CORE SAND MIXTURES

Though, the properties of moulding sand largely depend on the ingredients, correct quality of sand mixture can be arrived only by proper conditioning and preparation of sand which helps to develop the properties in the following ways:

(i) Moisture and binder are evenly distributed in the sand mixture and around the sand grains.

(ii) Sand grains are separated by aeration resulting in increased flowability of sand.

(iii) The foreign particles are removed from the sand.

3.7.1 Preparation of Naturally Bonded Sand

Naturally bonded sand having acceptable texture requires short period of milling and adjustment to achieve the required moisture content (generally 5 per cent moisture for green sand). After removal of all metallic particles and burnt sand from the used sand, it is reconditioned. In case of aluminum casting, contamination and deterioration of sand is not much and hence it may be sufficient to pass the sand through an aerator (Fig. 3.16). Regular control tests are performed to check the accumulation of fines and silt which will result in low green and dry strength and low permeability. The facing sand may require the addition of fresh materials to increase the refractoriness and texture of the mould surfaces.

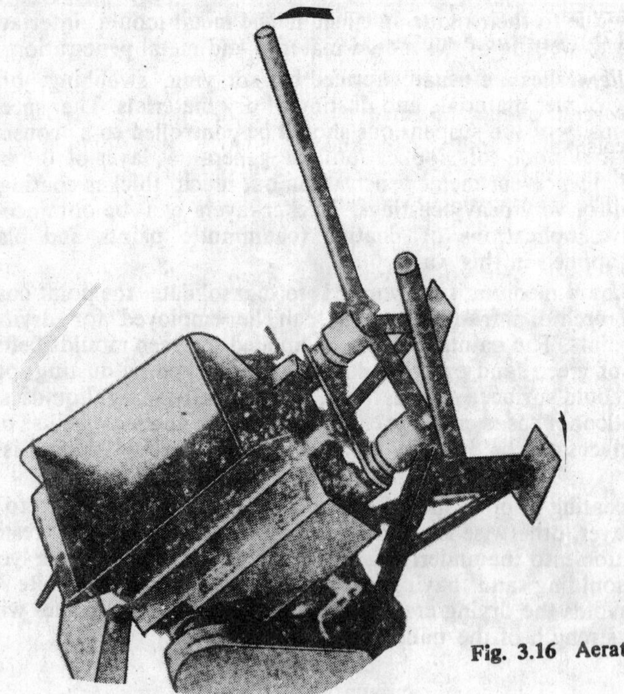

Fig. 3.16 Aerator

3.7 2 Preparation of Synthetically Bonded Moulding Sands

The oldest and simplest method of moulding sand preparation is by hand shovel which involves pouring of water over a heap of used or new sand followed by systematic working of moisture through out the mass. One of the first improvements on hand mixing was the use of screen (fore runner of modern aerator) through which the sand was passed before use. In early days, hand milling was successful because only naturally bonded sands were used. Synthetic sands widely used today cannot be properly tempered by hand mulling because bentonite clays require considerably more energy to develop plasticity than the fire clays associated with natural sands An improvement over hand mixing was the use of a rotary sand kicker which straddled long heaps of sand moistened with water and the mass was mixed with vigorously rotating blades of the machine.

Distribution of clay and water can not be accompalished by the gentle squeezing action of rollers which progressively expose more clay flakes to the adsorption of water. Modern sand mixers or mullers designed to mix the sand ingredients efficiently, are basically of following two types

(i) *Mullers having thick rimmed, heavy metal wheels* In this type, the wheels rotate with their axes horizontal and their full weight resting on the sand. The muller is provided with alternate rollers and ploughs for continuous agitation. A typical mixer of this type is shown in Fig. 3.17. Such mixers can be provided with automatic loading skips and discharge directly into the sand distribution system. Several variations of roller action mullers are manufactured mainly using the principle of either moving rolls and siaitionary pan (Fig. 3.17a) or rolls mounted rigidly with the pan rotating (Fig. 3.17b). For these mullers the mixing times are normally 3-6 minutes.

(ii) *Mullers having ligthter, thinner rubber covered wheels* In this type, the wheels spin rapidly ovre and through the mass with their axes vertical. The sand is stirred vigorously against the vertical sides of the muller shell. This type of muller is known as high speed mill which illustrated in Fig. 3.18.

Fig. 3.17 Sand Muller: (a) Stationary pan (b) Rotating pan

This can meet the demand for mare rapid batch mixing in mechansised systems. In this mill, circulation of sand is assisted by ploughs.

In some scases circulation of sand can be supplemented by an air blast which also provides cooling when processing hot sand. In this case due to high speed of milling, the preparation of sand, is completed in 1-1.5 minutes.

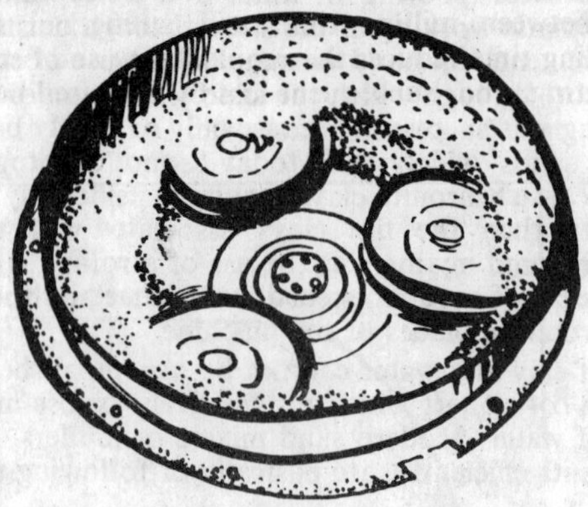

Fig. 3.18 High speed muller

Mullers may be supplemented by aerators designed to deliver the final mixture in light condition and free from lumps. Aeration can be achieved by power operated riddles or by passing the sand stream over toothed belts. This facilitates uniform density in ramming.

3.7.3 Preparation of Core Sands

The proportions of various constituents of core sand depend on various factors such as type of core, metal to be cast, method of core making, avilability and cost of the materials and equipment available for drying and curing. A general purpose core sand mixture may consist of 4.5 per-cent molasses (to provide green strength), 1-1.5 percent linsed oil (to provide dry strength) and balance dried silica sand.

The mixing of core sands is relatively simple. Oil, resin and other liquid binders can be distributed easily by the simple action of ploughs and rubb- ing plates. A typical core sand mixer is shown in Fig. 3.19. The mixtures using cold setting binders have short bench life and in such cases paddle or screw type mixers are used in which the constituents of the mixture are not brought together until ready for discharge at the moulding point.

Fig. 3.19 Core sand mixer

3.7.4 Effect of Milling Time on Propert'es of Moulding Sand

Milling should be carried out to ochieve complete and uniform distribution of bonding agent and water throughout the mixture.

Better mixing is caused by first running the mill containing sand and bento nite for some time before water addition. In the batch mixing of sand, pro petries improve as the ideal distribution of binder is approached. A general relationship between milling time and strength is shown in Fig. 3.20. Excessive milling time reduces the mouldability and bench life and leads to frictional heating and subsequent loss of moisture on standing.

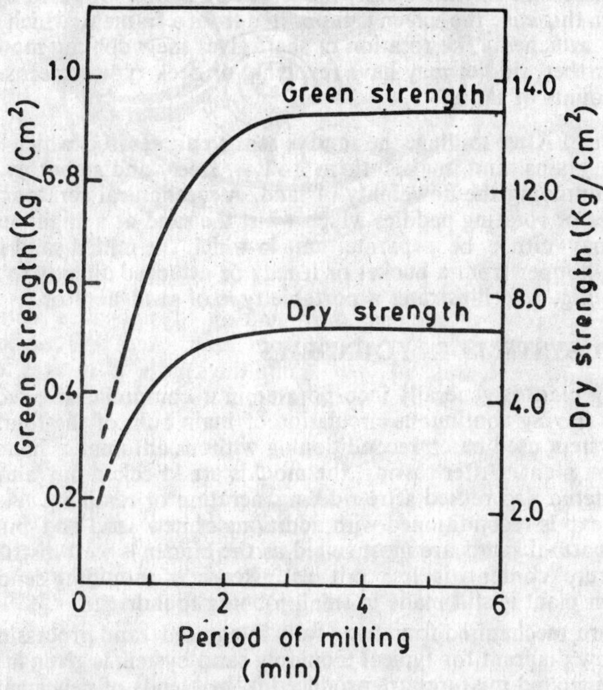

Fig. 3.20 General relationship between milling time and strength

3.7.5 Auxiliary Equipments in Sand Preparation

The equipments used to assist the performance of muller and to include convey the lump free sand to moulding section, the main equipments used include (i) magnetic separator, (ii) riddle and (iii) aerator.

(i) *Magnetic Separator* The raw sand as well as moulding sand coming from the shake out station should be freed of all iron particles, which otherwise may damage the machines used in sand preparation. The magnetic separator removes the iron pieces, wire nails, iron shots, etc. from the sand. The magnetic separator used for this purpose is simple and consists of a magnetic pulley over which a flat conveyor belt rolls. As the belt bends round the pulley, the sand slides off the belt, while magnetic particles go on moving with the belt due to magetisation effect and drop off only when the the magnetic effect ceases.

(ii) *Riddle* After the removal of iron particles, the sand is passed through the riddle or screen which separates out the lumps of dry sand cores, hard lumps of moulding sand, and other large size nonmagnetic pieces. The riddle may be either hand operated or mechanically operated which can screen the sand at much higher rate then hand riddles. Mechanical riddles can be operated by compressed air or electric motors. In the former case, the riddle consists of an air cylinder, a reciprocating piston and a screen connected to the end of the piston. The reciprocating action of the piston provides the screening action, whereas in the latter case, the screening action is caused by gyratory motion. In this case the screen is supported from a frame to which an electric motor is attached. The rotation of shaft gives the wobbling motion to the screen. Further, riddles may have revolving or deck type secreens to handle larger amounts of sand.

(iii) *Aerator* After milling, the sand is sent to an aerator which separates the sand grains and makes them to flow freely and smoothly. Aerating helps in improving the flowability of sand. A mechanical aerator consists of an impeller or rotating paddles which whirl the sand at a high speed. The aerator may either be separate unit in which the milled sand is charged through a hopper from a bucket or it may be attached directly to the outlet of muller. Eig. 3.16 illustrates a portable type of sand aerator.

3.8 SAND SYSTEMS IN FOUNDRIES

The mixing plant is generally incorporated in a comprehensive reconditioning system having continuous circulation of main bulk of the foundry sand. The old system used batch reconditioning without automatic handling and reclamation plant. After casting, the moulds are knocked out and the floor sand is watered and reused aftre riddling, aeration or remilling. At intervals the sand heap is reconditioned with additions of new sand and binder. For practice, natural sands are most suited as the binder is well distributed and this moisture content is less crit ical. Reuse of moulding sand without reclamation plant is still made in small jobbing foundries.

In modern mechanised foundries, fully integrated sand processing plant is used. A flow diagram for typical moulding sand system is given in Fig. 3.21. Closely controlled mixtures are produced from blends of new and reclaimed sand, binders and other additions. Sand systems of this type can be fully automated for the entire handling, mixing and distribution cycle.

3.9 RECLAMATION OE MOULDING AND CORE SANDS

The process of reclamation of the used sand consists of magnetic separation, disintegration, screening, and desilting. The processed sand is sent to mixing plant for the next cycle. The necessity of reclaiming the used sand is for the following reasons;

(i) To prevent pollution of the surrounding medium due to dumping of used sand.

(ii) To have more rational utilization of valuable sand deposits.

(iii) Properly reclaimed sand is as good as new sand in its performance.

(iv) Inventories and storage requirements are less.

(v) The troublesome work of unloading sand is reduced as the storage of reclaimed sand is automatic.

(vi) Sand is available at constant moisture content and of consistent quality.

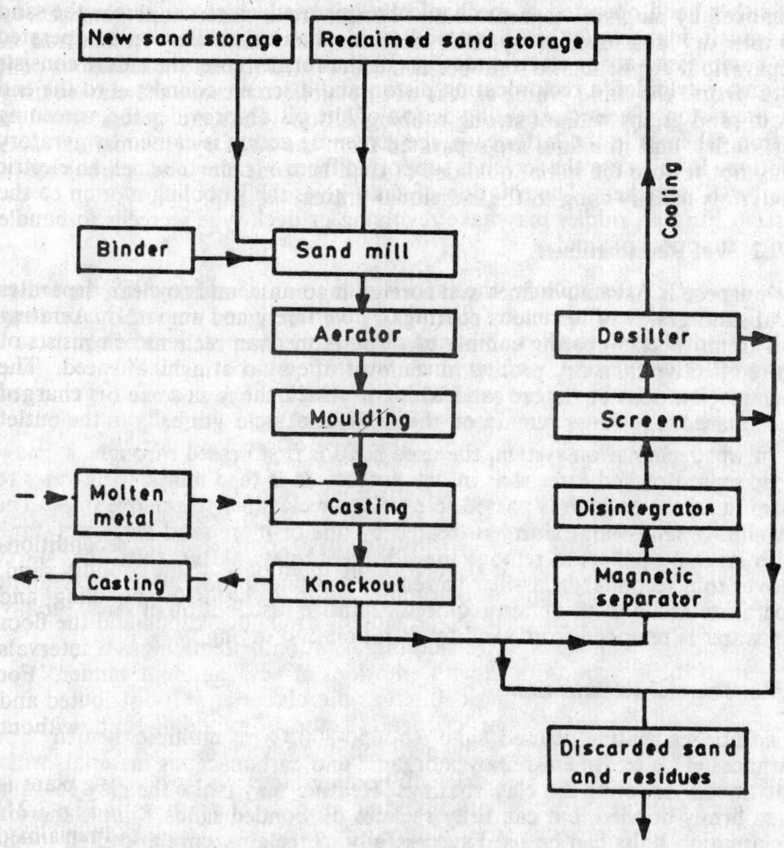

Fig. 3.21 Flowdiagram for typical moulding sand system

Intense heat at and near the mould-metal interface results in drying, burning, and baking of materials surrounding the sand grains into a hard, tightly adhering coating. Some of the sand grains may react with iron oxide at the surface of the casting to form fayalite (2 $FeO \cdot SiO_2$). Further, active clay is rendered inactive due to expulsion of crystallization water and this material adds to the fines which contribute nothing, rather require additional clay to bond these fines. There is also an inevitable accumulation of iron and steel and iron oxide particles of various shapes and sizes. Hence, the conglomerate mixture will not be suitable for reuse at the mould-metal interface due to (i) uneven composition and (ii) reduced refractoriness. The impurities reduce refractoriness and the fines reduce permeability. At the casting interface, always a new sand or reclaimed sand should be used. Reclamation of san l may be of different types, i.e. dry reclamation, wet reclamation, thermal reclamation, and combined system.

3.9.1 Dry Reclamation

This removes fines and iron oxide particles. Iron and steel particles are first removed by magnetic separators and lumps are crushed and screened. A simple dry reclamation system performs its function by screens and air separation. A pneumatic scrubber is incorporated to remove tenacious coatings from the sand grains as it is not possible to remove the clay coatings from sand grains withuot strong rubbing action. The sand grains travel to an outlet and tbe fines are separated from the air. Dry reclamation alone may not restore the initial qualities of sand as some clay and carbonaceous materials usually cling to the individual grains.

3.9.2 Wet Reclamation

Its purpose is to remove fines and foreign materials, and to clean the individual sand grains of tenacious coatings. A well designed wet reclamation unit can remove clay coating completely. Therefore, wet reclamation process is more effective than dry process in removal of clay coatings. However, wet reclamation does ot restore sand to its original quality as some organic and carbonaceous residues remain on the individual sand grains.

In wet reclamation system, the used sand is first passed through a magnetic separator and screens as in dry process. It is then mixed with water to form a slurry which is passed to a primary classifier to remove fines. The remaining sand-watar slurry is treated by one or more sand scrubbers (tank with large propeller) to remove the adherent clay. After this, the slurry moves to a secondary classifier to remove the fines produced in the cleaning operation and also to obtain required grainsize distribution of sand. Finally, the water is removed from sand by filtration and drying.

3.9.3 Thermal Reclamation

It comprises heating of used sand to 650-800°C in multiple hearth type furnaces. This treatment removes organic and carbonaceous materials without much affecting the clay coatings. Heating may make the clay coatings even firmly bonded, but can fully reclaim oil bonded sands. Simple thermal reclamation units can be used successfully to reclaim core and shell mould sands. Thermal reclamation should also be combined with screens and classifiers to ensure proper gain-size control of the used sand. However, thermal reclamation alone is not sufficient for most shops as it does not remove clay.

3.9.4 Combined System of Reclamation

In general, one single system of reclamation does not give the sand of original properties. The benefits of two or more systems can be realised and it is possible to reclaim used sands to their original quality by combining thermal reclamation with wet reclamation. First the used sand is treated by wet reclamation system and then after classification and dewatering, the sand enters the multiple hearth furnace fori the thermal treatment. The sand is finally cooled, reclassified and stored.

4

Moulding and Core Making Processes

4.1 INTRODUCTION

The mould forms the outside, whereas core forms the internal shape of the casting. The casting can be produced by a number of methods such as sand casting, permanent mould casting, centrifugal casting, etc. The sand casting techniques are most widely used and are flexible due to: (a) production of castings in almost all castable metals and (b) production of castings in sizes ranging from few grams to more than a hundred tonnes. Sand casting involves: (i) making a sand mould by ramming a moulding sand mixture (mixture of sand grains, clay, water and other additives around a pattern, (ii) removal of pattern from the sand, (iii) making the gating arrangement, and (iv) filling the mould cavity with molten metal. The various moulding materials are already discussed in chapter-3.

Since discussion of all the methods of metal casting will involve a good coverage, matter has been divided in two chapters i.e. chapter 4 dealing with only sand processes and Chapter 5 dealing with permanent mould castings.

The sand moulding started before 4000 B.C. and from that day to present, many methods of producing sand castings as well as special methods have emerged. However, the same basic principles involved in sand casting still hold good.

Sand moulding processes may be classified as following:

(i) Green sand moulding
(ii) Dry sand moulding
(iii) Skin dried sand moulding
(iv) Bench moulding
(v) Floor or pit moulding
(vi) Loam moulding
(vii) Shell moulding
(viii) CO_2 moulding
(ix) Air set or self curing moulding
(x) Investment casting and
(xi) Special methods of sand casting.

Sand moulding may also be classified as: (a) hand moulding and (b) machine moulding, depending upon the way of mould preparation. For hand moulding a number of hand tools are required.

4.2 HAND MOULDING AND TOOLS EMPLOYED

Hand moulding is the oldest method of mould prodution and relies mainly on the skill of the moulder. All the moulding operations such as ramming of sand, placing and withdrawing the pattern, turning over of mouldidg boxes etc. are performed manually. The moulding sand is placed around the pattern and compacted by the use of hand rammer. The finish and repair of the mould is carried out with the help of a number of tools. The main advantage of hand moulding is that the moulder can control the direction in which he applies the force to the sand to achieve a better degree of compaction on vertical faces of the pattern. However, the overall standard of hand ramming is not so high. Hand moulding suffers mainly due to low rate of production (as the hand ramming is time consuming). Production rates can be increased to some extent by the use of pneumatic hammer but the degree of compaction may vary at different places. In hand moulding a number of tools are required which are discussed below:

(i) *Shovel* A shovel is used to cut, mix, temper and handle the loose moulding sand. A shovel with a flat blade is often more convenient to let the sand slide off the end. This is particularly true when bench moulds or moulding flasks are to be filled with moulding sand.

(ii) *Riddleor foundry sieve* The riddle is used to remove lumps, shot iron, nails etc. from the sand placed next to the pattern or joint. The riddle has wooden frame with a screen of suitable mesh made of brass or galvanised iron. Power riddles are used for handling large volumes of sand.

(iii) *Rammers* Rammers are used for evenby and quickly packing the sand in the moulding box to make the mould sufficiently strong and dense. Rammers usually have two types of shapes separately or both in one. One is known as *peen rammer* (or peen end) which has a conical or wedge shape and other is known as *butt end* which has a round and flat shape. The peen end is useful to ram the sand in pockets and corners and hence it is used first. The butt end is used for final pressing of the sand after the moulding box (flask) is completely filled with sand. In the case of large moulds, pneumatic rammers are employed which save time and labour.

The pneumatic rammers are available in varying sizes. Large size rammers are used for ramming loam moulds in pits, medium size rammers for ramming large dry sand moulds, and small size rammers for ramming medium dry sand moulds and green sand moulds. Pneumatic rammers have the interchangeable butt and peen attachments, but in general butt ramming is employed.

(iv) *Trowels* A trowel consists of a steel blade (rectangular, triangular, or heart shaped) and a wooden handle at one end. Trowels are used to: (a) fill the mould (b) cut the ingates, (c) make the joints or partings (d) repair the mould, and (e) finish the flat surfaces of the moulds.

(v) *Slicks* These are double ended tools having one end flat (square) and a spoon on the other end. These are further designated by the shape of the blade and the width of the widest blade. Examples are heart and leaf, leaf and spoon, heart and square, spoon and bead, etc. (Table 2.2). These are used for rapairing and finishing the mould suface after the pattern has been with-drawn.

(vi) *Cleaners and Lifters* These are made of steel having one end flat and straight whereas the other end is bent at 90°. These are used to clean (to lift the sand particles) and finish the bottom and sides of deep and narrow openings in the moulds. The straight end helps in repairing and cleaning work, whereas the bent end helps in lifting sand particles.

(vii) *Gaggers* These are steel wires or rods (120-50 cm in length) bent at one end or both ends. These are used to reinforce the sand especially in the top portion of the moulding box (cope) and to support the hanging portion (unsupported) of the sand. The gaggers should be rough and coated with clay to provide good adherence of sand to them.

(viii) *Bellow* It is used to blow off loose sand particles from the pattern and mould cavity.

(ix) *Dust bag* It is used for dusting parting compound or plumb ago on the parting surface.

(x) *Brush* A brush is used to sweep away the parting sand and dust from the pattern or mould joint. Brush can also be used to apply the plumbago on the mould surfaces and to smoothen the mould surface by applying a paint.

(xi) *Swab* The sawb is a small brush made of hemp fibres. Swabs are used to moisten the sand around the edges of the sand about a pattern before its withdrawal from the mould. A floor swab should have a good body of fiber shaped to a point. A swab should be handled carefully because too free use of water around the mould may result in blow holes. A bulb swab used for the bench work consists of a rubber bulb to hold the water, a brass tube and a camel's hair brush on the end of the tube.

(xii) *Vent wire or rod* It is a steel (preferably of spring steel) wire piece having one end pointed and the other end fitted with a wooden handle. It is used to pierce small holes in the rammed sand connecting the mould cavity with outside air to allow the easy escape of water vapour and gases produced during pouring of molten metal.

(xiii) *Smootheners and corner slickers* These are employed to repair, finish and smoothen corners, edges, recesses bends etc. and are made in a wide variety of shapes (square corner slick, half round corner slick, pipe slick, etc.) and sizes. These are mainly used on dry sand and loam work.

(xiv) *Draw spike and draw screws* It is merely a steel rod with a loop at one end while the other end is either screw (draw screw) or pointed. It is used to withdraw the pattern from the mould. The pointed or screw end is driven into the top face of the pattern and then raised gently. In case of metallie pattern, an internal thread is cut in the pattern face which is engaged by the thread end of the draw spike. Wooden patterns may also have a lifting plate having a threaded hole in its top face.

(xv) *Rapping plate* It is made with nnthreaded separate holes, into which a pointed rapping bar (draw spike) is placed when rapping the pattern. It preserves the threads used for the draw-screw.

(xvi) *Runner pegs (sprue pin), sprue cutters and gate cutter* Runner pegs are made of wood having cylindrical, rectangular or square shape. These are used to form the sprue in the cope. Sprue cutter is a tube of steel which is forced into the rammed cope to trepan a hole for the sprue. A sprue cutter performs the same function as sprue pin. A sprue pin is placed in the cope before ramming the sand whereas a sprue cutter cuts a sprue af ter the cope has been rammed. A gate cutter is a shaped metal piece (usu ally U shape) and is used to cut the gate.

(xvii) *Straight edge or strike off bar* It is a wooden or metal bar of rectangular section and bevelled at one edge. It is used to remove the excess sand from the top of the rammed moulding box to provide a level surface.

(xviii) *Spray Gun* It is used to spray coatings on the mould or core.

(xix) *Sprigs and nails* These are used to: (a) reinforce thin projections of sand in moulds (b) fasten cores in moulds, and (c) reinforce sand in front of an ingate.

(xx) *Spirit Level* It is used to check the level of moulding box, i.e. whether it is horizontal or not.

4.3 MOULDING BOXES OR FLASKS

Moulding boxes or flasks are rigid frames of rectangular, square, round and irregular shapes (Fig. 4.1) made of iron, and aluminium alloys. Cast or welded steel flasks are considered to be better than iron boxes because they

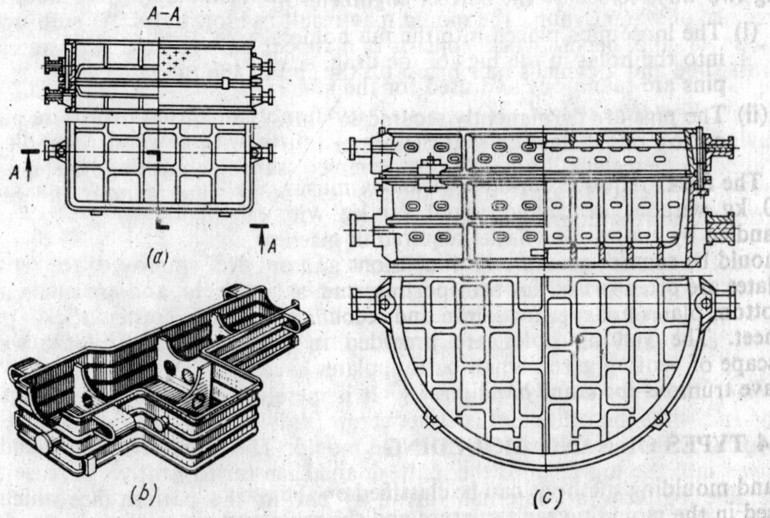

Fig. 4.1 Types of Moulding Flasks: (a) rectangular, (b) shaped, (c) round

are stronger. The moulding boxes impart rigidity and strength to the sand rammed in it. The flasks protect the moulds against damage during assembly, transportation and pouring.

The mould is generally prepared in two half flasks, the upper half is called the *cope* and the lower half is known as the *drag*. (Fig. 4.2). The surface of half flasks along the parting line is planed to secure a tight joint between the cope and drag. Some times the flask may be made in three parts and in this case the middle part is known as cheek.

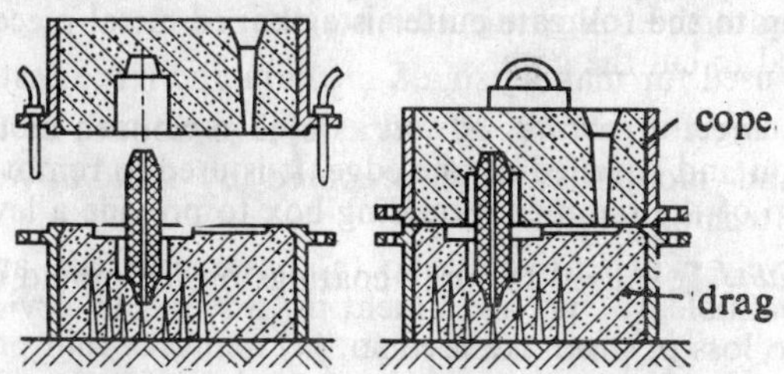

Fig. 4.2 Mould prepared in two half flasks

The grips are provided to small flasks for ease of transportation and handling during moulding, whereas lifting trunions are provided to large flasks for crane transportation. The flask walls are also provided with vent holes to facilitate the eseape of gascs evolved during pouring of molten metal into the mould. In large size flasks, ribs or cross bars are provided to hold the sand in place.

The location of the half moulds should be secured correctly which can be achieved by providing pins and holes in the lugs of flasks. There are following two ways to secure the correct alignment of half moulds.

(i) The loose pins placed into the pin holders of the cope are inserted into the holes in the lugs of the drag. After assembling the mould, the pins are taken out and used for the alignment of other half moulds.

(ii) The pins are permanently secured in the drag, and the cope is placed on top of the drag so that pins enter into the holes in the cope.

The flasks required for *hand moulding* must be as light as possible (about 30 kg without sand and about 60 kg with sand). The large flasks are handled by cranes. The flasks required in machine and automatic moulding should be rugged, accurate in dimensions and inter changeable. The bottom plates are oftenly used for transporting and assembling of moulds. These bottom plates are cast from iron and secondary alloys or welded from steel sheet. The venting holes are provided in the bottom plate to allow the escape of venting gases. Small battom plates have hand grips and larger ones have trunions for crane handling.

4.4 TYPES OF SAND MOULDING

Sand moulding methods can be classified on the basis of the sand ingredients used in the moulding sand mixture and the moisture content of the mould. These are green sand moulding, dry sand moulding, skin dried sand moulding, loam moulding, self hardening moulding, etc. These are discussed below.

4.4.1 Green Sand Moulding

Green sand moulding refers to the sand moulds made with moist sand mixture (composition discussed in chapter 3) and the moisture is present at the time of pouring the molten metal. Green sand moulding has been used for several decades and throughout that priod of time, the basic ingredients of the sand mixture have not altered significantly. Sand properties require ments have become more stringent and the tendency to use a fully synthetic sand has been added. Green sand moulding is by far the most popular and widely used sand moulding process (more than 90 pct. sand castings are produced) owing to the following advantages.

(i) It can be used for making small, medium and large castings.

(ii) It can be used for jobbing as well as high repetition moulding.

(iii) Green sand moulds can be prepared by hand as well as machine moulding techniques

(iv) It is relatively in expensive due to (a) basic materials are cheep and readily available, (b) no investment in stoves for drying operation, and (c) no loss of production time.

(v) The sand used can be reconditioned easity since there a is little dehydration of the clay bond.

(vi) Green sands have low compression strength and offer little resistance to contraction and thus there is less risk of hot tearnig.

(vii) Moulds joint closely and thus leave a little flash for removal by fettling

(viii) Moulds may be produced, poured and knocked out in a continuous cycle of short duration. This makes green sand moulding easily adoptable to mechanised and automated system.

However, green sand moulds suffer from the following disadvantages:

(i) Green sand moulds are not much strong and may be damaged during handling.

(ii) Due to the presence of moisture, the mould possesses low permeablity which can cause casting defects such as blow holes and pin holes.

4.4.2 Dry Sand Moulding

Dry sand mould refers to the sand mould made with a sand and a binder which does not require moisture to develop strength. The details of sand mixture used for dry sand moulds are given in chapter 3. The moulds are prepared in the identical manner as green sand moulds. The water content in the dry sand mix is kept high (6-8%) to give satisfactory green properties. The surface of green mould is sprayed with molasses water and then the mould is dried in an oven at 200-400 °Cuntil all the moisture is eliminated. Alternatively, the mould can be dried by circulating the hot air through it. Dry moulds offer the following advantages;

(i) Dry sand moulds are stronger and can be handled with less risk of damage to their shapes than green sand moulds.

(ii) Moisture related problems (low permeability) of green sand moulds are absent.

(iii) Dry sand moulds offer maximum resistance to distortion under metallostatic pressure and mould erosion and thus these are suitable for the castings of the lowest dimensions with high standard of accuracy.

(iv) In dry sand moulding, sands of lower permeability can be used which give impervious mould surfaces and thick mould coatings can be satisfactorily dried out. These result in better surface finish of casting.

(v) Moisture control is less critical as at the higher level of moisture, some moisture may evaporate on standing without loss of moulding properties

(vi) In dry sand moulds there is less tendency of surface chilling which facilitates metal flow in thin sections.

(vii) Moulds can be allowed to stand for some time before casting without the risk of drying out and surface friability.

Howeves, the costs for dry sand moulds run higher due to the cost of special binder and the equipment needed to dry the moulds. Thus the use of dry sand moulds is limited to the production of defectless large castings. Dry sand moulds are unsuitable for light and intrcate castings due to higher resistance of moulds to cooling contraction. This problem may be over come by the use of chills.

4.4.3 Skin Dried Moulds

This process refers to the drying of surface layer of the mould to a depth of about 25 mm by the use of gas torch, hot air or blowers fired with oil or gas. This pratice is intermediate between green sand and full dry sand moulds in advantages and disadvantages. A prior treatment of mould surface with sprays of water or dilute binder solutions can be given to assist surface hardening. The surface drying to a lower depth can also be achieved with the aid of inflammable mould coatings. A skin dried mould should be poured immediately after drying, otherwise moisture from green backing sand may penetrate the dried skin making it ineffective.

4.4.4 Air Dried Moulds

These moulds resemble skin-dried moulds in the sense that their skin is dried, but not by hot air or torch. The green moulds are allowed to stand in atmospheric air for certain period as a result of which moisture from mould surfaces gets evaporated resulting dried skin of mould. Large pit mould get dried in this manner, since they remain exposed to atmosphere for quite some time during and after their preparation.

4.4.5 Loam Sand Moulds

The loam sand mould is made of porous bricks which are cemented together with loam morter (a mixture of sand grains and clay in equal amounts and to the consistency of mud). The inner side of the brick structure which forms the rough contour of the castings is given 6-12 mm thick facing layer of loam sand. It is then finished with a strickle (sweep boad) to obtain the required shape. Finally the mould is dried by torches or hot air. The loam moulding is most suited for making large, heavy castings of a circular shape, e.g. large cylinders and paper machine rolls. In loam sand moulding, skeleton pattern or templates can also be used. This eliminates the need for pattern and thus reduces the costs of production considerably. However, skilled craftmen are required to make the loam moulds since the work is to be carried out by hand.

4.4.6 Cement Sand Moulds

This is one of the earlier moulding processes using chemical bonding. This was introduced in 1930's and is also known as the *Randupson Process.* The mould is prepared from a mixture of silican sand (80-90 pct) with an addition of cement (6-12 pct) and the required quantity of water 48 pct). This does not require hand ramming of the mixture. This produces a very rigid mould due to the setting action of cement and breakdown of the mould after casting is difficult.

In order to increase the setting rate of the sand, accelerators may be added. However, it is a normal practice to withdraw the pattern from sand mould only after its little setting and leave the mould for a minimum of 24 hours before pouring the molten metal.

Cement sand moulds offer many advantages such as ease of ramming, high dry strength of the hardened sand (3 MNm^{-2}), high permeability, superior surface finish and more dimensional accuracy, higher rigidity and capability to withstand metallostatic pressure and possibility of reclamation and reuse of sand. The process however, has the following limitation:

(i) Its use is limited to the production of large castings such as ships propellors required in small numbers.
(ii) Low refractoriness of the mould which renders the mould unsuitable for thick sectioned castings of high melting point metals such as steel.
(iii) Larger drying times needed render the process unadoptable for small castings.
(iv) Moulds are somewhat liable to erosion.
(v) Contraction of the casting is retarded due to high retained strength in the mould.

4.4.7 Sodium Silicate Processes

There is a whole range of sodium silicate binders varying in $Na_2O : SiO_2$ ratio and a foundry can have the choice of particular sodium silicate to meet its requirements. In general higher the ratio, the more reactive is the binder requiring a greater degree of control for proper mould and core quality. The popular choice is a 2 : 1 ratio silicate but use of higher ratio silicates of 2.2 : 1 to 2.5 : 1 is also common. Sodium silicate process is a compromise between green sand and dry sand practice for mould hardening.

Common advanrages of sodium silicate processes are as following:

(a) Moulds and cores produced are much more rigid and hence the process is suitable for core assembly mould production.
(b) The sand mixtures possess high fiowability and hence the compaction of the sand is much simpler.
(c) No baking ovens or driers are needed and thus more capacity is obtainable from the same floor space.
(d) The castings obtained have high dimensional accuracy and good sur-

face finish. Thus this processes can be used as a substitute for dry sand moulding.

(e) The moulds have greater resistance to friability and erosion and hence less chances of sand inclusions.

(f) Gas evolution on casting is low.

(g) The process is adoptable to either short run or large run production.

(h) The process facilitates the withdrawal of pattern of intricate shapes, since the mould is hardened before withdrawing the pattern.

(i) The process can be mechanised for mass production.

(j) Cracking and deformation of moulds is prevented as no drying is needed.

The main disadvantages of sodium silicate processes are:

(a) There is much greater difficulty in knocking out of the moulds, particularly it is severe in the steel castings where a large volume of metal surrounds a small volume of sand.

(b) The bench life of sand mixture is shorter and hence it can not be left for long periods.

(c) Reclamation of sand is much more expensive because of the build up of soda and fines.

(d) Need for hardening agent (CO_2, ester, or dicalcium silicate) adds significantly to the cost of the process.

(e) Stricter technical control is necessary.

(f) The requirements of the pattern equipment are different and stricter control is necessary for the same.

The sodium silicate process is adoptable to all the conventional techniques for mould and core production. However, machine moulding is not an entirely suitable due to incorporation of gassing techniques or the waiting period for the self hardening systems to begin. It can be used as facing sand and thereby compete with clay bonded systems. Therefore, sooium silicate process is more suitable in jobbing rather than repetition moulding and for longer cores in small numbers.

The various processes using sodium silicate binders include CO_2 process, ferro silicon process or Nishyama Process (N. Process), die-calcium silicate process, calcium sulphate process and processes using organic hardening agents. These processes are described below:

(i) *CO_2 Process* This is the principal sodium silicate process in which rigid mould or core is produced by the application of CO_2 gas. This process was introduced in the foundry industry during 1950s. The process consists of producing a mould with silica sand mixed with sodium silicate and passing CO_2 gas through the porous sand mouid (patten remains in position) for the predetermined length of time. The reaction between CO_2 and sodium silicate ($Na_2O\ mSiO_2 \cdot xH_2O$) produces SiO_2 which contains a certain number of water molecules ($SiO_2 \cdot xH_2O$) called silica gel. This silica gel gives the necessary strength to the mould.

The reaction proceeds rapidly in the early stages of gas passage and the compressive strength of the sand mixture reaches to a maximum value on passing a certain critical amount of CO_2 gas. If passage of CO_2 gas is continued, the strength of the bond gets affected adversely. Further, it is also observed that a given quantity of gas produces higher strength when passed at low pressure and for a longer duration. The volume of CO_2 required can be computed for a known amount of sodium silicate present in the sand mixture. In general, 0.5—0.75 kg of gas is needed for every 1 kg of sodium silicate present in the sand mixture.

The gassing can be carried out from CO_2 cylinders or storage tanks through a probe having a number of holes at the base. For larger moulds, several probes con be used but the flow rate in all of them should be the same as in a single probe. Gas leakage should be prevented by frequent testing. The flow rate of CO_2 gas will depend on the desired depth of penetration.

For economic reasons, CO_2 is generally supplied in liquid state, but in advanced countries it is all supplied in solid state which requires a pressure of 58 kg/cm^2 at room temperature. At atmospheric pressure, a temperature of—78.5 C is required to maintain CO_2 in solid state. It changes directly to the gaseous from above this temperature without passing through a liquid phase.

The hardening reaction is affected by three main factors, i.e. temperature, moisture and sand quality. The hardening reaction does not proceed below 10 °C and thus control of temperature is essential, best range being 20-30 °C. Higher temperatures will cause problems with premature drying out of sand mixtures giving friable moulds and weakly bonded cores. Damp sand causes difficulty in process control of gassing time for optimum strength development. Silica sand used should have consistent properties. Finer sands give faster bonding action. Coarse sands have a minimum requirement of about 2 pct. water increasing to 4 pct with finer sands.

The CO_2—silicate process suffers from two major disadvanthges, i.e. low tensile strength (<345 kNm^{-2}) and poor collapsibility. The ease of knockout (callapsibility) depends largely upon the amount of silicate present in the binder. For this reason, sugar-silicate and other silicate organic systems generally have improved collapsibility. In many cases, cores are not easily removed after casting due to sintering of the bonded sand.

It is generally necessary to apply a refractory coating (foundry paints) to the hardened surface to prevent metal burn on and penetration. These coatings are alcohol based, since small amounts of water accidentally present in the coatings can give severe scabbing defects.

(ii) *Ferro Silicon Process or Nishiyama Process* In this processes (developed in Japan) sodium silicate is mixed with powdered ferrosilicon (about 2 pct) and the mould is made by usual techniques. Hardening and bonding of sand are achieved by the silicagel formed by the exothermic reaction of silicon with the sodium hydroxide produced by the hydrolysis in the solution of sodium silicate. The reaction involved are

$$Na_2O \cdot mSiO_2 + H_2O \rightarrow 2NaOH + mSiO_2$$

$$Si + 2NaOH + H_2O \rightarrow Na_2O \cdot SiO_2 + 2H_2$$

The above reactions take place with temperature rise (reaching a boiling condition at about 90° C) and evolution of steam and hyrogen, i.e., setting is accompanied by dehydration.

The reaction will continue till silicon and water are present. If silicon is added in slight excess, almost all the water is expelled by decomposition and evaporation. As soon as the temperature rises to some extent, reaction proceeds at a faster rate and a hard spongy mass is obtained. The preparation of mould consists of following steps:

(a) Mixing of sand with about 2 pct finely powered (3-3.5 microns) ferrosilicon (75-80 pct Si)

(b) About 5 pct. sodium silicate of correct specifications is added and the mixing is continued.

(c) Suitable additives to impart collapsibility and bench life are added.

(d) Moulding is completed as early as possible within the period of bench life.

(e) Moulds are coated with suitable wash.

In addition to the usual advantages of sodium silicate processes, ferrosilicon process possesses the advantage of very low residual moisture in the moulds. The main disadvantage of the process is the evolution of hydrogen gas presenting a risk of explosion.

In this process, moulding sand mixture can be made to a flowable slurry by the adjustment of basic ingredients and mixing a foaming agent. The flowable slurry can fill the mould uniformly and ramming is avoided to a greater extent.

(iii) *Dicalcium silicate or fluid sand processes* This process was developed in Russia during the early 1960's. In this process sodium silicate (5 per ecnt) is added to the sand together with a suitable hardener, dicalcium silicate (2-3 per cent) being the most common. The sand mixture is made fluid by adding water and a foaming agent and the mixing should be quite vigorous to attain the best fluidity of the sand mixture. The mix so obtained can be poured into the moulding box eliminating necessity for ramming. How ever, vibration can be used to aid compaction of the sand mass. The sand then hardens *in situ* and then the pattern is withdrawn to make the mould ready for pouring.

In addition to the usual advantages of sodium silicate process, this offers the following advantages:

(a) The processes is attractive as the sand mix can be prepared at one end and then piped to various moulding station in the foundry to produce the necessary moulds.

(b) Sand behaves as a fluid to give a mould which varies very little in hardness from one point to another.

(c) Slags from certain melting or reduction processes such as basic lined hot blast cupola, basic open hearth furnace, ferro chrome furnace, etc. contain appreciable amount of dicalcium silicate and hence can be used in place of synthetically produced dicalcium silicate.

The rate of hardening depends on the fineness of the silicate (should not be less than 70 microns) and temperature of the sand (higher temperature increases the rate of reaction and shortens the bench life). The mixing time is usually 3-5 minutes and the life at moderate climate (about 25 °C) is 25-30 minutes. The sodium silicate used in the process should have $Na_2O:SiO_2$ ratio of 2:3 to 3:3 and sp. gr. 1.48-1.50.

In this process, the permeability of sand is higher because of the foamed condition and the compression strength after curing tends to be much lower (about 1 mNm^{-2}) than that obtainable by other sodium silicate processes. Therefore, a thick coating of suitable mould dressing (zircon based for larger sections and graphite for smaller sections) is applied to prevent metal penetration.

The fluid sand process is adopted in heauy and medium castings, both in grey iron and steel (such as ingot moulds, steel mill rolls, castings for cement and mining industry, etc.). The process has become quite popular in many European countries, the USSR and Japan and presently it is also finding its place in Indian foundries.

(iv) *Cement moulding process* Though, dicalcium silicate has found most application influid sand proeess using sodium silicate as binder, cement can be used along with sodium silicate to achieve a good combination of strength, permeability, and flowability. A typical sand mix consists of about 2 pct cement, 4-5 pct sodium silicatc, and 1 pct pitch or molasses. Similar to the dicalcium silicate process, flowable cement slurry can be produced by using a foaming agent. The sand mix has a useful bench life of about 15-20 minutes and thus can be employed for large sized moulds for ferrous castingr with good shake out properties.

(v) *Process using organic hardener* Organic hardeners can also be used along with sodium silicate for the silica gel formation. Organic hardners of sodium silicate process are easter compounds such as monoacetin, diacetin, triacetin ethylene glylol, diacetate, etc. This process is also associated with the usual advantages and limitations of sodium silicate processes.

4.4.8 Moulding Processes Based on Organic Binders

These processes were initially developed for core making as the cost involved is high. However, due to superior quality and shorter production cycle of castings obtainable from mould prepared by using organic binders, the process is adoptable for preparation of moulds. Various types of organic binders have long been in use and many other are developed. The use of organic binders has become very popular for the production of high quality castings of small and medium size.

In the last 35 years, there has been a tendency to replace oils, silicates and clays with resins for foundry binders. There are many resins used in the foundry which are from the simple urea and phenol formaldehyde resins to the complex furan and polyal resins. Even the same ingredients when react in different proportions and to different can produce differing mould core properties. The main advantages of these processes degrees, are:

(a) Organic binders (resins) give in box curing.
(b) Do not have the breakdown limitations of silicates.
(c) Require less skill in the production of moulds as in case of silicate moulds.
(d) The cured strengths are very much greater facilitating the production of complex moulds and cores, and handling problems are also minimised.
(e) The process can be used for mould production without the constraints of moulding boxes, when as series of cores are produced and assembl-

ed to produce the mould cavity.

(f) The sand can be reclaimed fairly simply and amount of resin and catalyst can be reduced using reclaimed sand and thus expenses can be reduced.

Processes based on organic binders can be broadly classified as

(i) Shell moulding
(ii) Hot box moulding,
(iii) Cold box moulding or self setting moulding, and
(iv) Gas setting resin process.

(i) *Shell moulding* This process was first developed during World War II (1940s) by Johannes Croning in Germany and still retains the name 'Cronig' or 'C' *process* in parts of Furope. Now it is known as *shell* process. The basis of the process is to make moulds and cores as relatively thin shells from a mixture of fine refractory material (70—100 microns) such as silica sand or zircon sand and thermosetting resin. The resin coated sand is then placed in contact with a heated metal pattern (first sprayed with liquid parting agent) to produce a hardened sand retaining its shape without distortion on withdrawal of the pattern. The layer of the hardened sand produced is referred as 'shell' and hence the process is known as shell moulding process. The thickness of shell to be produced mainly depends on particular casting requirements, e.g., a relatively thick shell is required to support the metal when poured into the mould cavity or the shell has to be supported by a backing medium such as sand, metal shot, etc. The shell moulding offers the following advantages:

(a) Tolerance as close as 0.002 cm/cm. can be held
(b) Less pattern draft is required for shell moulding than ordinary sand moulding.
(c) Castings can be poured with thinner wall section and with metal at a lower temperature than is possible in green sand.
(d) Small cored holes (0.6 cm in dia and 1.2 cm deep) can be cast accurately.
(e) Much smoother surfaces of castings are obtained due to the use of finer sand.

The disadvantages associated with shell moulding are

(a) The pattern equipment being of metal, is more expensive.
(b) Additional capital outlay is required for shell moulding machine.
(c) Large castings can not be produced conveniently.
(d) The resin bond adds to the cost of the moulding mixture. To some extent, this may be balanced off by less quantity of sand mixture required in shell moulding.
(e) A reasonable high volume of castings is required to justify the use of the process.

Shell moulds are produced in halves which are brought together along with cores if necessary before pouring. These can be assembled with a resin based adhesive. Shell moulding can be accompanied by two methods i.e. (a) dump method and (b) blowing method, each has its own field of use.

Dump method of shell moulding This method consists of dumping the resin mixed sand under the influence of gravity on to a heated pattern plate, leaving for a specified period and then allowing the excess sand to fall away from the pattern leaving behind a shell of hardened sand on the pattern. The shell so formed is released by suitable means. This method does not produce a shell of controlled thickness. The thickness of shell will depend on the temperature of pattern and length of time for which sand is allowed to remain in contact with the pattern.

The process of dumping is very simple and shell moulds can be produced with a simple equipment (Fig. 4.3). The steps involved are

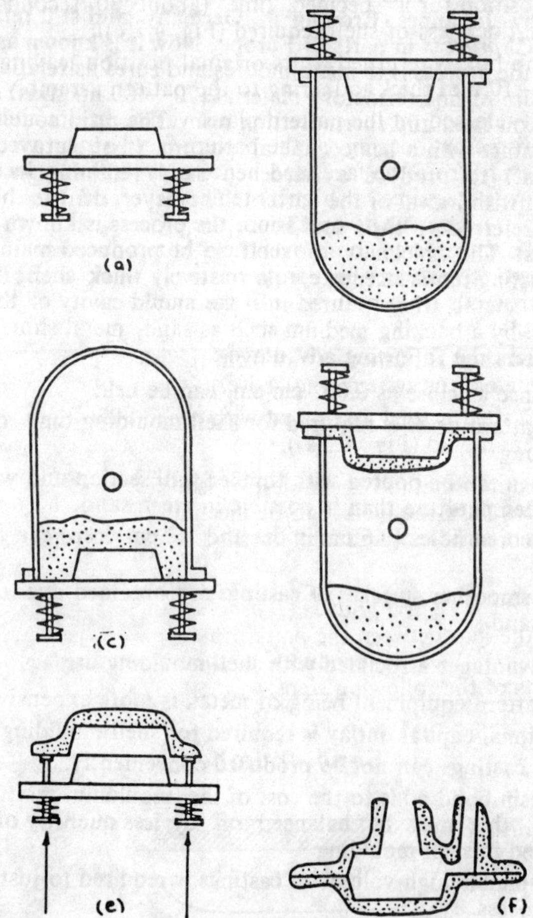

Fig. 4.3 Sequence of shell moulding: (a) pattern plate, (b) pattern plate mounted on dump box, (c) and (d) investment, (e) ejection (f) shell assembly

(a) Plate mounted metal patterns (made of cast iron or light alloys) are employed which are fitted with integral gating and feeding systems Each plate is also provided with ejector pins (spring loaded or simple) which are distributed round the peripheny of the pattern (Fig. 4.3(a)).
(b) The complete pattern assembly is preheated to 180—200 °C and its surface is treated with a parting medium (usually silicone emulsion). The heated pattern is then clamped against the open end of a dump box containing the sand resin (resin is usually phenol novolaks containing hexamine as a hardener) mixtures (Fig. 4.3(b)).
(c) The dump box is inverted to allow the moulding mixture to fall into contact with the heated pattern. The assembly is allowed to remain in this position for a specified time (about 30 seconds) depending upon the thickness of shell required (Fig. 4.3(c)).
(d) The dump box is returned to its original position leaving the bonded layer (5—10 mm thick) adhering to the pattern asrembly (Fig. 4.3(d)).
(e) The sand retained on the pattern plate varies in its degree of cure from satisfactory curing at the pattern face to relative uncured sand farthest away from the face, and hence curing is required. Full and rapid curing requires a higher temperature than that of preheated pattern plate. For curing operation, the pattern assembly along with the shell is transferred to an oven held at temperature of 300—450 °C or exposed to radiant heat from burners or electrical elements. Hardening proceeds from both sides of the shell and is generally complete in 2 minutes.
(f) The cured shell is pushed away from the pattern assembly by actuation of the ejector system (Fig. 4.3(e)).
(g) the cured shell is then assembled with other shells and cores to form a complete mould (Fig. 4.3(f)).

Though the dump method is simplest, it is a slow and laborious process. However, it does have the facility to be mechanised to quite a high degree. The main disadvantage of dump method is relative lack of control over shell thickness.

Blowing method or Dietert process (D process) In this process, sand mixture is blown into the gap between the pattern plate and a profiled back plate or carrier (Fig. 4.4). This produces a shell of controlled thickness. The process consists of the following steps:

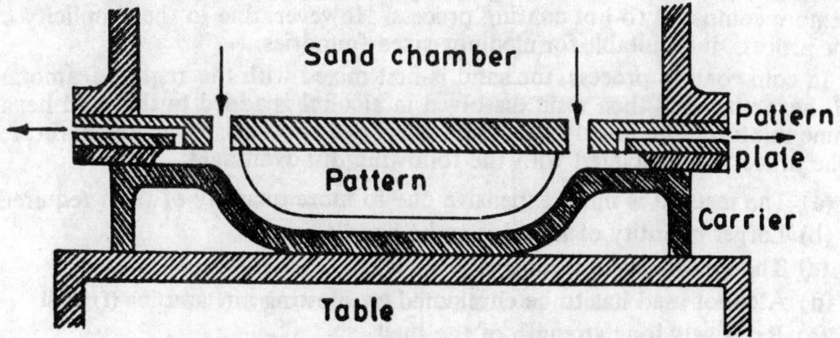

Fig. 4.4 Principle in D-process of shell moulding

(a) The metal pattern assembly complete with gating and feeding systems is mounted on the blow plate of the sand (core) blowing machine.

(b) The lower half of the blowing assembly is formed by a separate carrier or cradle (profiled back plate) having the general shape of the pattern which shapes the back of the shell. The gap between the two is maintained about 10 mm (equivalent to thickness of shell). The carrier can include slots etc. to strengthen the shell. The pattern assembly is provided with blow holes and vents to obtain uniform density in the thin shell.

(c) The carrier is preheated and brought upto the blowing head and the moulding mixture is blown into the enclosed space.

(d) The carrier is withdrawn along with moulded shell and transferred to the stove for a brief cure.

(e) After curing, the shell is ejected and clamped to a further shell for casting.

The main advantage of blowing method is the productlon of shell of controlled thickness. The process is capable of very high rate of production and can be applied to heavier castings. The process is also adoptable to CO_2 sodium silicate process.

Preparation of sand resin mixture The sand resin mixture can be prepared by three different methods, i.e., hot, warm, and cold coating process. The choice of the process depends on the type of equipment available and the type of resin.

In hot coating process, dry silica sand is first heated to a temperature of 120—175 °C and then hot sand is mixed with resin and hexamine to coat the sand grains with resin uniformly. The sand mix is screened to eliminate umps and agglomerates produced during mixing. The sand mix is then cooled to about 45 °C to prevent lump formation and to impart flowability. Flowability is further improved by aeration of cooled sand. The hot coating process is employed for mass production of castings, since it is the most economical method due to less amount of resin required and free flowing of mixture. The process is also safe as no alcohol (explosive in nature) is used.

In warm coating process, the catalyst is used with a different resin formulation (in the form of liquid solvent solution) so that curing occurs only by using warm sand at about 80 °C. The process has certain disadvantages such as (a) the preparation of the sand mix takes more time, (b) the mix is prone to lump formation during storage, and (c) quantity of resin required is more compared to hot coating process. However, due to the simplicity of the process, it is suitable for medium sized foundries.

In cold coating process, the sand is first mixed with the required amount of hexamine and then resin dissolved in alcohol is added to the sand hexamine mixture. The mixing operation is carried out at room temperature. The process is associated with the following disadvantages:

(a) The method is more expensive due to more quantity of resin required.

(b) Larger quantity of liquid is to be handled.

(c) The mixing cycle is longer.

(d) Alcohol used has to be eliminated by blowing air, and

(e) Relatively long strength of the shell.

However, with the development of new resins, the process is gaining importance as small quantities of sand mix can be prepared as and when required without the need of any special equipment.

Precoated sand is also available in a ready mix form which can be stored for a long time (upto about 3 months) and thus can be used whenever required. These sand mixtures are particularly preferred in small foundries which can not afford to instal the equipment for sand coating.

(ii) *Hot box moulding* This process is an improvement of shell moulding. The resin sand mix is blown over the heated pattern or corebox and the mould of sand is heated and allowed to form a solid mass. In this care pattern is turned over as in the case of shell moulding. It offers the advantages:

(a) The cores or moulds are relatively much more free from distortion and shrinkage,

(b) Dimensional accuracy is greater than in the case of shell moulding, and

(c) Method is largely adopted for small sizes with high production rates.

The resins used in the hot box process were originally sugar based and urea formaldehyde. Presently, phenol formaldehyde, urea furan, phenol furan and urea phenol furan are commonly used. The urea formaldehyde is the cheapest followed by phenol formaldehyde with the furans being the most expensive. The urea formaldehyde resins result in rapid breakdown on casting and high nitrogen which has limited their unblended use. The furan resins give much faster setting higher strengths, and decreased gas evolution. Zero nitrogen furan resin is excellent for steel and S.G. iron castings, whereas a straight furan is quite suitable for nonferrous castings.

(iii) *Cold box process* This is one of the most recent methods based on organic binders. In this process, no heating is required and hardening is obtained with the help of a catalyst. The cold box process is now tending to replace even the hot box process as it is excellent for mass production. Though, cold box processes are available under many properietary names, three methods are usually employed in foundries, i.e., amine gas method, alkyd or phenolic isocyanate process and furan process.

Amine gas method This process involves mixing fine dry sand with a poly isocyanate (1 pct.) and alkyd phenolic resin (1 pct.), blowing the mix into a sealed core or mould box and injecting tri-ethylamine or dimethylamine gas (TEA or IMA) with CO_2 or air respectively as the carrier. Usually 0.2—0.5 pct amine in the mixture is used at 200 KN/m^{-2}. The mixed sand has a bench life of 2—3 hours, but gassing with the amine hardens the mould or core in 10—30 secs. when the compression strength will be 2000 KN/m^2. The amount of amine required is 0.01—0.02 parts by weight of the core sand (1 – 0.3 ml amine per kg of core).

The mixture should be free from water, since isocyanate reacts preferentially with water instead of resin. Excess vapours purged from the hardened mould should be abscorbed or incinerated in suitable equipment. The limitations of the process are: (a) an expensive equipment is required for closing the catalyst gas through the core/mould box, and (b) the gas is toxic and poisonous.

However, the process is extremely suitable for small sized castings required in large number.

Alkyd or phenolic isocyanate process This method involves mixing fine dry, clay free sand with an alkyd resin and polyisocyanate and a liquid catalyst such as cobalt oxide or magnesium oxide. The mix is filled in the moulding or core box and compacted lightly. When the mould is left out in air, it gets hardened. The setting time is usually 20—40 minutes which can be controlled by adjusting the ingredients. The process is very efficient and suitable for large castings requiring accuracy and good finish, Wooden patterns can be employed and no special equiment is required.

Furan process The common furan resins include urea and phenol furfuryl alcohol formaldehydes. The catalyst commonly used for hardening is phosphoric acid (70—90 pct. strength depending on climatic conditions). The resin added to the sand varies from 2.0 to 2.5 pct. and the catalyst used is 30 pct. of resin. The bench life of sand mix is about 30 minutes and hardening is obtained in 1—2 hours which mainly depend on temperature of the environment. This process offers the advantages such as

(a) High collapsibility of mould and thereby fettling cost is lower,

(b) Low moulding costs as no baking is required and ramming is substantially less

(c) The process in non toxic and non inflammable.

However, the furan process is expensive and thus usually restricted to core making and castings requiring good finish. Large moulds can be prepared in composite form, i.e., using furan sand for the facing and other process for the backing. The urea type furan sand is cheaper and quite satisfactory for grey cast iron but it is unsuitable for steel castings as the mould gets affected by high temperature and metal penetration. Further, presence of nitrogen in the resin may produce pin hole porosity. Though phenol type furans are more expensive, they are suitable for steel castings.

(iv) *CO_2 gas setting resin process* This method has been developed in Japan as a substitue of CO_2-sodium silicate process. This process involves mixing dry sand with saponified ammonium acrylic ester or water soluble phenolic resin as a binder. After the moulds are prepared in the usual way, CO_2 gas is passed through the mould to acelerate the hardening action. Some amount of lime is also mixed along with the resin to act as binder cum activater. This process produces moulds of reasonably good strength and overcomes the problem of collapsibility. This process can be used for heavy and simple castings of all cast metals.

(v) *SO_2 gas setting resin process* It is possible to produce acid *in situ* by reacting SO_2 gas with a suitable oxidising agent and this forms the basis of SO_2 process (developed by SAPIC in France). The process involves mixing of dry sand with 1.5 pct. resin (preferably high furfuryl alcohol treated with silane) along with a peroxide catalyst. The peroxide catalysts are potentially dangerous and thus specially formulated materials such as material based on methyl ethyl ketone peroxide (MEKP) solution in plasticiser is used. SO_2 gas as a dry air 10 pct. SO_2 mixture is blown into the compacted sand moluds. Excess purged gas is absorbed in specially designed scrubber units using caustic soda mist in tower containig polypropylene packing. The process has been adopted for jobbing work by using specially designed gassing chamber. A further development of this process allows boxless shell type moulds to be produced competing very favourably with shell moulding techniques. This process has found particular application in production of high quality aluminium alloy engine cylinder blocks and aerospace castings.

4.5 MOULDING TECHNIQUES

In general, a prerequisite of moulding technique is to bond together sand particles in such a way that the mould profile may be formed and retain its form during various activities such as core laying, closing, pouring of molten metal, and finally the casting should be knocked out with minimum efforts. The choice of moulding technique is governed by the basic factors such as casting quality, production rate required, size of casting, shape of casting, number of castings required, surface finish, dimensional accuracy, composition of the alloy to be cast and the overall cost of production. Basically depending on the method of preparation, the processes conducted with hand tools are referred to as hand *moulding* and those requiring the help of machine are referred to as *machine moulding*. Hand moulding may be further grouped as floor moulding and bench moulding. There are some special methods of moulding also such as investment casting, full mould tedhnique, etc. These different mould techniques are discussed below:

4.5.1 Floor and Pit Moulding

Floor moulding can be carried out in two ways, i.e., open sand type and one box type. In the former case, the mould cavity is prepared in the floor and the molten metal is poured directly in the cavity without providing any passage in the sand for the metal. Open sand type can be used for the castings that do not require good surface finist on the upper face such as floor plates, weights, mould boxes, manhole covers and drain covers. One box method is employed to overcome the draw back of open sand type. In this case, one part of the flask is placed at the top of the floor mould where the flask acts as a cope and carries sprue and risers. For easy escape of gases from beneath the casting, the pattern is bedded on to a previously rammed sand base laid over a coke bed for venting. Vent pipes may also be embeded into the floor beneath the coke bed to facilitate the escape of gases.

Pit moulds are used to produce castings too large for a flask. In this case concrete lined pit equivalent to the size of the mould is made in which a bed of charecoal, cinders or coke is placed to facilitate the escape of gases. The pattern is suspended in a pit in the position and sand is rammed under and around the sides of the pattern. Moulding sand used is of special grade as moulding takes several days to make the mould. Chamotte and compo can be used for this purpose. The coke for the complete mould may rest on the drag (pit) at or above floor level and may be bolted down to prevent run out of molten metal at the parting plane. If the design of the casting is such that a pattern can not be drawn out of the mould. the entire mould cavity can be made with cores. Large castings should be cooled slow to prevent the formation of excessive internal stresses.

4.5.2 Bench Moulding:

Bench moulding refers to the moulding carried out fully in flasks (moulding boxes). Bench moulding is used for small and medium sized castings which can be easily handled. Bench moulding includes two box moulding, three box moulding, moulding with false cheek, plate moulding, stack moulding and odd side moulding.

A brief description of these techniques is given below:

(i) *Two box moulding*; This technique makes use of two flasks, the upper one is called *cope* and the lower one *drag*, which are fitted with a suitable clamping arrangement.

The clamping arrangement is required to: (a) prevent the cope from lifting due to the pressure of molten metal on pouring and: (b) maintain proper alignment at all times. This is the simplest and most commonly used technique for simple and castings.

(ii) *Three box moulding*: This technique makes use of three flasks with pattern made in parts as required. This technique is used for flanged type pattern where flanges are to be cast horizontally. This type of mould can not be easily prepared in two box system. The additional box in the middle is referred as 'cheek'. All the three boxes are clamped together while pouring. This technique is illustrated in Fig. 4.5

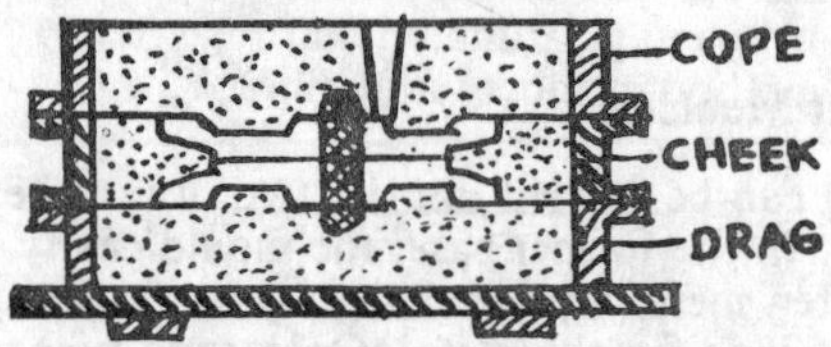

Fig. 4.5 Three box mould

(iii) *Moulding with a false cheek*: When three boxes are not available or it is not convenient to use three boxes, the mould can be prepared in two boxes with a sand core known as false cheek. This is illustrated in Fig. 4.6 This type of mould is prepared in following manner:

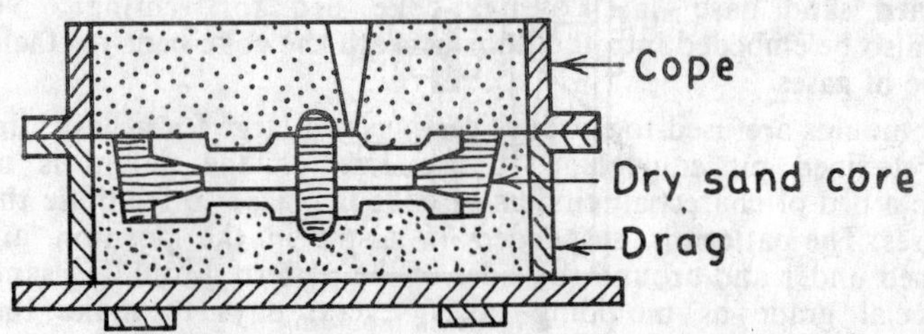

Fig. 4.6 Moulding with a dry sand core

(a) Ramming the cope half of the pattern in the cope and rolling over the assembly.

(b) Formation of a parting around the pattern in the cope.

(c) Preparation of false cheek with the green sand in the cavity formed.

(d) The drag is then rammed ovar the false cheek

(e) The drag is then drawn, the pattern removed and the drag is again replaced.

(f) The cope is drawn off and remaing pattern half is removed.

(g) The mould is finally closed for pouring.

When the castings are required in large number, dry sand cores prepared separately can be used in place of green sand cores.

(iv) *Plate moulding*: This method is used to produce simple castings several at one time. The form of a iven part is divided into two halves which are attached respectively to each side of a board. One side of the broad is moulded with drag and after turning it over, the cope side is moulded. The cope and drag are then seperated and patterns are removed. The cope and drag a reasse mbaled to mach the moulded impressions to make the mould ready for pouring. This method is usually employed with machine moulding where all the operations are carried out mechanically.

(v) *Stack moulding*: When small castings having on surface flat are to be produced, it is possible to use both the sides of one half of the mould. This can be obtained by stacking the moulds in such a way that the lower face of each part is a cope face and the upper face is a drag face as shown in Fig. 4.7. A common passage running through the stack is made for the molten metal. This method produces a number of castings in a single pouring operation.

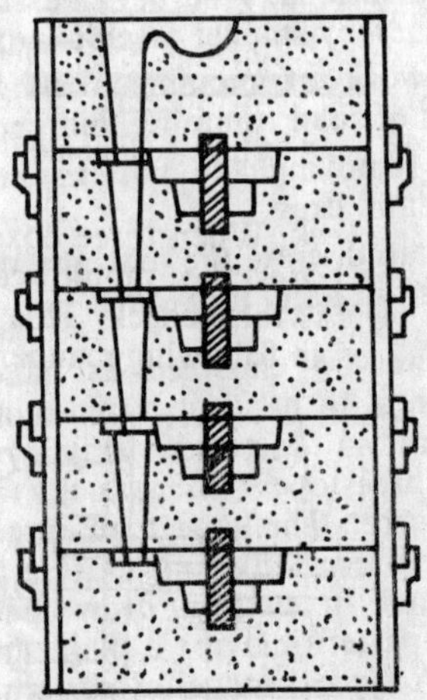

Fig. 4.7 Stack moulding

(vi) *Odd side moulding*: When one piece of pattern does not have flat face and can not be made into the split type an odd side moulding is employed by bedding the pattern in mouding sand. The method entails the following three steps.

(a) Bottom part is rammed on odd sid (Fig. 4.8 (a))

(b) Top part is rammed on bottom part after turn over and removal of odd side (Fig. 4.8 (b))

(c) Mould is completed by separating the boxes, withdrawal of pattern and finally closing the mould for pouring of molten metal (Fig. 4.8 (c))

Fig. 4,8 Basic sequence in loose pattern moulding

4.5.3 Machine Moulding:

Machine moulding enables moulds to be produced at high rates. The principal feature of machine moulding is the use of power operated mechanism for most of the operetions of mould preparation (compaction, pattern withdrawal, etc.). Machine moulding is distinguished from hand ramming by the reliance placed upon flowability of the material to achieve the desired packing density. Machine moulding sands are thus selected for high flowability instead for maximum green strength. The main advantages of machine moulding are:

(a) It enables higher out put and thus affords great saving in time, particularly when large number of identical castings are required.

(b) It gives increased dimensional accuracy of castings (due to less tapered patterns used) and thereby smaller machining allowances are required.

(c) It eliminates hard work of moulders owing to mechanisation of laborious operations of sand ramming and pattern removal.

(d) It uses vibrators in removing the pattern and secures more exact location of the cope and drag.

(e) Less skilled worker can operate the machine compared to the skilled craftmanship required in hand moulding.

Moulding machines can be classified as hand operated and power operated moulding machines. In hand operated moulding machine, one or more operations, such as ramming, pattern drawing, and mould rolling-over are performed by the machines operated manually (power is not used) either by hand lever or pedal control. The hand moulding machine may be of the pattern draw type or the pattern draw and squeeze type depending on the type of operation performed. In pattern draw machines, the mechanism used for withdrawing the pattern may be a plain stripper type, a pin lift type or a roll-over type. A typical model of pattern draw and squeeze type hand moulding machine is shown in Fig. 4.9

Power operated moulding machines make use of external power and thereby drastically reduce the mannual labour and fatigue and the production rate is much higher. These machines make use of hydraulic or pneumatic action to perform various operations such as raising and lowering the table for pattern withdrawal, ramming the sand and rolling over the moulding boxes. The power operated machines can be classified based on their principal functions. In most cases, sand is filled with the help of compressed air, the mould is inverted by rolling it over and the pattern is

Fig. 4.9 A typical model of pattern-draw and squeeze type hand moulding machine

drawn out mechanically. Various operations required to produce the moulds are performed outomatically one after the other in proper sequence. The moulds ready for assembly are transported from the machine to the mould assembly and pouring floor by a conveyor. The pattern is drawn by the similar mechanisms as in case of hand mouldidg machines.

In general, the moulding machines are classified depending on the method of ramming, i.e. squeezing, jolting slinging or the combination of two. Different moulding machines are discussed below:

(i) *Squeez Machine*: This is one of the simplest machine which compacts the sand by squeezing it around the pattern. The various steps involved in making a mould are: (a) Placing a box around the pattern, (b) filling the box with sand and: (c) applying pressure through a squeeze board by means of levers (in hand moulding machine) or peumatically or hydraulically (in case of power operated moulding machine) to compact the sand to the desired density. Sometimes squeeze action may be applied by means of electromagnets. A simple hand squeze machine is shown in Fig. 4.10. The squeeze pressure can be applied by keeping the machine table fixed and lowering the squeeze board on to the sand or vice versa.

The main limitation of squeeze method is that the sand is packed more densely at the top whcre the squeeze board presses against the sand and the density decreases as the depth of box increases. This results in the lowest density at parting plane. Therefore, this method is restricted to moulds having less than 150 mm depth. Depending on the size of machine, squeeze pressure may vary from 3000 kg to 20,000 kg,

In squeeze moulding it is not possible to eliminate differences in bulk density arising from large disparities in the depth of sand below the squeeze

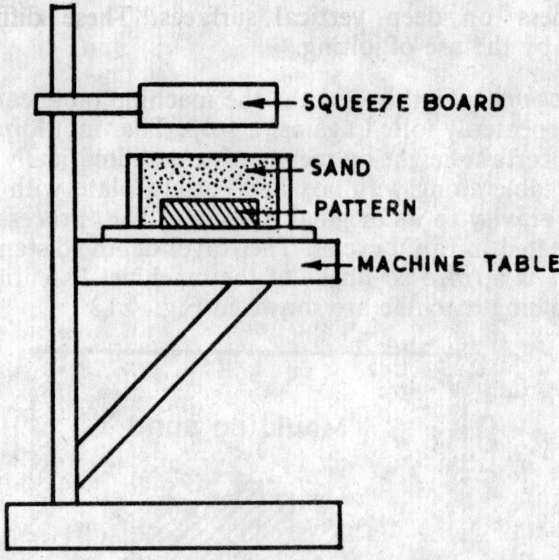

Fig. 4.10 Simple hand squeeze machine

head. Therefore, to overcome this problem, *contour squeeze* or *flexible diaphragm* method is required which enables more uniform compression to be achieved despite uneven depth of pattern. Different contour squeeze heads and flexible pressure diaphragms are illustrated in Fig. 4.11 These

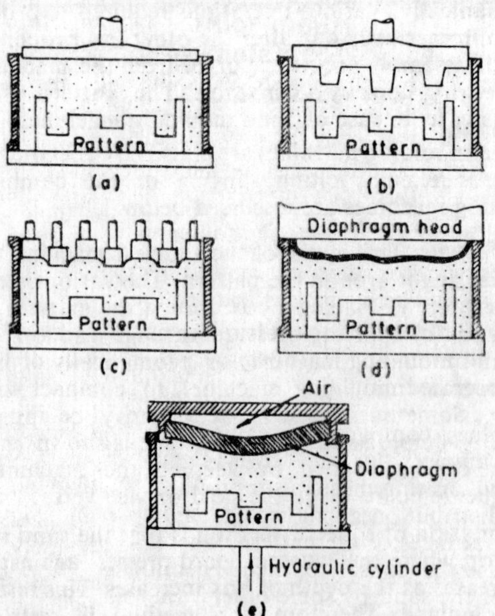

Fig. 4.11 Different contour squeeze heads and flexible pressure diaphragms: (a) conventional suqeeze (b) profile suqeeze (c) equilising piston squeeze (d) diaphragm squeeze and (e) high pressure hydropneumatic squeeze

techniques require less extreme pressure as the lateral thrust produce high mould hardness on deep vertical surfaces. These difficult cases becan further aided by the use of jolting.

(ii) *Jolt Machines*: In this case the the machine table carrying the pattern assembly is repeatedly iolted against a stop. The moulding machine table is raised to a certain height by means of a pneumatically operated piston and then the table along with box and pattern plate with sand is allowed to fall under gravity to its original position. The process of raising and falling is repeated in rapid cycles. The travel depth (distance) of moulding machine table is termed as stroke of the machine. Essential components of jolt type moulding machine are shown in Fig. 4.12

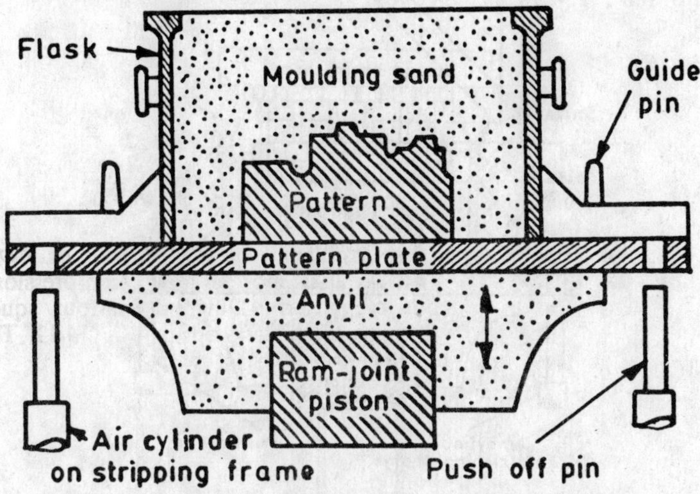

Fig. 4.12 Essential components of jolt type moulding mechine

The jolting action may be one of the following two forms

(a) A straight forward or plain jolt against a fixed part of the machine (a fixed anvil) in which shock is transmitted to the foundation.

(b) A shockless jolt in which the jolt takes place against a spring supported by a floating anvil. In this case shock is mainly absorbed within the box assembly.

In jolting machines, compaction is caused by the impact of falling sand, i.e., partly by the impact due to its own velocity and mainly by the impact of the sand column. Jolting mechanism produces the lower horizontal faces a mould density distribution giving maximum hardness at the joint line and the lower horizontal faces of the pattern, whereas the sand at the back of the moulding box is very loosely packed. Therefore, it is customary to finish the compaction of the sand by some other means such as hand ramming or squeezing. Jolting action is used on a wide range of machines. The jolting pressure varies from 200 kg to 1000 kg depending on the size of machine.

The density and hardness of the moulded sand will depend on weight of the stroke, the amount of sand heaped above the mould and the number of strokes which can be controlled to give the required density and hardness.

(iii) *Jolt Squeeze Machine*: The draw back of both squeeze and the jolt action can be overcome by the use of combination of jolting and squeezing which give uniform density and hardness in all portion of the mould. The machine using the combined action is referred as jolt squeeze machine. It uses the same pattern equipment as jolt machine, but after jolting operation, the sand is squeezed by air pressure to produce more uniform and harder mould. Jolt squeeze machines can take up the flasks upto approximately 100x125x50 cm and are widely used for their simiplicity and low capital cost. The essential components of a jolt squezze moulding machine are shown in Fig. 4.13

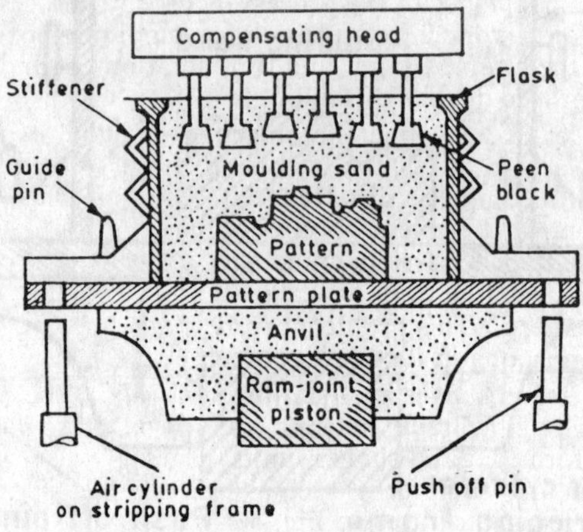

Fig. 4.13 Essential components of jolt squeeze moulding machine

In jolt squeeze machine, both squeeze and jolt actions can be obtained one after the other. The table is attached to the ram-jolt piston which is raised and dropped in the jolt cylinder by the action of compressed air. During jolting, the squeeze piston rests on base of its cylinder and the lift of the jolt piston raises the table. During squeezing, the jolt piston and cylinder move along with the squeeze piston and the table is raised through the required height. A typical jolt squeeze moulding machine accomodating moulding box of maximum 100x75 cm may have the specifications as given below:

Squeeze pressure	10,000 kg.
Jolting load	500 Kg.
Pattern lraw	40 cm
Table size	110x80 cm
Squeeze stroke	11cm
Air pressure	7 kg/cm^2

There are two types of jolt squeeze operation i.e. (a) consecutive jolt squeeze and (b) simultaneous shockless jolt squeeze.

The latter operation gives better compaction if a preliminary free jolt is given to the sand before applyilg the simultaneous jolt squeeze. In the consecutive jolt squeeze method, a better mould hardness distribution can be achieved by moving the sand about by hand during jolting stage before squeeze operotion.

(iv) *Slinging Machine or Slingers*: In these machines compaction of sand is obtained due to its impact with the pattern. In this method, a stream of loose sand is thrown downard through a slinging head (high speed rotor with blades) on the pattern at high speed. Due to the rapid impact, the sand particles settle down instantly and get rammed. The density of packing can be controlled by variation of the speed of blades and velocity of throw of the sand. The advantages of the process include: (a) process is suitable for jobbing or repetition work in relatively large moulding boxes, (b) unifrom bulk density of the rammed sand and (c) aeration of sand during the fall and there by improved quality of sand.

The sand slinger may be either stationary or travelling type. In order to accommoedate largest boxes, the traversing ability of the slinger should be rapid and multiple impeller tips are used. The sand used should be low in temper water content to achieve good flowability of sand otherwise sand may stick to the impeller system. Sand slingers have some limitations such as difficulty in compacting deep or vertically walled patterns, and excessive wear of the pattern. This problem can be overcome by applying a facing sand and precompacting in those areas (deep pockets, etc.).

In general, the sand slinger is of stationary type and the moulding boxes are conveyed along a track beneath the slinger head and the necessary operation is performed. Sand slinger units generaly work in conjunction with complementary equipment such as rollers pattern draw machines. A typical sand slinger is illustrated in Fig. 4.14.

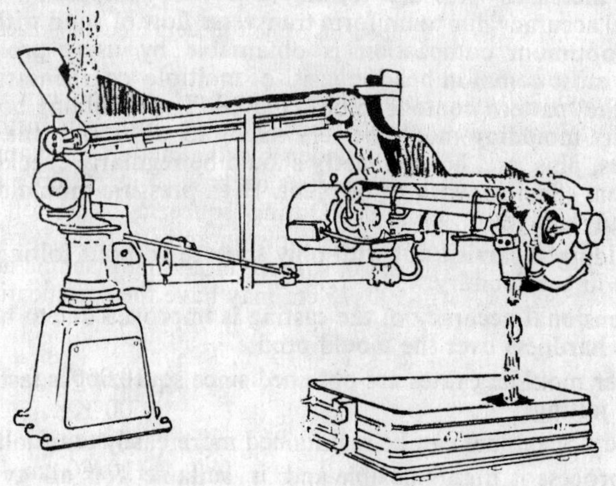

Fig. 4.14 Sand slingers

(v) *Blow squeeze machines* A further development in machine moulding is the boxless squeeze moulding. The sand is first injected into the moulding chamber by compressed air from the air receiver into the sand hopper (Fig. 4.15). The sand is then campacted by movement of the squeeze plate (rear pattern plate attached to squeeze plate). These machines can produce horizentally as well as vertically split moulds which are placed on a moving track for pouring. However, there are two limitations of the process, i.e., (a) coring should be simple because very little time is available in the machine cycle for caring up and (b) low sand permeability can cause bloweing problems.

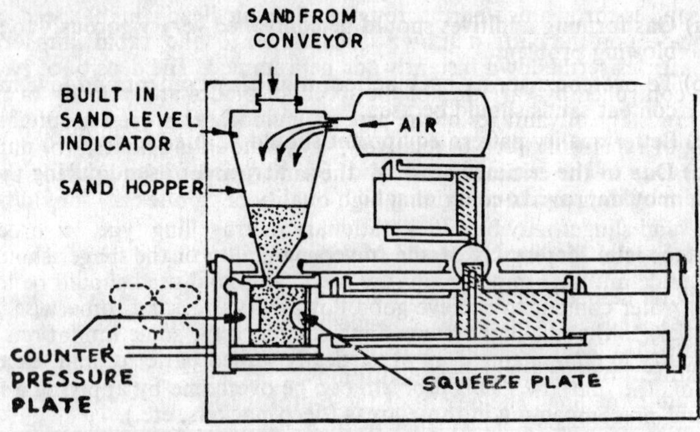

Fig. 4.15 Simplified diagram of boxless squeeze moulding machine

(vi) *High pressure moulding machines* When the compaction (squeeze) perssure is much higher, i.e., 5-10 times than that for conventional moulding machine work (about 3 MNm^{-2}), the process is referred as high pressure moulding. Increased pressure results in better compaction with better dimensional accuracy due to uniform transverse flow of sand without jolting. However, optimum compaction is obtainable by using profiled squeeze heads. The most common heads consist of multiple compensating cylinders to enable the pattern contour to be achieved. The moulding boxes used in high pressure moulding should be very strong to withstand the high pressure. Pushes, pins, and box geometry should be regularly checked to ensure the optimum effective working of plant. High pressure moulding offers the following advantages:

(a) Moulding is carried out with only squeeze without jolting and much noise in the foundry.

(b) Dimensional accuracy of the casting is improved due to high and uniform hardness over the mould profile.

(c) H igher moulding rates are obtained since squeezing is faster operation than jolting.

(d) Squeeze machines can be maintained more easily than jolt squeeze.

(e) The procese is highly flexible and is suitable for all types of cast metals.

Inspite of the above advantages, the high pressure moulding is suitable only for mass production work using cycle operation due to (a) high initial cost of the equiment and (b) precise control requirements for sand. Other disadvantages of high pressure moulding are (c) high degree of compaction results in low sand permeabilty and thus castings may be susceptible to blowing defects, (b) higher thermal conductivity of hard rammed sand may also give rise to chilling of cast iron and (c) spring back effect on releasing pressure, i e., expansion of the mould cavity.

Therefore, while using high pressure moulding, following adjustments should be made:

(a) Gas forming additives should be controlled very vigorously to prevent blowing problem.

(b) To overcome spring back action, high clay content and low moisture content sands should be used.

(c) Better quality pattern equipment should be used.

(d) Due to the critical nature of the sand properties required, synthetic moulding sands containing high quality bentonite clay should be used.

4.5.4 Special Methods of Mould Making

Many moulding processes significantly differ from conventinoal moulding process. These processes find considerable scope for special applications required in the modern industries. The important ones are precision or investment or lost wax process, full mould process, plaster moulding and vacuum moulding. The details of these processes are discussed below:

(i) *Precision (Investment) casting* Investment casting uses a pattern of an expandable material such as wax or plastic (polystyrene). The investment mould is prepared by surrounding an expendable pattern with a refractory slurry that sets at room temperature. The pattern (usually of wax or plastic) is melted out or burned off leaving the mould cavity. Investment casting is also known as lost wax process, precision casting or the *Cire Perdue* process. The investment process developed from an art dating back as early as 1776-1122 B.C. during the early Egyptian and later Shang Dynasty periods. Earlier, jewelers and dentists were the principal commercial users of the process until world war II. The commercial viability of the investment casting industry developed from the early 1940's with the inceasing demand for complex close dimensional tolerances parts required for the aerospace industry. Since then investment castings are gaining importance in the general engineering industries. The process of investment casting comprises the following steps which are illustrated in Fig. 4.16.

(a) *Pattern making* An expendable pattern is first produced by injecting a wax under pressure into a split die (die of aluminium alloy, cast steel, etc.) Patterns can be made from variety of materials such as waxes containing addition of resin and filler materials. Sometimes, polystyrene and frozen mercury are also used for pattern making. Silicon rubber dies are also used for small quantities and intricate shapes. If the pattern is small in size, several such patterns are joined together about a common wax sprue through ingates and then investment is prepared about the assembly of patterns. For pattern work special wax blends are formulated in order to have good properties of pattern such as good hardness, good tensile and

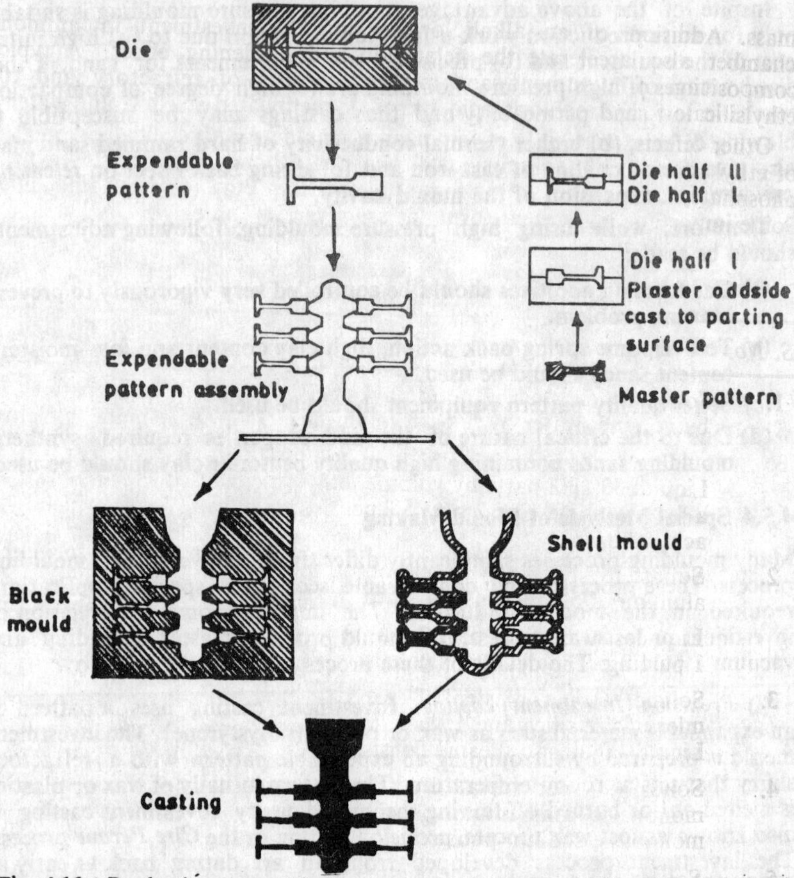

Fig. 4.16. Production squence in investment casting (expendable pattern technique)

impact strength, solubility in specific solvents, resistance to the action of chemicals used in investment sand, and low shrinkage during solidification.

(b) *Investment preparation* Investment moulds can be produced by block process or solid investment or ceramic shell investment.

In solid investment, a precoating of binder mixed with fine refractory sand (silica or zircon) is applied to the pattern which is then placed and complete with gating in a metal flask and is filled around with refractory slurry. The binder used for the primary coat and investment slurry is ethyl silicate. The reaction involved forms the silica gel in hydrolysed solution of water which imparts a permanent silicious bond.

$$\underset{\text{Ethyl silicate}}{Si\,(OC_2H_5)_4} + 4H_2\,0 = \underset{\text{Ethyl alchol}}{4\,C_2H_5OH} + \underset{\text{slica Acid}}{Si\,(OH)_4}$$

Since ethyl silicate is insoluble in water, methylated spirit may be used as a mutual solvent or dilute hyrochloric acid (10 pct.) also serves as a catalyst and stabilizer. Silica gel thus on heating produces hard, and permeable mass. Addition of an alkali or keeping the investment in ammonia chamber also accelerate the gelatation and hardening effect. A typical composition of investment sand mixture consists of refractory sand 100, ethylsilicate 5, hydrochloric acid 10 pct. 1, methylated spirit 1, and water 9.

Other binding materials of importance inculde sodium silicate, a mixture of ethyl silicate and sodium silicate, phosphate bonded materials (gypsum phosphate, ammonium dihydrogen phosphate, etc.) and colloidal silica. Some important investment formulas are given in Table 4.1.

TABLE 4.1 Coating and Investment moulding mixtures for Investment castings

S. No.	*Materials and their proportion*	*Uses*
1.	Solids: Silica sand 67, tetraethyl silicate. 8 parts by volume, water by 1 part volume. Liquid: 33 (1-2 parts by volume ethyl alcohal, a few drops to 1 or 2 pct of 3 pct hydrochloric acid solution	Precoating for high melting alloys
2.	Solids: 187 parts (94 silica of 50 microns, 56 alumina of 50 microns, 37 silica of 400 microns. Liquids: 80 parts (4 part 2 pct sodium silicate, 1 part 2 pct, polyvinyl alcohal)	-do-
3.	Solids: (60 pct plaster of paris, 25 pct. 300 micron size silica and 15 pct. talc.) Liquid: Water to creamy consistency	Plaster moulding for non-ferrous alloys
4.	Solids (90 pct silica, 6 pct magnesia, 3 pct monobasic ammonium phosphate and 1 pct monobasic sodium phosphate	Precoating for high melting alloys
5.	Solids: (95 pct sand, 5 pct alumina cement as binder) Liquid: water 27-31 pct	Investment moulding
6.	Solid: sand Liquid: 3 pct or more ethylsilicate or sodium silicate	Investment moulding and for ethyle silicate precoat
7.	Solids: sand 91.2 pct, 6.5 pct calcium phosphate, 2.3 pct MgO (50 microns) Liquid: water 33.8 pct.	For ceramic and investment moulding
8.	Solids: sand 93.3 pct, 5.17 pct primary ammonium phosphate	-do-

The ceramic shell proces consists of dipping the pattern assembly into a refractory slurry, the excess slurry is allowed to drain from the pattern assembly and then a refractory aggregate is applied. The refractory particles attach themselves to the wet slurry layer. The resulting coating is then dried by heat or by chemical means and the cycle of dipping, application of refractory particles and drying is repeated until an adequate thikness of material is obtained (5-15 mm).

Selection between the solid investment and shell investment process mainly depends on factors such as casting size, required tolerances. composition of metal poured, cost, and availability of equipment.

(c) *Dawaxing* After the investment mould is ready, dewaxing is necessary in each case (sheel or solid investment) to remove the pattern material before pouring molten metal. In case of block moulds, dewaxing can be carried out by heating the mould to a temperature slightly above the melting point of wax and main bulk of the wax is allowed to drain out. This is followed by firing to volatilize and burn the residual wax. In large scale production, the moulds are placed in a gas fired or electric furnace, usually of tunnel type. The moulds are introdueed at the low temperature end (about 400°C) and are slowly moved through the furnace to higher heat zones (temperature may be as high as 1000-1100 °C). The wax melts out of the mould and is burned off as the mould advances through the furnace and the mould emerges from the hot end of the furnacc preheated and ready for pouring.

Shell moulds cane be dewaxed by a number of techniques, e.i., flash dewaxing, steam pressure treatment, solvent method and heating with support of refractory grannules. In flash dewaxing, the temperature is raised rapidly to about 1000 °C by placing the mould assembly directly into the furnace. In this case a steep temperature gradient produces superficial melting of the wax before the appreciable expansion of main bulk and thus perventing the cracks in the mould. Dewaxing and firing both are completed in a single operation of about 2 hours period. In steam pressure treatment, the mould is placed in an autoclave and the steam pressure is developed to complete the dewaxing in a few minutes with high wax recovery. In this case temperature involved is lower and thus thermal shcock is much less than in flash dewaxing method.

The solvent method employs hot vapour of a solvent such as trichloroethylene. The vapour permeates the porous shell and immediately dissolves the wax faces adjoining the investment, before the heat of the solvent vapour expands the wax. Equipment for this process consists of a conventional trichloro ethylene vapour unit of the type used for vapour degassing.

In the support technique, the ceramic shell is supported in hot refractory granules so as to maintain the mould under compression during the critical stage of dewaxing temperature.

Polystyrene patterns can be eliminated by preliminary drying below 70 °C (to avoid mould cracking with a sudden expansion of polystyrene at about 82 °C) following by firing at high temperature.

Firing of moulds other than those based on plasters is normally completed at about 1000 °C. Firing brings about final dehydration and full development of dry strength eliminating traces of organic matters. Preheating of mould also facilitates metal flow and reproduction of mould details. The less refractory plaster moulds used for non ferrous alloys are fired at lower temperature (550-650 °C). Ignition of wax is avoided to prevent carbonacious residues.

(d) *Casting* As soon as the dewaxing is complete, the moulds are removed from the furnace and poured immediately. After pouring, the moulds with metal are allowed to cool down at a slow rate, keeping the mould out of contact with open air.

Some Variations/modifications of investmeut casting The Shaw Process employs ceramic moulds in conjunction with permanent patterns to produce a limited range of castings such as tools. The *Antioch Process* makes use of plaster moulds for casting low melting alloys with improved finish and dimensional accuracy.

The most important modification of investment casting is *Mercast Process* which uses frozen mercury as pattern material. Mercory patterns are prepared by pouring mercury into pattern dies maintained below-50 °C to allow the solidification of mercnry. The investment mould is then prepared at about-38 °C using frozen mercury pattern in the conventional way. When the investment is brought out at room temperature, the mercury melts and drains out. The moulds are finally fired and poured. The most important advantages of Mercast process include. (a) highest dimensional accuracy as the mercury does not expand or shrink during change of state from liquid to solid or vice versa and (b) injection equipment is not required as mercury can be poured by gravity.

The main disadvantages of the process are: (a) expensive equipment needed for low temperature work (b) high cost of mercury (c) handling problems, and (d) poisonous nature of mercury.

Investment casting possesses the following advantages:

(a) Castings produced have high degr. e .f accuracy (± 0.1 mm) and fine surface finsh 1-5 microns) and thus eleminate or reduce machining
(b) The process is adoptable for all types of metals and alloys.
(c) Complex and difficult shaped castings can be conveniently produced.
(d) High qulity castings free from external and internal defects are obtained on account of close control of mechanical properties (grain size, grain orientation, directional solidification, etc.).
(e) Process is adoptable for mass production.

The main limitations restricting the use of investment casting for high quality and small sized casting are:

(a) Physical and economical considerations limit the casting from few grams to 5 kg.
(b) All the stages of production require precise control and the process requires special equipment for preparing patterns, making investment, etc.
(c) The raw materiales, special tooling, equipment required etc., all are expensive.

(ii) *Full mould process* In this method the pattern is produced complete with gates and risers from expanded polystyrene having a density of about 15-35 kgm^{-3}. The pattern can be easily made by carving or cutting with hot wires or suitable instruments. The pattern is embeded in a suitable mould system and it need not be extracted from the mould (Fig. 4.17).

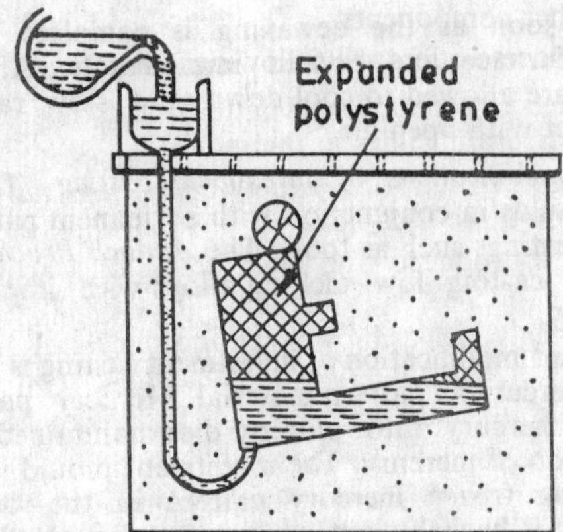

Fig. 4.17 Full mould process

The main variations of full mould processes include (a) no bake moulding sand process (b) nonbinder process and (c) magnetic moulding process.

(a) *No bake moulding sand process* In this process the pattern with gates and risers is coated with suitablele mould wash and is placed in the mould box. A facing sand such as organic no-bake sands (furan alkyd isocyanate bonded), CO_2 sand, ferrosilicon sand, or fluid sand is used over the pattern and then it is backed up with green sand. The facing sand is used to prevent distortion of the mould and pattern. While the pattern is inside the mould, molten metal is poured, the heat of which is sufficient to gassify the pattern and progressive displacement of pattern material by the molten metal. The small amount of gas produced can easily escape through the sand without causing any back pressure. In order to compensate the heat required to gassify the pattern, molten metal is poured at about 30° C higher than that required for sand casting.

No binder process In this process, dry silica sand is poured around the polystyrene pattern. In this case due to the absence of bond, the sand grains flow relatively easily into all recesses of the pattern to take up the correct impression of the pattern profile. Flow of sand can be aided by vibratary technique if required. The molten metal is then poured into the mould, the polystyrene vaporises in front of the advancing molten metal until it occupies the complete space previously taken up by the pattern.

Magnetic moulding process In this method, the polystyrene pattern is embeded in iron sand containing fine iron shots, iron oxide power, or grinding swarf. The moulding box carrying the pattern is subjected to magnetic field which causes easy flow of moulding material into and around the pattern recesses and results in a fully consolidated mould. The molten metal is then poured into the mould vaporising the polystyrene pattern. After the solidification of the casting, the magnetic field is removed by switching off the current to electromagnet. On removal of magnetic field, the mould disintegrates instantaneously and the casting becomes available without requiring any fettling or cleaning. The iron sand is collected, cooled, and recirculated to the moulding section for further mould production. The process has been successfully used in some foundries for the production of automobiles and other components.

The full mould process offers the following advantages:

(a) There is no need to produce split patterns and thus the process eliminates complicated joint lines in the mould.
(b) The process can be used for a variety of castings ranging from small to large and-simple to complex.
(c) The castings are produced with high dimensional accuracy and metallurgical quality.
(d) The casting defects such as blow holes and pin holes are prevented due to absence of air in the mould at any time.
(e) The method is most suitable for one off casting required at short notice or for prototype castings.
(f) This process in an economical substitute for investment casting as the pattern for full mould process is made from pre-expanded polystyrene granules.
(g) The rate of production is high as a large number of patterns can be moulded together.

(iii) *The vacuum sealed moulding proeess* This unique technique was invented in Japan in 1971. The process does not use any bonding medium between the sand grains. The conventional pattern is first covered with a plastic film and then it is placed in moulding box which is filled to the top with dry sand. The whole system is sealed and air tightened by plastic films (Fig. 4.18). The air is sucked out of the mould via a vacuum pump. The moulding boxes are specially designed with cavity walls to facilitate the extraction of air. In case of large moulds, pipes are laid through the sand mass to enable a more efficient extraction of air. Under vacuum, suction pressure, the mould becomes stiffened and the suction is maintained until the casting has solidified.

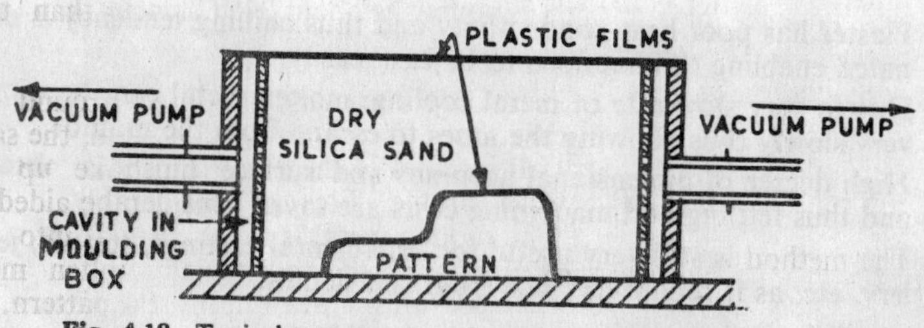

Fig. 4.18 Typical arrangement of vacuum sealed moulding process

The best material as plastic film is ethylene vinylacetate co-polymer con taining about 15 pct. of vinylacetate. The film of this in thickness of 0.05 – 0.1 mm has good forming properties to be stretched without tearing over the pattern and joints profiles. Under the application of vacuum the moulding box is removed from the pattern with a film supporting the sand at the top and bottom of the box and hence the shape of pattern is retained. The top box is then prepared in the identical manner and finally the mould is closed. The molten metal is poured into the mould cavity with maintenance of vacuum until the casting solidifies. On solidification of casting the vacuum is released and the sand falls away from the casting. The pouring rate should be as fast as possible to prevent destruction of the film layer and there by exposing loose sand leading to sand inclusions in the finished castings.

Vacuum moulding technique offers the following advantages:

(a) Surfaces of castings are similar to that obtainable by investment casting methods with accuracy in dimensions.
(b) Process is much cheaper with low capital investment and maintenance cost
(c) Sand recovery is very high (upto 99%)
(d) Foundry hazards are minimum with low environmental pollution.
(e) Neither binder nor water is required and hence clean environment is possible.
(f) Sand conditioning and fettling problems are eliminated.

In addition to the powerful vacuum unit, vacuum moulding process requires special pattern plates and double flasks for effecting vacuum, efficient venting system in the inner face of moulding box (to prevent the sand and particles from being sucked by the vacuum pump) and a device for strething and heating plastic sheet.

(iv) *Plaster moulding* The mould is prepared in plaster of Paris (gypsum plaster) which is mixed with suitable quantity of water (by adding plaster to water) and additives (such as asbestos, talc, silica flour, etc.). The additives impart more strength, permeability and the refractoriness. The slurry so prepared is poured over the pattern placed in the moulding box. The mould is vibrated to allow the proper settling of slurry. The setting is usually complete in 30 minutes after which the pattern is withdrawn and the mould is dried by slow heat, to about 200 °C. Molten metal is poured after the moulds are completely dehydrated. The process offers the following advantages:

(a) Plaster has poor heat conductivity and thus chilling tendency is eliminated enabling thin sections to be cast easily.
(b) Due to very slow rate of metal cooling, molten metal can be poured very slowly thus allowing the gases to escape from the mould.
(c) High degree of dimensional accuracy and surface finish is obtained and thus fettling and machining costs are saved considerably.
(d) The method is still very useful for ornamental castings, statues, jewellery, etc. as it reporduces fine from and details.

However, some limitation of plaster moulding are:

(a) Plaster has low permeability and thus use of plaster mould requires preparation of mould and pouring of metal under vacuum. Alternatively metal can be fed under pressure by using compressed air.
(b) Owing to the low refractoriness of plaster, plaster moulding is limited to the casting of non ferrous metals.

(v) *Ceramic moulding* This process uses chemical and ceramic slurries to produce metal castings of highest precision and very fine finish. A refractory slurry prepared by mixing specially developed ceramic aggregates and a liquid chemical binder (alcohol based silicon ester), is poured over the pattern placed in a flask. The slurry fills up all the cavities and recesses by itself and sets in 3-5 minutes time. The pattern is then withdrawn and the ceramic mould removed from the flask is treated with catalyst or hardener to produce full chemical stabilisation. The mould is then heated at 900 – 1000 °C in a furnace to expel the binder completely. Heating time varies depending on the size and sectional thickness of the mould. In the heated mould, the metal is poured and allowed to cool down slowly as in the case of investment casting. This process is also known as Shaw Process and Unicast Process. The processes offer the following advantages.

(a) The process is suitable for all types of cast metals including reactive metals, such as titanium and uranium.
(b) In general there is no need of risers, venting, or chilling as the cooling rate is very slow.

The ceramic moudling process is specifically suitable for making forging dies, tyre moulds, dies for plastic moulding, dies for drawing and extrusion, patterns for shell moulding, impellers of pumps having very narrow passages, parts of atomic reactors, etc.

4.6 CORES AND CORE MAKING

The cores are separate shapes of sand and are essentially a mould component of sufficient strength. The cores are placed in the mould to provide castings with contours, cavities, passages, etc. that are not otherwise practical or physically obtainable by the mould. Many castings are produced at reasonable cost only due to the use of cores. In general, cores may be considered as the shapes of dry mass prepared separately (by baking, air setting, CO_2 setting, etc.) and then placed in the mould. Most of the principles that apply for preparation of moulds, apply also to preparation of sand cores.

The principal functions of the cores are summarised in Fig. 4.19. These functions include the following:

(i) To provide cavity in hollow form (Fig. 4.19(a))
(ii) To provide under cut features (Fig. 4.19(b))
(iii) To provide deep recess (Fig. 4.19(c))
(iv) To provides cover core to close the mould cavity and thus dispensing with a conventional top part (Fig. 4.19(d))
(v) To provide full core assembly as block mould (Fig. 4.19(e))

The cores should have core prints or projections locating in corresponding mould recesses to support the core in the correct position.

4.6.1 Requirements of Cores and Core Sands

The cores should fulfill the following requirements:

(a) During the preparation of core, it should be strong enough to retain its shape without deforming.
(b) After baking or drying the core should be strong enough to withstand handling and to resist erosion and deformation by metal during the filling of the mould.

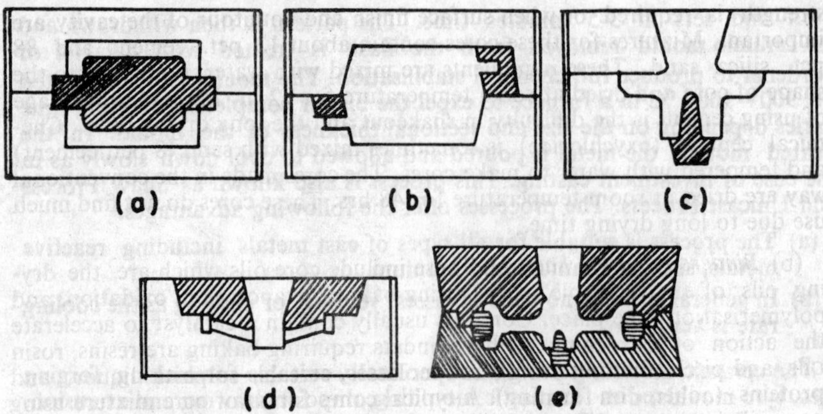

Fig. 4.19 Functions of core: (a) cavity in hollow form, (b) under cut, (c) deep recess (d) cover core (e) core assembly

(c) The core should be stable with minimum of contraction and expansion in order to produce a true form of the casting.

(d) The core should be sufficiently low in residual gas forming materials to prevent excess gas from entering the metal.

(e) The core should collapse after solidification of metal to minimise strains on the casting and to facilitate removal of the core from the casting during shake out.

The composition of cores varies considerably although sand is the basic ingredient. Binders and fillers are the other ingredients. In general, a high silica sand containing very little clay (if any) is best suited due to its low cost and ability to withstand high temperature, but zircon, olivine, carbon, chamotte sands can also be used in some specific cases. Coarer sands permit greater permeability and are preferred for cores. In selection of core sand, it must be considered that most of the sand from burned out cores enters the moulding sand system and may represent a significant portion of the foundry sand. Hence, a sand selected for core should be suitable for both moulds and cores.

Core binders most commonly used can be classified in the following four general types i.e.

(a) Inorganic binders that harden at room temperature

(b) Binders that requirebaking to harden

(c) Organic resin binders and

(d) Clays.

In order to meet the specific requirements of cores, more than one binder may be employed.

(a) *Binders that harden at room temperature* These include sodium silicate (used in CO_2 process), portland cement, and chemical cement such as oxychloride 20 pct. $MgCl_2$ and 80 pct. dead burned magnesite). The details of sodium silicate—CO_2 process have been already decsribed under the section 4.4.7. Portland cement can be used for heavy castings when extra core strength is required or when surface finish and countour of the cavity are important. Mixtures for these cores contain about 12 pct. cement and 88 pct. silica sand. These ingredients are mixed with water, rammed into the shape of core and dried at room temperature for 72 hr. Main disadvantage of using cement is the difficulty in shakeout and the long drying time. Chemical cement (oxychloride) is sometimes mixed with sand (8 pct. cement) and tempered with water to make cores. The cores made in the conventional way are dried at room temperature for 48 hrs. These cores do not find much use due to long drying time.

(b) *Binders requiring baking* These include core oils which are the drying oils of the vegetable type. During baking, evaporation, oxidation, and polymerisation take place. Core oils usually contain a catalyst to accelerate the action of hardening. Other binders requiring baking are resins, rosin oils, and pitch (thermosetting), and molasses, cereals, sulphite liquor, and proteins (adhere on heating). A typical composition of core mixture using core oils is given in chapter 3.

(c) *Organic binders* All the organic binders used for moulding sand mixture can be employed for core making also. The details of these binders and the processes used are already given in section 4.4.8.

(d) *Clays* Bentonites and fire clays also are used as binder materials, but mainly to increase green strength. The clays are used in combination with core oils or some other type of binders.

Other core materials include siltca flour (to increase the hot strength of the core) and saw dust or wood flour (added as filler to increase collapsibility of the cores).

4.6.2 Types of Cores

The cores employed in foundries can be classified as following depending on their shape and position in the mould (Fig. 4.20).

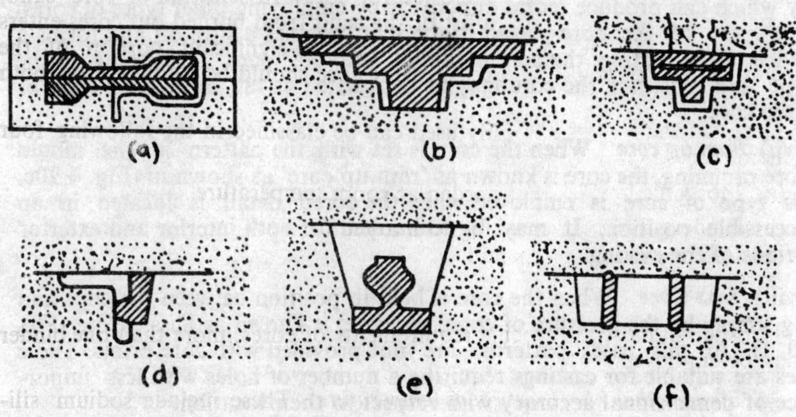

Fig. 4.20 Different types of cores: (a) balanced eore, (b) cover core, (c) hanging core, (d) wing core, (e) ram up core, (f) kiss core

(i) *Horizontal core* This is the most common type of cores which is usually in cylinderical form and is laid horizontally in the mould. The core ends rest in the seats provided by the core prints on the pattern. Horizontal cores are used more frequently and are kept ready in various diameters and lengths and thus termed as "stock cores"

(ii) *Vertical Core* This refers to the core placed along a vertical axis in the mould. The top and bottom ends of the core fit into the seats provided in the cope and drag halves of the mould. These cores are also used more frequently and are also termed as 'stock cores' as horizontal cores and are kept ready in various diameters and lengths.

(iii) *Balanced core* When there is an opening in the casting only one side and only one core print is available on the pattern, a core known as balanced core (Fig. 4.20a) is required. In such cases, the core print should be quite large to support the weight of the core extending into the mould cavity. The core should be able to withstand the force of buoyancy of the molten metal surrounding it. Sometimes chaplets (small pieces of same metal as of casting) are inserted to support the core in the mould cavity.

(iv) *Cover core* A cover core is shown in Fig. 4.20b. It stretches vertically down words instead of extending horizontally as in the case of balanced core. A cover core is required in the situation when the entire pattern is rammed in the drag and the core is required to be suspended from the top of the mould.

(v) *Hanging core* The cores hanging from the cope without any support at the bottom of the drag are known as 'Hanging Core' (Fig. 4.20c). In this case, the core may be fastened on the top side of the cope with a wire or rod extending through the cope.

(vi) *Wing core or stop off corc* This type of core is shown in Fig. 4.20d. This type of core is employed when a hole or recess is to be obtained in the casting either above or below the parting line. The use of this type of core becomes essential when pattern can not be placed in the mould in such a way which can produce recess directly or by employing other types of cores. As a part of the core placed in the seat becomes a stop off and forms a surface of the casting, the core is called "stop off" core. Depending on the shape and position, the core may be designated as tail core, chair core, or saddle core.

(vii) *Ram up core* When the core is set with the pattern in the mould before ramming, the core is known as 'ram up core' as shown in (Fig. 4.20e. This type of core is employed when the cored detail is located in an inaccessible position. It may be employed for both interior and exterior portions of the casting.

(viii) *Kiss Core* When the core is held in position between the cope and drag simply by the pressure of cope, the core, is known as Kiss (Core Fig. 4.20 f). In this case, patterns are not provided with core prints. These cores are suitable for castings requiring a number of holes with less importance of demensional accuracy with respect to their location.

4.6.3 Core Boxes and Ramming of Cores

Cores may be rammed by hand, jolting, squeezing and blowing operations analogous to those in moulding. As the moulds are produced in moulding boxes, the cores are produced in the core boxes, the details of which have already been discussed in chapter 3. Special core box features are required for special techniques of core making such as core blowing, hot box, cold box techniques, etc. When small cores are required in large quantity porduction times can be greatly reduced by the use of multigang core boxes having several core impressions, This system enables a number of cores to be rammed or blown simultaneously and thus reducing the time in stripping and reassembly. In addition to the use of core boxes, standard core sections can be produced hy extrusion of sand mixture through an orifice using a simple machine.

There is no need of description of those operations parallel in the moulding field such as hand ramming hot box process, CO_2 process, cold box process, etc., major difference is that most core making is based on sand mixtures of low green strength, needing little ramming. In case of very large cores, partial or complete hardening can be accompolished within the core box. However, the most distinctive technique of all core making methods is the use of core blowers for the mass production of small and medium sized cores.

Core blowing This is accompalished by filling the core box cavity with sand mixture at high velocity by the use of compressed air (560 – 700 kN/m^2). The air is exhausted through the vent holes in the core box and core plate, while the sand is retained by the cavity walls. Core blowing is a high speed operation and under proper conditions. a blown core ean be delivered within 2-6 seconds, depending on the size of the machine.

Core blowing machines have several forms ranging from bench models producing small cores, to heavy floor models with sand capacities of several hundred kilogrammes. A simplest type of bench blower consists of a separate sand cartridge (manually filled tube) which is inserted between blowing head and core box. The whole assembly is clamped together and the main air valve is opened to propel the sand slug into the core box.

Heavier machines (horizontal or vertical in construction) or known as pneumatic core blowing machines consist principally: (a) a sand reservoir into which air is introduced directly at mains pressure through a quick active valve, and (b) blow plate arranged to travel back and forth from the filling position to the blowing position.

This type of machine may also incorporate agitator blades to prevent channelling in the sand chamber. A typical vertical pneumatic machine is illustrated in Fig. 4.21. In the filling position, the upper throat of the sand reservoir is open to receive sand from thc hopper. After filling, the reservoir and blow plate are moved into position directly over the core box. The machine table and core box are raised to form a seal between the sealing ring in the head of the machine and the upper throat of the reservoir, and between the blowplate and the core box itself. When the elements of the machine are brought to this above position, compressed air is applied.

After the blow has taken place, the table is dropped and the reservoir is returned to the filling position The squence of operation may be controlled by a combination of hand and foot operating valves or by the use of a fully automatic operating valve. In core blower, compressed air consumption and core box wear are quite high.

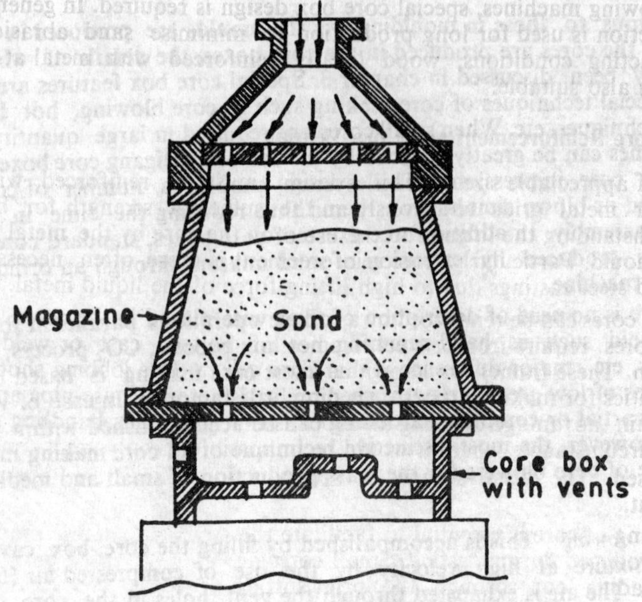

Fig. 4.21 Section through core blower

The other important machine is the core shooter (Fig. 4.22). In this case blowing is accompalished by the admission of a fixed volume of compressed air from a separate machine reservoir to the sand chamber through a large capacity quick acting valve (mean while the valve to the main air line is kept closed). The rapid pressure build up in the blowing chamber results in effective sand propulsion. In core shooter, air consumption and core box wear are greatly reduced. In core shooter, sands of appreciable green strength can be employed as flowability is less important.

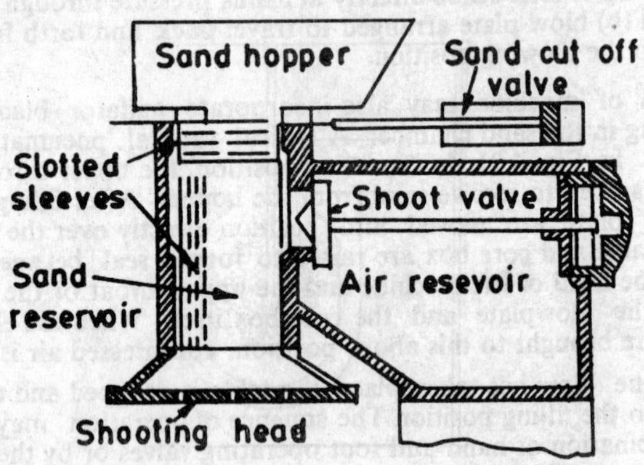

Fig. 4.22. Section through core shooter

In blowing machines, special core box design is required. In general metal construction is used for long production to minimise sand abrasion. For less exacting conditions, wood locally reinforced with metal at wearing points is also suitable.

4.6.4 Core Reinforcement and Core Venting

Cores of appreciable size or fragile design must be reinforced with core irons or metal grids to give them the suffieient strength for handling and withstanding the lifting force exerted on the core by the metal under it in the mould. Particularly, core reinforeement is more often necessary for iron and steel castings due to high lifting force of the liquid metal.

Small cores can be reinforced by a rod or wire placed in the cores, whereas large cores require iron or steel frames of bolted, cast or welded construction. These frames are known as *Core arbors*. In a jobbing shop having no facilities for making arbors, it is more satisfactory to use pipe and rods. The main advantage of using an arbor is that it permits handling of core under circumstances where it would be impossible if only rods and pipes were used. The disadvantage of arbor is the difficulty in its removal at shakeout.

Venting of core is essential to facilitate the escape of volatile matters and gases produced during casting. Small cores are usually vented with pierced holes leading out through the print surface or alternatively, waxed cord, string, or woven nylon vent tubing can be incorporated during core making. In case of large cores, pierced metal pipes can be moulded round which serve both to strengthen and as conduits for gases. Coke centres can also be used with vent tubes to permit the escape of gases through the core print. Examples of core vents are illustrated in Fig. 4.23.

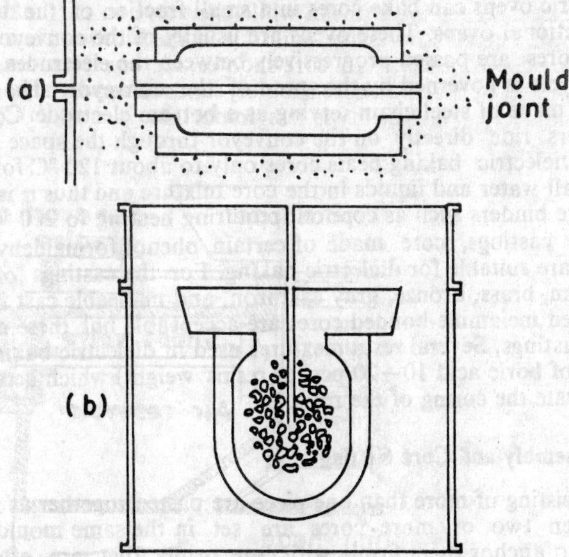

Fig. 4.23 Core venting: (a) central vent in small core, (b) coke centre in heavy core

4.6.5 Core Baking

The cores are heated to remove moisture and to develop the strength of the binder. For baking of cores, various sizes and designs of the ovens are available. The temperatures for baking may vary from 150° C to 400° C. The type of the oven and temperature of baking employed depend on the size of the cores, duration of baking time and binder used. The heating of ovens may be carried out by coal, coke, oil, gas or electricity depending on the availability, cost and quality of cores required. The ovens are usually of the air circulated type which distributc the heat uniformly all over and carry away the moisture liberated from cores. Dielectric heating is employed in modern ovens for the production of high quality cores made from resin binders.

The core baking ovens can be classified as batch type or continuous type. Batch type ovens make use of portable racks which are loaded with cores and transported to the oven by suitable mechanical means (trucks, trolley or conveyor). The batch type ovens may also be of drawer type in which the cores are placed in a number of sliding drawers which are operated by an overhead pulley. The batch type ovens are usually suited in jobbing foundries and production of castings in batches.

Continuous or a vertical type ovens are preferred for the high rate of production of small and medium size cores. In continuous type the oven is operated steadily at a fixed temperature and all cores remain in the heated zone for the same length of time. This restricts the use of continuous type ovens for only a fairly narrow range of sizes. In continuous type ovens, the loading and unloading of cores is continuous and the baking time is controlled by the rate of travel of conveyor. The temperatures of the various zones of the oven and the speed of the conveyor are so coordinated that the core emerges from the oven as baked and cooled.

Dielectric ovens can bake cores in a small fraction of the time required in conventional ovens. These ovens are usually of the conveyor type wherein the cores are passed progressively between the electrodes, and thus the heating cycle is governed by the speed of the conveyor. The conveyor is generally made of steel chain serving as a bottom electrode. Core plates and core driers ride directly on the conveyor through the space between electrodes. Dielectric baking heats cores only to about 120 °C for the time to vaporise all water and liquids in the core mixture and thus it is not possible to use core binders such as core oil (requiring heating to 2C0 °C or more). For steel castings, core made of certain phenol formaldehyde resin sand mixtures are suitable for dielectric baking. For the castings of aluminium, magnesium, brass, bronze, gray cast iron, and malleable cast iron, dielectrically baked melamine-bonded cores are acceptable but these cause porosity in steel castings. Several resin mextures used in dielectric baking require the addition of boric acid 10–20 pct. of resins weight) which acts as a catalyst to accelerate the curing of the resin.

4.6.6 Assembly and Core Setting

Core consisting of more than one piece are pasted together as a sub assembly. When two or more cores are set in the same mould and it is not possible to anchor them firmly with core prints, they are often pasted in place. Sometimes, two cores are pasted together in the assembly to seal the joint against metal penetration. Though a ready mixed paste can be used for sealing, core paste can be prepared by mixing talc, dextrin, flour and molasses.

Patching of slightly damged core can be done by the core sand mixture as used in making of core. Generally, patching should be done before baking. However, patching can also be done after core assembly in which case the patches can be dried by baking or torching before setting the core into the mould. Talc sealers are used between moulds and cores to prevent runouts. A satisfactory sealer is made by mixing 4.5 kg. of talc and 6 kg of core oil. Use of sealer also prevents blow holes caused by cores which are surrounded by metal. *Core mudding* is carried out by using proprietory mudding compound to smooth rough surfaces and fill small holes and seams. They must be dried thoroughly, otherwise casting defects will be produaed.

5

Permanent Mould Casting Processes

In general the permanent mould casting processes refer to metal mould casting processes which differ from sand casting processes. Metal moulds have the following advantages over sand moulds.

(a) Metal moulds produce castings with comparatively closer tolerances and smooth surface finish.

(b) Metal mould can be housed repeatedly to obtain a large number of castings.

(c) High thermal conductivity of the mould material greatly facilitates the solidification and cooling of the casting, resulting in improved mechanical properties.

(d) High casting yeld is obtainable owing to smaller machining allowances, which results in saving of metal.

(e) The metal mould processes can be mechanised and automated with improved working conditions.

(f The production rate is very high. (One cycle takes only abuot 15 seconds). However, there are also some unfavourable points

(i) Cost of metal moulds is high.

(ii) The process is uneconomical for production of small number of castings (not less than 200 for small castings and 50 for large castings.)

(iii) The process tends to form hard spots in iron castings, the elimination of which may require annealing of castings.

In general, the metal mould casting process involves the following steps.

(a) Preparation of moulds including cleaning and heating operations and application of facings and points to the working surfaces of moulds.

(b) Assembly of moulds by setting the cores in the mould, closing the mould and fasting the complete mould assembly.

(c) Pouring of mollen metal into mould and allowing sufficient time for the solidification of the castings.

(d) Removal of castings after cooling by opening the mould mechanically or mannually.

(e) Fettling and heat treatment of castings.

Metal mould casting processes include permanent mould casting or gravity die casting, pressure die casting, vaccum diecasting, low pressure die casting, centrifugal casting, continuous casting, semi continuous casting, squege casting, slush casting, shot casting and flus casting. Non-metallic permanent moulds include rubber moulds, graphite moulds and ceramic baked moulds.

5.1 GRAVITY DIE CATING OR PERMANENT MOULD CASTING

In this process molten metal is fed into moulds by the force of gravity without any external pressure. The castings obtained are much thicker in cross section (more than 6 mm) since gravity alone does not provide adequate pressure to force the molten metal into metal moulds. The process is suitable relatively for fluid alloys due to high freezing rates in metal moulds. The moulds need repair and renewal only after long period of usages.

5.1.1 Dies for Gravity Die Casting

The dies are usually simple in construction which are in pair and produce a cavity for the complete casting when the two halves are placed together. The dies are provided with the following arrangements.

(a) Feeding of molten metal,

(b) Vents for escape of gases, and

(c) mechanism for clamping and ejection.

The joint on the dics may be straight (vertical or horizontal) or curved depending on the shape of the casting.

The die may also consist of several pieces with numerous joints for producing complexe shapes. A generous taper allowance is given to (facilitate) the removal of castings and to reduce the wear and tear to the die surface. In sand casting, no allowance for thermal expansion of mould is given, whereas in die casting an allowance for thermal expansion of the dies is given in addition to the normal contraction allowance for the casting. The castings should be ejected soon after pouring to prevent hinderance to contraction on further coling. In order to increase the production rate (to deuce the cycle time), the dies may be either water cooled or air cooled. In rgeneral following four types of dies are in common use:

(i) *Two part die* A typical two part die is split along a vertical joint line passing through the die casting (Fig. 5.1).

(ii) *Recessed die* The die is made hollow to decrease its weight and improve the thermal balance. In the hollow portion ribs may be provided to increase the strength and stiffness of the die.

(iii) *Hinged die* In this case two halves of the die are hinged to facilitate their assembly and mutual location. This is particularly advantageous in case of mannual operation of small dies.

(iv) *Multi staged die* This type of die is made up of several pieces which are either joined together firmly or kept loose to facilitate the production.of complex shapes not obtainable from simpler shapes.

5.1.2 Die Materials

The dies are usually made of high qualty cast iron or die steel, graphite, copper, and aluminium have also been used. for some specific applications When using dies of cast iron or die steel, they are preheated (to a temperature of 200 – 500 °C depending on the alloy to be cast) to reduce the cooling rate. However, only initial heating is necessary and later on the equilibrium temperature of 300 – 400 °C is maintained by a regular time cycle for casting and ejection. All the dies are coated with insulating refractory coatings to prevent welding of the casting to the mould and to provide more uniform rate of cooling of the casting. The coating material may be different for die and the feeding system. The usual coatings applied for aluminium and magnesium alloys are shown in table 5.1.

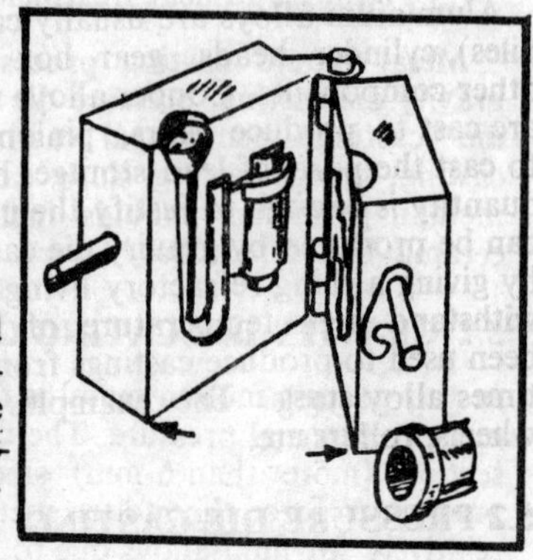

Fig. 5.1 Two part die for permanent mould casting

TABLE 5.1 Composition of Coatings for die Costings

Alloy	*Part of mould*	*Composition in gms/litre of water*				
		Zinc oxide	*Sodium silicate*	*Absestos*	*Talc*	*Boric Acid*
Aluminium Alloys	Dies	75	50-60	—	—	—
	Feeding system	130	60	25	—	—
Magnesium Alloys	Dies	—	35	—	100	65
	Feeding system	—	30	100	—	—

5.1.3 Cores for Permanent Mould Castings

Cores are used to obtain (holes) cavities and under in the castings and may be of sand, cuts plaster or metal. Sand and plaster cores offer greater flexibility of design. The metal cores may be either collapsible or heavily tapered to facilitate their removal while the castings are still hot. The cores may be solid or hollow. The solid cores are easy to prepare and are suitable for small castings. Hollow cores weigh less, get heated and cooled easily and thus are preferred for large castings. For higher cooling rate, the cores may be water cooled.

5.1.4 Metals and Alloys Cast by Permanent Mould Casting

Permant mould casting process has been successfully used for the commercial production of all the common metals and alloys such as tin, zinc, aluminium, magnesium, copper, their alloys, cast iron and steel. The process is applicable for the light and medium weight range (0.5—50 kg) which are roughly equivalent to those produced as repetition sand casting by machine moulding.

Aluminium alloys are usually cast to produce pistons (for light automobiles), cylinder heads, gear box (housings) compressor parts, and similar other components. Copper alloys which are fluid enough in molten state, are cast to produce gears, pinions, impellers, etc. lead alloys are used to cast the parts of lead storage batteries and other components. When quantity is enough to justify the use of pressure die casting, zinc castings, can be produced by graviry die casting. Cast iron castings can be produced by giving a thing refractory living or coating to the mould surfaces (to withstand high temperature of molten metal). Permanent moulds have been used to produce castings from low carbon, medium carbon and sometimes alloy steels. The examples of steel castings are car wheels, gear wheels, rollers etc.

5.2 PRESSURE DIE CASTING

This is modified permanent mould process in which the molten metal is injected into the die cavity at high pressure instead of flowing by gravity and allowed to solidify under externally applied pressure. The maintenance of high pressure during solidification results in high quality castings. Pressure die casting process offers the following advantages:

(a) Intricate components can be cast at relatively low cost and high production rate (100-400 castings per hour).

(b) The castings are produced with low surface roughness and can be buffed directly.

(c) Dimensional accuracy of castings is high (+0.025 to 0.075 mm) and thus requiring practically no machining.

(d) The molten metal injected under pressure quickly takes the shape of the mould and thus very thin section (down to 0.8 mm) can be cast.

(e) Due to saving in labour and time, the process is very economical for the production of large number of castings.

(f) Rapid cooling rate results into casting of high strength and quality.

(g) Dies retain their accuracy and reproducibility for a long time. For example, dies for zinc alloys may produce as many as 5 million castings.

The process of pressure die casting is however limited to certain non ferrous metals and small sized castings. The process is further suited for produciug large number of costings, since the cost of dies and die casting machines is quite high.

5.2.1 Dies and Gating System for Pressure Die Casting

The dies employed in pressure die casting are similar to those employed for common type of permanent moulds. For pressure die casting, machinery is used for exerting pressure, opening and closing of dies, ejection of casting, etc. In the machine, one die half is stationary ealled the cover die whereas the other half opening and closing is known as ejector die.

The dies are generally made of an alloy steel (5 pct. Cr, 1 pct. Si, 1 pct

Mo, 1 pct. V, 0.37 pct. C hardened to about 450 B.H.N.) which is (i) stable to dimensions (ii) resistant to heat, (iii) resistant to soldering with metal and (iv) resistant to erosion. Other materials for dies idclude cast iron, heat resisting tool steels, nickel and cobalt base alloys, and refractory metals. The die cavities are machined very accurately.

The die cavity is usually designed with a taper of 0.2 to 1.0 per cent with the joint face. Greater allowances are provided for core surfaces and shallow walls. The taper on the vertical faces (draft) minimises the damage to the die surface during ejection and there by increases the die life.

Water cooling channels may be provided in the die blocks to prevent local overheating in the region of the gating system. The dies should be pointed or lubricated to prevent excess rise in surface temperature which should be maintained within an optimum operating range depending on the metal/alloy cast viz. about 200 °C for zinc base alloys and about 300 °C for aluminium base alloys. The moving parts of die assembly also need frequent lubrication. Usually, a thick suspension of graphite in oil is used as a lubricant.

The dies made of conventional material (die steel) generally have a working life as follows

Copper base alloys	<10,000 castings
Zinc base alloys	>100,000 castings
Aluminium or Magnesium base alloys.	20,000—100,000 castings

In individual cases much longer working life can be obtained.

5.2.2 Cores in Pressure Die Casting

Metal cores are provided to form the desired cavities in the casting. The cores can be retracted as part of the ejection sequence. The movement of the core is accomplished with the help of a cam, rack and pinion or piston mechanisms actuated by the opening stroke of the machine or as a separate movement. The cores may be fixed or loose which can be actuated automatically (Fig. 5.2) or set by hand for each operation in case of deep holes.

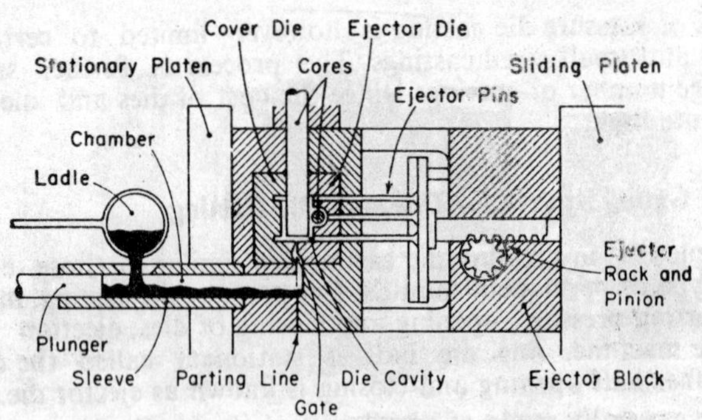

Fig. 5.2 Fixing of cores in pressure die casting

5.2.3 Die Casting Machines

The die casting machine should perform the following two main functions:

(a) Clamping the die and actuating its movements during the production and ejection of a casting.

(b) Injecting the molten metal into the die.

In addition to performing the above two functions, the machine should be sturdy in design to support the die halves in correct allignment. Further. the clamping force needeed to hold the die halves in position should be greater than the maximum force developed by the molten metal to ensure no leakage at the parting line. In many cases, the clamping force may be as high as 1000 tonnes, depending upon the die size and the pressure of molten metal and is a crucial feature for a die casting machine. This is determined by the projected area of the casting and gating in the plane normal to the direction of closing. The movements of die may be actuated by hydraulic and mechanical, or purely mechanical mechanism.

The basic principle of a typical machine is illustrated in Fig. 5.3. Now a days, very large die casting machines are in use which have locking forces exceeding 2000 tonnes. A large machine employed for the production of aluminium alloy castings for automobiles is shown in Fig. 5.4. The modern machines are provided with automatic control of operational phases.

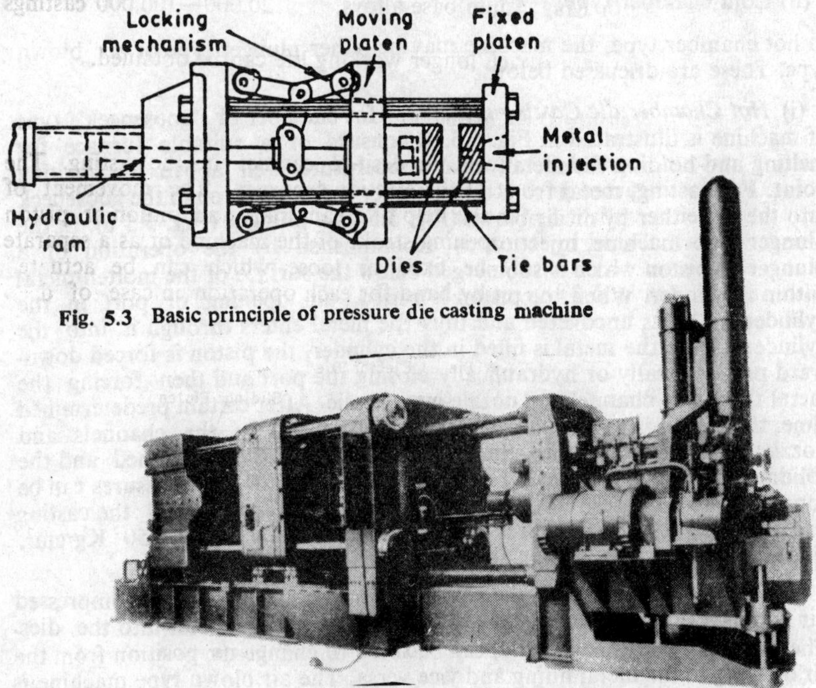

Fig. 5.3 Basic principle of pressure die casting machine

Fig. 5.4 1500 ton cold chamber die casting machine

Closing force of die actuating system, P_{cl} can be represented by the following formula

$$P_{cl} = KP_{pr} + \frac{m_{pr}^2 W_n^2}{2\Delta l (m_{pr} + m_m)} \qquad (5.1)$$

Where $m_{pr} = G_{pr}/g$ = mass of the plunger and attached parts in kg. Sec²/m, G_{pr} = weight of the plunger, m_m = Gm/g = mass of the mobile part of the moved punch, W_n = rate of plunger movement at the final stage of pouring, Δl = gap between the plunger and die, K=a factor determing the displacement of the plunger in lhe machine.

$$K = \frac{\text{Area of casting projection}}{\text{Cross sectional area of nozzle}} = \frac{Ac}{An}$$

From equation (5.1), it may be observed that the main effect on closing force is due to the mass of plunger (m_{pr}) and the velocity of the pressure punch (W_n) If Δl and m_n are higher, the closing force required will be less The closing force is also affected by the casting projection area (A_c) and. cross sectional are of nozzle (A_n). Increase in A_c results in increase of closing force, whereas reverse is true for A_n.

Basically, the pressure die casting machine can be classified in two categories i.e.

(i) Hot chamber or gooseneck and

(ii) Cold chamber type.

In hot chamber type, the machine may be either plunger type or air blown type. These are discussed below:

(i) *Hot Chamber die Casting machine*: Hot chamber or 'gooseneck' type of machine is illustrated in Fig. 5.5. It consists of a suitable furnace for melting and holding the metal, at a temperature well above the melting point. For casting, metal from the chamber is forced through the gooseneck into the die either by or air blown (Fig. 5.5 a) a plunger (Fig. 5.5 b). In the plunger type machine, injection is accomplished by the operation of a plunger or piston which is submerged below the surface of the molten metal within a cylinder. When the piston is raised, an opening or port in the cylinder wall gets uncovered and thus the metal enters through it into the cylinder. After the metal is filled in the cylinder, the piston is forced downward pneumatically or hydraulically closing the port and then forcing the metal through a channel and nozzle into the die. After certain predetermined time, the piston is raised allowing the excess metal in the channels and nozzle to drop back into the cylinder. The die is then opened and the solidified casting is removed. The metal feeding speeds and pressures can be controlled to any desired levels depending on the metals and the casting size and shape. The pressures employed are usually below 150 Kg/cm², since higher pressures do not give any advantage.

In case of air blown type die casting machine (Fig. 5.5 b), compressed air is blown (air pressure is raised) to force the molten metal into the dies. The chamber (gooseneck bottle) can be tilted to change its position from the air blowing to the metal filling and vice versa. The air blown type machine is much simpler in operation and construction than the plunger type, but the

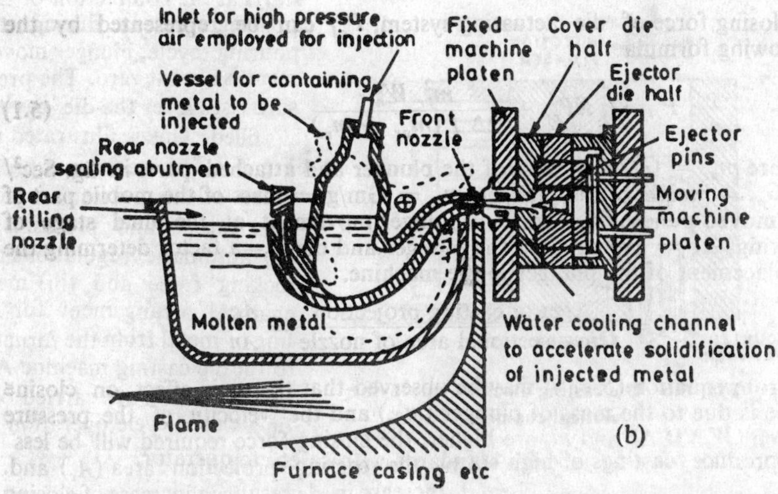

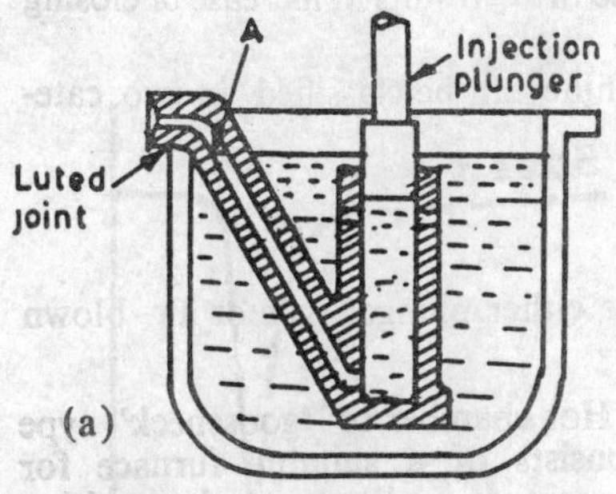

Fig. 5.5
Hot chamber die casting machine:
(a) Operated by direct air pressure
(b) Operated by air actuated arm

former is restricted to pressures in the region of 25 – 30 kg/cm^2. Air blown type machine requires more attention as the most of the work is carried out mannually. Therefore, plunger type hot chamber machines are largely preferred due to much higher rate of production and easy adoptability to mass production.

(ii) *Cold chamber die casting machine* The general principle of a cold chamber die casting machine is illustrated in Fig.5.5. It consists of a cylindrical chamber through which molten metal is introduced. After feeding the chamber with slightly more metal than is needed to fill the die, the piston is advanced to force the metal into the die. After allowing some time for solidification, the die is opened and the piston pushes out the flat piece or excess metal from the cold chamber. Finally, the casting is removed. The whole operation is carried out automatically and completed within few seconds.

In cold chamber machines, much higher injection pressures (300—1600 Kg/cm^2) can be employed which permit the use of lower metal temperatures. The casting of greater intricacy can be obtained with better quality (less entrapped air and less shrinkage effect). Most commonly used cold chamber die casting machine embodies a horizontal chamber as shown in Fig. 5.5(a) but a vertical chamber (Fig. 5.6) can also be used.

Modern cold chamber machines consist the provision for multiphase injection of molten metal to ensure complete and uniform filling of die cavity, and thus preventing porosity in castings. The pressure build up

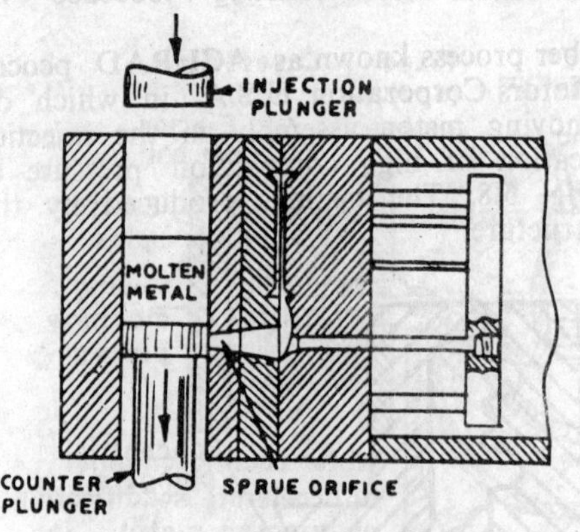

Fig. 5.6 Die casting machine employing vertical chamber

starts at the completion of the pouring cycle i.e. during the pouring cycle, plunger movement is almost zero. The pressure rises after the die cavity filled. This is illustrated in Fig. 5.7 The machine is provided with (i) variable injection speeds and presusres, (ii) automatic regulations of locking force and (iii) mechanical arrangement for ling of metal from the furnace to the die casting machine. All these function are carried out on a closed cycle operation to produce castings of high standards with safety to operator.

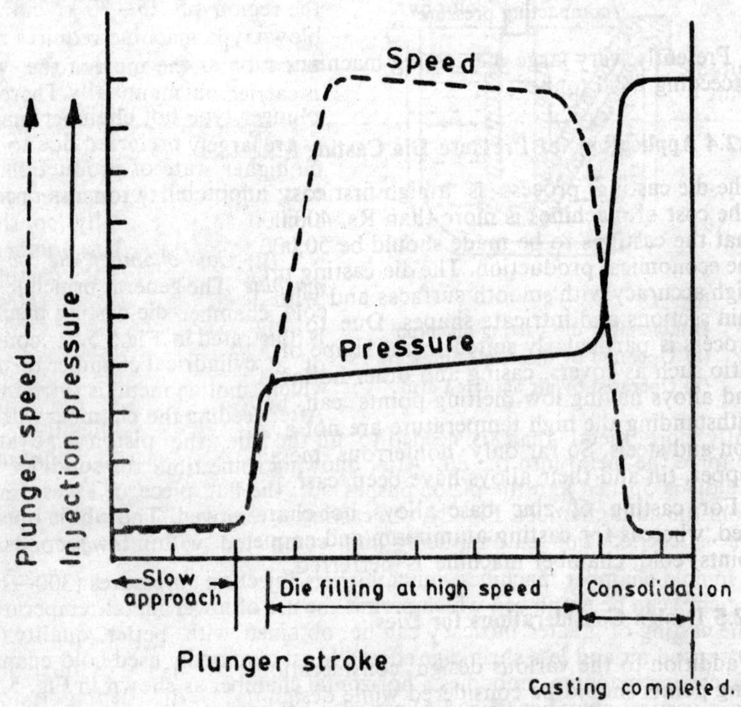

Fig. 5.7 Changes in plunger speed and pressure during metal injection in

A modified form of cold chamber process known as ACURAD pcocess has been developed by General Motors Corporation U.S.A. in which die cavity is first filled slowly by a moving piston assembly in the injection cylinder. It is followed with application of high compaction pressure by secondary plunger as shown in Fig. 5.8. The castings produced by this process are accurate with dense structure.

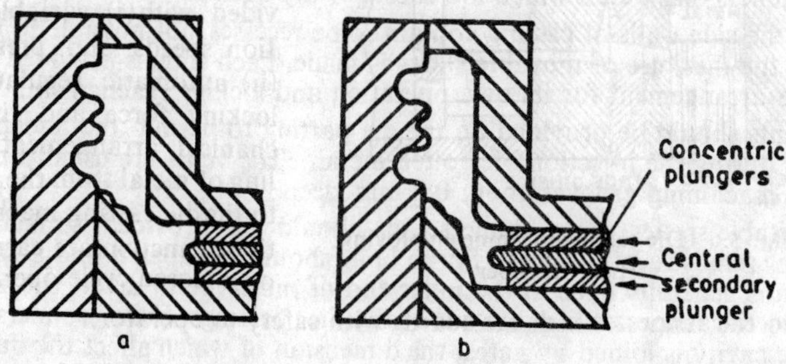

Fig. 5.8 ACURAD Process: (a) slow filling of die cavity, (b) using high compacting pressure

Presently, very large die casting machines are in use with lokcing forces exceeding 2000 tonnes.

5.2.4 Applications of Pressure Die Casting Machines

The die casting process is a high first cost, high production rate process. The cost of machines is more than Rs. 400,000. It is generally considered that the castings to be made should be 50,000 to 100,000 in number for the economical production. The die casting process produces the castings of high accuracy with smooth surfaces and with design capacity for extremely thin sections and intricate shapes. Due to high rate of metal flow, the process is particularly suited to the shape of high sueface area to volume ratio such as covers, casing and other hollow forms. However, only metals and alloys having low melting points can be cast or since die materials withstanding the high temperature are not available. It is difficulty to cast iron and steels. So far only nonferrous metals such as zinc, aluminium, copper, tin and their alloys have been cast.

For casting of zinc base alloys, hot chamber machines are extensively used, whereas for casting aluminium and copper alloys (of higher melting points) cold, chamber machine is perferred.

5.2.5 Design Considerations for Dies

In addition to the various design considerations for sand castings, following points should be considered while designing the die casting dies.

(i) Fixed cores can be placed either in the movable or fixed portion of the die. The cores should have their axes parallel to the direction of motion of the die.

(ii) Movable cores should be provided with means for moving and locking in place.

(iii) A shape having an internal undercut or recess can not be produced by a permanent core and thus for such cases, a loose core or knock out core essential'

(iv) The castings should cling to the ejector portion and the ejector pins should not leave the objectionable marks on the casting. As far as possible, the pins should be made to bear on the other projections (runners, flash etc.) which are subsequently cut off.

(v) If the side walls of casting contain some recesses not possible to form by the dies, use of movable slides is made. Each slide is provided with the arrangement for its own operation and locking macnanism.

(vi) Vents should be provided on the die parting to permit the escape of air during the injection of molten metal. The vents are usually made by machining grooves about 0.2 mm deep.

(vii) Suitable sprues, gates, runners, etc. should be provided to facilitate the removal of casting. Sprue holls should be tapered and a spruee pin is generally fixed at the inner end of the sprue to divert the metal into the runners which are cut at the die parting. From runners, the die cavity is joined by gates, the dimension of which affect the quality of the casting.

(viii) Sometimes overflow wells should be included to collect the extra metal and to facilitate the escape of air.

(ix) Water cooling should be provided for large dies and cores needed for mass production. For water circulation, the holes are drilled through the dies.

(x) When the shape and size permit the provision of several die cavities in the same die, it should be incorporated to inerease the productivity.

5.3 VACUUM DIE CASTING

In pressure die casting, the air and other gases tend to be entrapped in the casting in the form of fine bubbles which result in low ductility and low impact strength. The gas holes can further expand during subsequent heat treatment causing blistering. These problems can be overcome by vaccum die casting process in which the die cavity is evacuated before each shot. Evacuation of die cavity reduces the porosity, assists flow in thin sections and gives better surface finish, with comparatively low injection pressures. Evacuation may be accomplished either by sealing the die halves (after clos ing) and then applying a vacuum directly to the die cavity, or enclosing the entire die assembly in vacuum tight container.

A typical system for vacuum die casting is illustrated in Fig. 5.9. In this case the die is surrounded by an airtight chamber which is evaquated before each injection. The vacuum may also be used to feed molten metal into the chamber. In this process, the reduced level of porosity in the castings reduces the risk of blistering and surface blemishes. However, except some special cases the process is not universally acceptable. It is expected that vacuum die casting process will make a major contribution in future pressure die castings.

5.4 LOW PRESSURE DIE CASTING

This technique is comparatively recent one and enables the production of

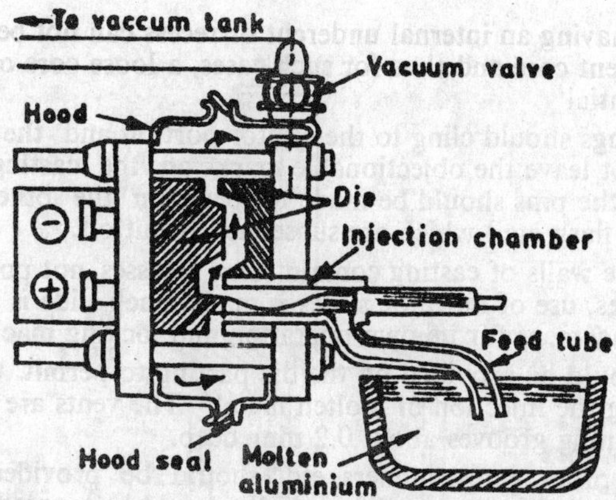

Fig. 5.9 A typical system for vaccum die casting

the castings of very thin sections and free from flaws. This process gives the yield approaching the 100 per cent even in case of aluminium and magnesium. This process is illustrated in Fig. 5.10. The metal mould or die (usually of cast iron) is filled by upward displacement of molten metal from a sealed melting chamber or bath. The displacement is achieved by applying relatively low pressure of dry air (0.5-1 kg/cm^2) at the surface of molten metal bath. The metal rises through a central metal tube and gets filled into the die cavity. Sufficient venting is provided into the die to allow the escape of air. The pressnre is maintained upto the solidification of metal and then pressure is releasad to allow the excess metal to drain back into the bath through the connecting tube. Displacement of metal into die can also be effected by evacuation of the die. In this process the metal is filled upward which avoids the use of runners and risers. There is hardly any wastage of metal and the castings produced are of very high standard in surface finish and soundness. The use of low pressure eliminates turbulance and air suction. The process also ensures smooth and progressive filling of the die cavity and the development of proper thermal gradients for directional solidification.

Low pressure die casting is suitable for the manufacture of a large number of components, such as automobile and air craft components, impellers for pumps, rain water fittings etc.

5.5 CENTRIFUGAL CASTING

Centrifugal casting is essentially the method of producing castings by pouring the multen metal into a rapidly rotating mould (metal or refractory material mould) and allowing it to solidify. The centrifugal force plays an important role in shaping and feeding of castings. The centrifugal force produced by rotation is quite large compared to usual hydrostatic forces. The centrifugal force is used for making castings in two ways, i.e. (i) distributing the poured molten metal over the outer sufaces of a mould providing the means of forming hollow cylinders and other annular shapes, and (ii) developing a high pressure in the casting during solidification.

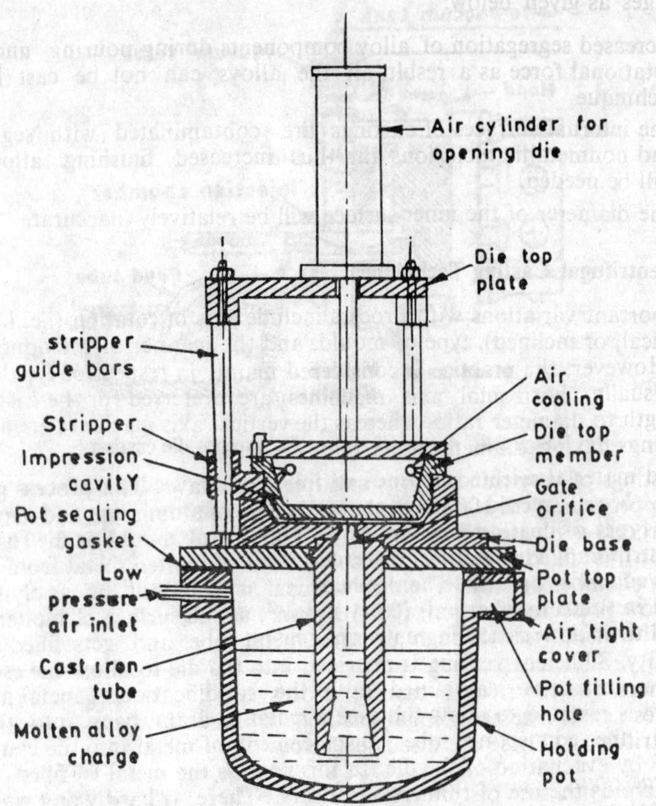

Fig. 5.10 System for low pressure die casting

The mould can be rotated about a vertical, horizontal or an inclined axis or simultaneously about its horizontal and vertical axis. The centrifugal casting process offers the following advantages:

(a) Centrifugal force in conjunction with directional solidification assists feeding and accelerates the separation of nonmetallic inclusions and precipitated gases. Thus the casting is produced free from gas and shrinkage cavities and porosity.

(b) The process is suitable and economical for producing high quality cylindrical forms.

(c) Hollow interiors in cylinders are produced without cores, since the rotating mould causes the liquid metal to move to the periphery by the rotational force.

(d) The cast tubes produced by centrifugal casting are superior in quality

(e) Practically there is no wastage of metal in gates.

(f) Production is rapid with low production cost.

The centrifugal casting method is however associated with certain disadvantages as given below:

(i) Increased segregation of alloy components during pouring under the rotational force as a result, all the alloys can not be cast by this technique.

(ii) The internal surfaces of castings are contaminated with segregates and nonmetallic inclusions and thus increased finishing allowances will be needed.

(iii) The diameter of the inner surface will be relatively inaccurate

5.5.1 Centrifugal Casting Techniques

The important variations with process include axis of rotation (i.e. horizontal, vertical, or inclined), type of mould, and the manner of introducing the metal. However, the practice is considered mainly in respect of type of product. Usually, horizontal axis machines are preferred for the castings of high length to diameter ratio, whereas the vertical axis machines are adopted for castings having diameter more than the length.

Mould material selected depends mainly on the shape and number of castings. Metal moulds (steel or cast iron) are used for simple shapes. Various types of coating are applied to moulds for their protection and ease of stripping. The coating materials include refractory slurries and dry powders containing graphite, ferrosilicon, etc. For complex shape or castings required in small number, refractory moulds are preferred. Graphite moulds also find their application in centrifugal casting for some special applications.

In general the centrifugal casting processes can be divided into three categories i.e. (i) true centrifugal casting, (ii) semi centrifugal casting, and (iii) centrifuging. These are discussed below:

(i) *True centrifugal casting* In this process the mould rotates about its (horzontal, vertical or inclined) and the metal poured is to form the internal shape by the centrifugal action. In this case no risers and no central cares are required to form the hollow shape. However, the end cores may be used to prevent the escape of metal. Plain pipe or tubes are made by this process. In this case hollow shape is produced only by centrifugal force and the wall thickness is determined by the rate of metal introduced and the rate of mould rotation.

The method of true centrifugal casting is ideal to produce hollow castings, such as gun barrels, cast iron pipes, bushings, and hollow propeller shafts. The positioning of the axis of rotation depends on the length of the casting andspeed of rotation. Usually the axis of rotation is kept horizontal for long castings (length is more than 4 times of diameter) in which case the central cavity formed is a true cylinder. Tubular castings having upto 1200 mm diameter and 4800 mm length have been produced successfully with a wall thickness as low as 6 mm.

An inclined axis is preferred for the castings having the length less than four times the diameter but more than the diameter. The vertical axis is used for very short castings. When the axis is kept either inclined or vertical, the resulting cavity is not true cylindrical but a paraboloid which can be approached to almost cylindrical shape by employing fairly high rotational speeds. The slight taper obtained can be removed by machining. The use of inclined or vertical axis is more convenient due to ease of pouring the metal and ease in removing the casting. Principle of vertical and horizontal centrifugal casting is illustrated in Fig. 5.11.

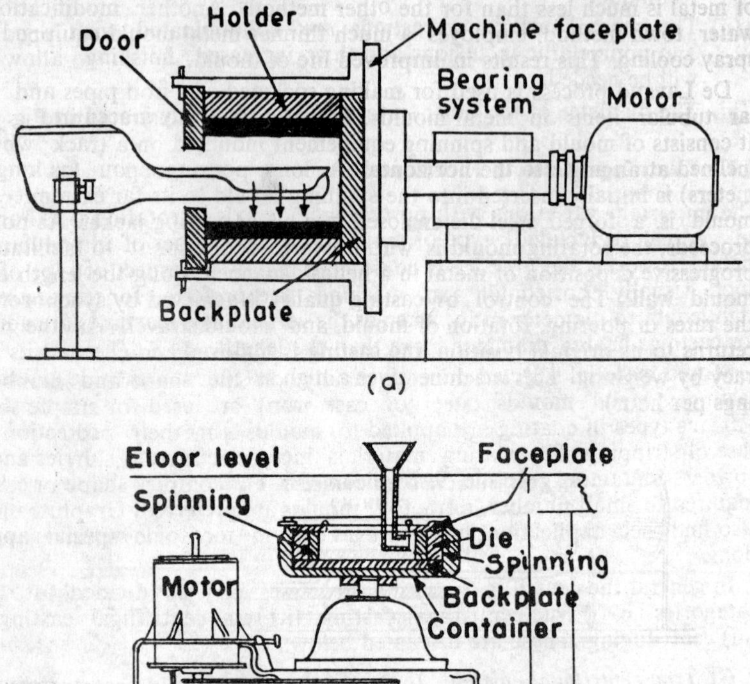

Fig. 5.11 Principle of centrifugal casting: (a) horizontal (b) vertical

Several techniques have been employed for true centrifugal casting, namely, Watertowns method, DeLavaud method, sand spun method, vertical axis centrifugal casting method, etc.

In Watertoun method, a heavy cast iron rotor is used as a chill mould, the inner diameter of which is machined to produce a gun tube of definite outside diameter. The mould is given a coating of refractory wash. The mould is rotated at speeds of about 1500 r.p.m. and a weighed amount of metal is poured into a pouring basin which is provided with a short spout at the large end of the bore. The feeding rate of metal is accurately adjusted with spinning speed by controlling the spout diameter to provide an even distribution of molten metal without causing any tear of solidified hot-metal. Slow feeding rates will cause surface laps resulting from premature freezing, whereas fast rates of feeding will result in craks due to the presence of excessive forces of molten metal on the solidified thin skin just next to the chill mould. Another important feature of the process is the application of pressure (about 5 kgs/cm^2) against the small end of the tube just after solidification. This prevents the mould restraint and consequent tearing during the lengthwise contraction in the mould.

In one modification of Watertown method known as sand spun method, a sand lining is rammed inside the steel mould. In this case the chilling rate of metal is much less than for the other methods. Another modification of water town method employes a much thinner metal mould equipped with spray cooling. This results in improved life of mould.

De Lavaud process is used for making socketed cast iron pipes and similar tubular items in metal moulds. This process is illustrated in Fig. 5.12. It consists of mould and spinning equipement mounted on a track which is inclined at an angle to the horizontal. A long pouring spou (as long as 5 meters) is initially inserted into the spinning mould to its far extremity. The mould is a forged steel die enclosed in a water cooling jacket. As pouring proceeds, the rotating mould is withdrawn over the spout to facilitate the progressive deposition of metal in a helical manner along the length of the mould wall. The control of casting quality is achieved by synchoronising the rates of pouring, rotation of mould, and mould travel. As the mould returns to its original position, the casting is removed and checked its accuracy by weighing. This machine gives a high rate of out put (30—100 castings per hour).

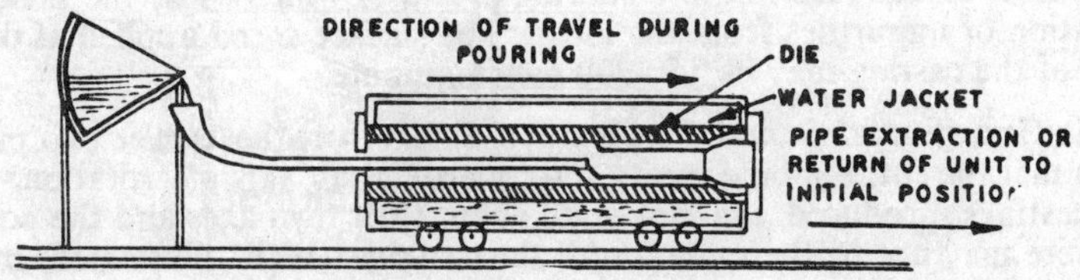

Fig. 5.12 Simplified diagram of de Lavaud pipe casting machine

(ii) *Cemicentrifugal* This technique is employed for making heavily shaped components which are symmetrical about their own axis. The examples are pulleys, spoked or disked wheels, gears and propellers. The provisions are made for directional solidification towards a central source of feed metal taking the measures as employed in static casting, such as selective chilling and padding. If the components are too small or irregularly shaped for centrifugal casting about their own axis, these are pressure cast by spacing the moulds about a central sprue which may carry out the function of a common feeder head.

Semicentrifugal casting is generally carried out on vertical axis machines to facilitate mould assembly. It is illustrated in Fig. 5.13. The system consists of a steel flash having a base plate, rings and top cover. The mould may be prepared either directly in the flash or built from high strength cores coated with refractory wash. The complete assembly is attached to the turn table by means of a fixture provided in the baseplate. The metal is poured into a central sprue with rotation of mould about its vertical axis. The metal from central sprue enters the hub and then is forced outwords to the rim by centrifugal force. For a central base in the casting, a dry sand core is best. The central sprue also functions as riser for the hub portion.

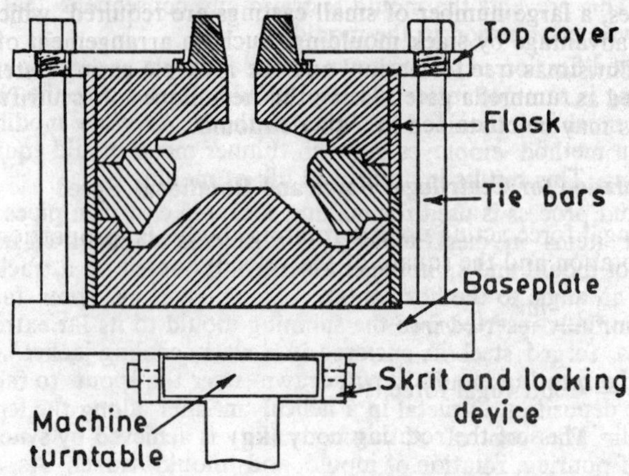

Fig. 5.13 Semicetrifigul casting

The spinning speeds in this case are much lower than in case of true centrifugal casting due to the tendency of metal to spurt out of the mould joint. Low speeds result in low pouring pressures which cause the inefficient sepration of impurities from the metal. The surface speed applied at the out edge of the casting may be 150-200 meters/minute.

(iii) *Centrifuging method* This method differs from the earlier two processes in that the entire mould cavity rotates around an axis of rotation. Thus the castings produced are not spun about their own axes and the pouring pressure may not be the same for all the castings. It is illustrated in Fig. 5.14. Several identical or nearly similar mould cavities are symmetrically arranged around the axis of rotation of a central sprue. The molten metal is fed into the central sprue which feeds the metal into cavities through a number of radial gates.

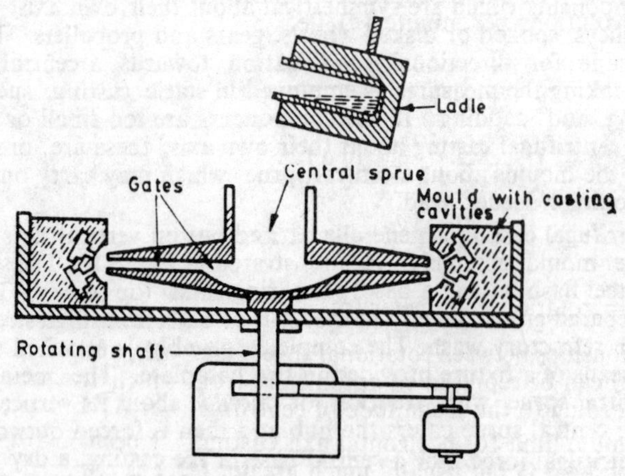

Fig. 5.14 Schematic diagram for centrifuging casting

Sometimes, a large number of small castings are required which can be made with advantage by stack moulding. Such an arrangement of casting is known as Christmas tree formation' and the multiple gating system employed is referred as 'umbrella gate'. Using the technique of ceutrifuging, even 150 castings may be contained in a single mould.

5.5.2 Calculation for Centrifugal Force and Rotational Speed

The centrifugal force acting upon a rotating body is proportional to the radius of rotation and the square of the velocity, i.e.

$$F_c = \frac{mv^2}{r} = mr\omega^2 \tag{5.2}$$

Where F_c = Centrifugal force (N)

m = Mass of the rotating body (kg)

r = radius of rotation (m)

ω = angular velocity (rad/sec)

v = peripheral speed (m/sec)

and the gravitational force (Fg) on the same mass will be

$$Fg = mg \tag{5.3}$$

Where g = acceleration due to gravity (m/sec^2)

A factor known as G factor by which the gravity force is multiplied during rotation is given by

$$G = \frac{F_c}{Fg} = \frac{r\omega^2}{g} \tag{5.4}$$

The equation (5.4) can be expressed more conveniently in terms of speed units, i.e. revolutions per minute (n), i.e.

$$G = \frac{r}{g}\left(\frac{\pi}{30}\right)^2 n^2 = \frac{0.011\,rn^2}{g} \tag{5.5}$$

or

$$n = \left(\frac{G \times g}{0.011r}\right)^{1/2} = 29.9\left(\frac{G}{r}\right)^{1/2}$$

$$= 42.3\left(\frac{G}{D}\right)^{1/2} \tag{5.6}$$

Where D = rotational diameter (m)

The relationships between rotational speed, diameter, and centrifugal force (or G factor) can be represented graphically from which the speed can be selected according to the magnitude of centrifngal force required.

The minimum limit for horizontal axis casting has been reported to be in the region of 3-4.5 G but usually much greater force is required in practice to take the full advantage of the process.

In vertical axis casting the required minimum speed is governed by the tendency to form a parabolic bore due to gravitational component of the force acting on the metal. In this case the speed of rotation in revolutions per minute (n) can be given by the following expression.

$$n = \frac{30\,(2\,g)^{1/2}}{\pi}\left(\frac{L}{r_t^2 - r_b^2}\right)^{1/2}$$

$$= 42.3\left(\frac{L}{r_t^2 - r_b^2}\right)^{1/2} \qquad (5.7)$$

Where L = axial length (m)

r_t = top radius of bore (m)

r_b = bottom radius of bore (m)

5.5.3 Mouldidg Materials in Centrifugal Casting

Metal as well as sand moulds are employed for centrifugal casting. Sand moulds may be of dry sand, green sand or skin dried sand. Cluster castings are usually made in dry sand. While employing green sands, special bonding materials are used to impart strength. Sometimes graphite moulds are also used, particularly when the metals show a tendency of wetting the mould surfaces. Metal moulds are used in the following instances:

(i) To produce large number of identical castings.
(ii) For the castings of fine grain structure.

The metal used for moulds may vary from grey cast iron to alloy steel depending on various factors such as size and composition of metal, temperature, and number of castings.

Sand moulds are preferred in the following instances:

(i) When the chilling tendency is to be prevented.
(ii) When the number of castings required is small.
(iii) Shape of casting is typical and requires destruction of the mould for the removal of casting without causing any damage to the casting.

5.5.4 Effect of Process Variables on Quality of Castings

The main variables in centrifugal casting are (a) speed of rotation, (b) pouring temperature (c) pouring speed, and (d) mould temperature.

(i) *Speed of Rotation* In case of true centrifugal casting, the important consideration is the retention of bore shape against gravity without causing any longitudinal tearing, which is much dependent on the feeding pressure (function of speed of rotation). In semicentrifugal casting, the speed of rotation is the main criterion. The structure of cast metal is also influenced by the speed of rotation. Increased speeds of rotation promote refinement. Therefore, highest possible speeds should be used with prevention of tearing to obtain the maximum benefit from centrifugal casting.

(b) *Pouring Temperature* Pouring temperature influences the mode of solidification and thus the temperature required depends on the type of structure needed in the cast metal. Low temperatures result into maximum grain refinement and equiaxed structures, whereas higher temperatures favour columnar structures. However, the range of pouring temperature is limited by other practical considerations. In general, the pouring temperature should be sufficiently high to ensure good metal flow and freedom from cold laps, and at the same time, course structures and hot tearing should not occur due to excessive superheat.

(c) *Pouring speed* The pouring speed mainly depends on the need to finish casting before the metal becomes sluggish. If the pouring speed is too high it can cause excessive turbulence and ejection, whereas slow pouring speeds offer many advantages, such as improvement in directional solidification and less risk of tearing.

(iv) *Mould Temperature* The mould temperature is of secondary importance in relation to structure of cast metals. Its major influence is on the degree of expansion of die with preheating, since the expansion reduces the risk of tearing in true centrifugal casting.

5.5.5. Applications of Centrifugal Casting Techniques

The centrifugal casting has been used as the principal method for the production of cast iron pipes and cylinder liners. The process has also been used commercially for making tubes and annular components in most casting alloys (carbon, low alloy and high alloy steels and copper and nickel base alloys). The process is of great importance for the shaping of alloys which are unsuitable for forging or manufacture of wrought tubes. The special products made by centrifugal casting include bimetal tubes (obtained by successive pouring of two alloys under controlled conditions), composite products (lining of structural outer shell with an alloy of special properties), hollow blanks for further mechanical working, and many other items. Some typical applications of centrifugal casting techniques are given in Table 5.2.

5.6 CONTINUOUS CASTING

This is the technique used for casting metals directly into simple shapes such as billets, rounds, bars, tubes, etc. but the technique can not be employed for shaped castings. Therefore, this process is not much relevant to foundry practice and thus will be discussed in brief. The process is illustrated in Fig. 5.15. The process consists of continuous pouring of molten metal in a water cooled vertical copper mould (stationary or rotating) through a heated basin (called tundish) from the crucible. The mould is open at both ends and is usually 300—400 mm long. The internal shape of the mould represents the cross section of the casting needed. By the time the metal leaves the bottom end of the mould, a solid crust is formed, whereas the interior metal is still liquid. The solidified section then passes vertically downward with the help of a set of water cooled rollers. The casting is further cooled by compressed air. The casting may be carried away either fully along the vertical axis or partly along a vertical and partly along the horizontal axis. The cast metal block is cut into required size pieces either by gas cutter or a circular saw. Continuous casting process is used for casting nonferrous metals and alloys as well as steel.

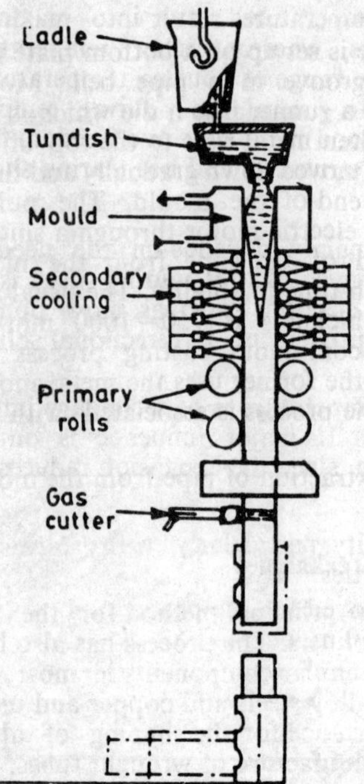

Fig. 5.15 Principle of continuous casing

TABLE 5.2 Important Applications of Centrifugal Castings

Method	*Objects*	*Metal/Alloy*
1. True centrifugal casting		
(a) As cast	Pipes for water, gas, and sewage and rain water pipes	Cast iron
	Tubing for reformers and radiant tubes	Heat resisting steel
(b) Machined form small cylinderical castings	Piston rings and cylinder liners.	Cast iron Copper alloys
	Bearing bushes and pape r making rollers Gas turbine rings	Heat resisting steel and nickel base alloys
	Runout rollers	Carbon steel
2. Semi-centrifugal	Pulleys,	Steel
	Nozzle boxes	Heat resisting steel
	Gears	Steel
3. Centrifuging	Hinges and brackets	Steel
	Dental casting and Jewellery	Cobalt-Chromium alloy

5.7 SEMICONTINUOUS CASTING

This is illustrated in Fig. 5.16. A sand core is set up on a bottom plate with a dummy slab. This core serves to form a groove in a pipe bell. Molten metal is poured from a ladle through a runner into a die which is thin walled water cooled mould. When the molten metal rises to the top of the mould, the plate carrying the sand core is moved down gradually and thereby a solidified pipe is pulled out from the end of the mould. The pulling of the tube is achieved by means of an electric motor through a suitable transmission mechanism. The pulling speed of the pipe from the mould should be such that the metal gets enough time to solidify. The usual pulling rate for tube with 10 – 12 mm wall thickness is 750 – 1000 mm per minute. The main difference between the continuous casting process and semi-continuous casting process is that the former uses the metal mould, whereas later employes the sand mould. The process is associated with the following advantages:

(a) As the process permits progressive extraction of pipe from the mould, the output is very high.

(b) The pipes produced are of high quality particularly with respect to surface finish and mechanical properties.

(c) The process can be successfully employed for the production of large pipes (800 mm diameter and up to 10 m length)

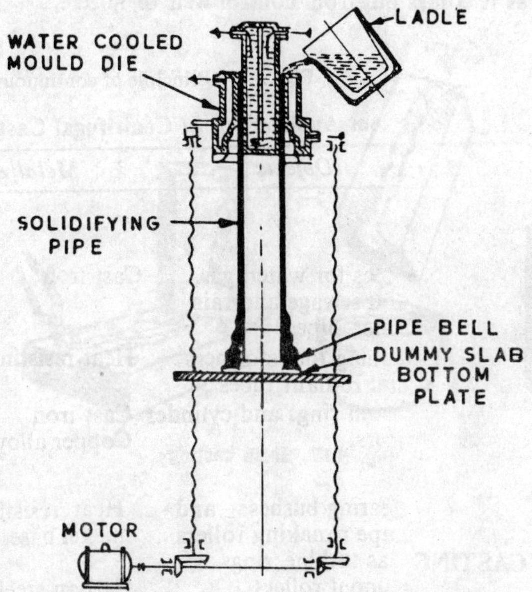

Fig. 5.16 Semicontinuous casting of bell type pipes

5.8 SLUSH CASTING AND COTHIAS PROCESS

These are essentially the forms of die casting used for the production of hollow castings without the use of cores. In the former process, the outer surface is formed by a shaped die, whereas the inner surface is produced by decanting out the excess metal after a thin skin of solid metal is formed against the mould walls. In the latter process the metal is filled in the original die cavity only partly and then the metal is displaced partly to fill the entire cavity by inserting a metal core to form the internal surface.

This process is not much applicable to engineering items. However, the process is important for casting where the wall thickness is not important. The main applications of the process are in making models and statues in low melting metals and alloys (tin, lead bearing alloys, aluminium alloys, zinc alloys, etc.). Fig. 5.17 illustrates the process of slush casting.

5.9 SHOT CASTING

This is the technique used for the production of shots required in shot blasting equipment. The shots are produced by dropping molten metal from some height in a pool of water. The molten metal is allowed to flow over a control weir at a constant rate so that the thin stream of molten metal gets separated into droplets. The droplets become spherical in shape as they fall freely through air and freeze immediately as they strike the water. The size of shots so produced depends on type of metal, temperature of molten rate of falling of metal stream, distance of fall of droplets, add the mechanism (air jet or mechanical device) used to disperse the stream of molten metal as it comes out from control weir or nozzle.

Fig. 5.17 Slush casting

5.10 FLUSH CASTING

This is another technique employed to produce hollow castings with making use of cores. This process is illustrated in Fig. 5.18. The required quantity of molten metal for one operation is poured into machine keeping the piston in top position. The piston is then lowered such that uniformly thick hollow costings are produced. The left over metal on the piston surface may produce the bottom of the casting.

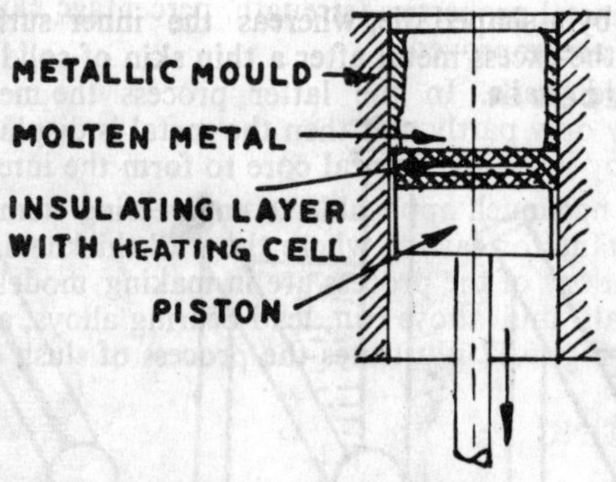

Fig. 5.18 Flush casting

5.11 SQUEAZE CASTING AND LIQUID FORGING

These may be considered as a variation of die casting in which two parts of casting die are held apart in the beginning, but brought together after pouring to squeaze or forge the solidifying casting to get the beneficial effects of forging.

In squeeze casting the liquid metal is poured into a receiver (Fig. 5.19a) of a special casting machine. It is followed by turning the movable half mould (Fig. 5.19b) to file the metal between the mould walls by squeazing action. As the mould closes completely, the distance between the half moulds becomes equal the casting thickness. The excess of metal is squeezed out into a ladle (Fig. 5.19c). After the casting is solidified the movable half mould turns back to the initial position and the casting is extracted from the stationary half mould. This technique is used for the production of thin walled large size articles of panel type and shells measuring 3 m × 2.5 m × 1.0 m (used in refrigerators, heat exchangers etc.) with the wall thickness of 2.5—5 mm. Common metals cast by this technique include aluminium and magnesium alloys. The machines employed are complex in nature and are operated automatically. The two half moulds can be made either of metal or sand depending on the design of casting and the alloy used. Squeeze casting process possesses the following advantages.

(a) The technique cuts down sharply the time and labour required in assembling of similar articles by rivetting, stamping, welding etc.

(b) Various properties such as strength, rigidity, resistance to vibration, etc. are improved.

The principle of liquid forging is the same as of squeeze casting, except that in the later case, high piston pressures (500—1500 kg/cm) are employed. The technique gives the following beneficial effects:

(a) Crystallization of metals and alloys under high pressures of piston.

(b) Plastic deformation of the solidfying casting which closes the intercrystalline pores, compresses the gas—air voids, and prevents the entrapment of gases during the crystallization of metal.

(c) The castings obtained are of high density.

(d) Fines grained structures are obtained in casting due to high rates of crystallisation.

(e) Most of the metal properties (strength, percentage elongation, impact strength, etc. are improved.

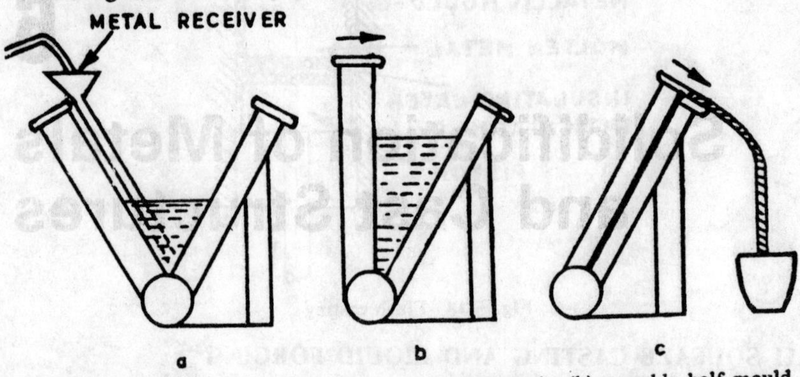

Fig. 5.19 Scheme of squeeze casting: (a) metal poured, (b) movable half mould turned, (c) excess metal squeezed out

The process may employ split (Fig. 5.20 (a)) as well as unsplit (Fig. 5.20 (b)) moulds. The molten metal is poured into a mould (split or unsplit) to a certain level and then squeezed with a piston at a speed of 100—500 mm per minute to fill the mould cavity with the metal. The quality of the castings produced is dependent on various variables such as pressure applied, compression time, temperature of die and punch and type of lubricant used. For a given casting and its composition the values of these variables should be decided carefully.

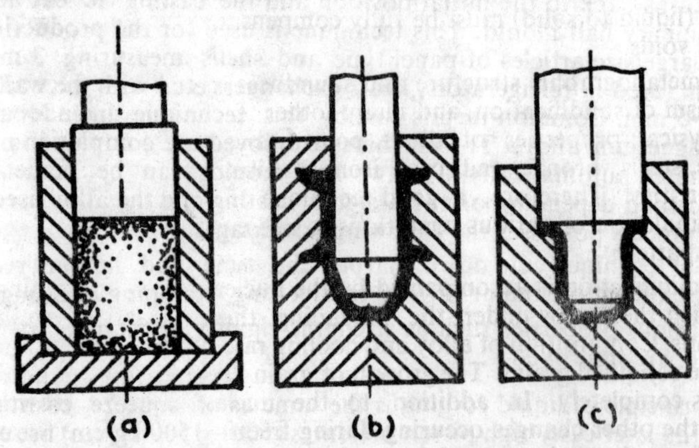

Fig. 5,20. Scheme of liquid metal forging: (a) Closed forging, (b) Open forging (e) Metal poured to a certain level and then squeezed.

The liquid forging employing piston pressure can be used only for the mass production of ingots and thick walled castings of various non ferrous as well as ferrous alloys due to high cost of moulds and requirement of high energy. The operating conditions of liquid forging depend on the composition of an alloy, size of the casting, and shape of the casting.

6

Solidification of Metals and Cast Structures

In casting processes, the liquid metal is poured into a relatively cool mould which initiates the processes of solidification. During the stage of solidification the cast from develope cohesion and acquires structural properties. The mode of solidification influences the final properties of castings in two ways, i.e. (i) metallographic structure, and (ii) soundness which is a degree of true metallic continuity.

The metallographic structure determines many of the inherent properties of castings. The metallographic structure includes grain size, grain shape, orientation of grains, distribution of alloying elements, crystal structure, and imperfections in crystal structure. The soundness of casting is also established during solidification since the volume shrinkage during the change of state (liquid to solid) must be fully compensated by liquid metal to the internal voids.

Both metallographic structure and soundness of castings depend an the mechanism of solidification, and many other factors such as constitution and physical properties of alloys, pouring and mould conditions, etc. For example, steels, bronze, and cast irons exhibit entirely different feeding characteristics. Therefore, a good technique of founding is concerned with the manipulation of various conditions to achieve full control of the solidification pattern

The soliaification is accompanied by the nucleation of tiny grains or crystals, which then grow under the prevailing thermal and crystallographic conditions. Composition of alloy and cooling rate control the size and character of the solidified grains. The growth of grain ceases when the liquid metal solidifies completely. In addition to the nucleation and growth of the grains, the other changes occuring during freezing process are (i) extraction of heat from the molten metal soon after the entry of metal into the mould, referred as superheat which should be removed before the solidification starts, (ii) extraction of latent heat of fusion which must be transferred to the surrounding mould before the casting is completely solidified, and (iii) transfer of heat from the solid metal to the mould and then to the atmosphere. During these three stages of cooling, the metal contracts resulting into shrinkage of casting.

The major processes thus covered under solidification are (i) crystallisation from the liquid metal, (ii) heat evolution and transfer, and (iii) dimensional changes. In addition, various factors influence the solidification process. These factors are (i) mould material and its thickness, (ii) mould geometry, (iii) thickness of casting, (iv) metal properties, (v) heat transfer from the mould to the atmosphere. (vi) use of in-oculants etc. to control grain size, and many others. The properties having relationship with solidification process are also covered in this chapter.

6.1 CRYSTALLISATION FROM THE MELT

The crystalline form of matter is chracterised by regularity of atomic spacings over considerable lengths. The metallic crystals deposited from the liquid metal possess internal symmetry but are usually irregular in external form due to uneven growth rates and mutual constraint in the last stages of freezing. Thus the general erystalline unit in the cast metal structure is allotriomorphic grain, although the eutectic grain represents two or more crystal structures resulting by simultaneous growth of separate phases. In the solids the crystal lattice represents a more closely packed state of matter than the liquid and thus freezing of melt will invariably be accompanied with contraction in volume. Further, liberation of energy in the form of latent heat of crystallisation results reduction in molecular motion which affects markedly the rate and mode of crystal growth. Latent and specific heat values for important common metals are given in Table 6.1.

TABLE 6.1 Latent and specific heat values of important metals

Metal	*Latent heat of fusion* Cal/g	KJ/Kg	*Specific heat* Cal/g deg C	KJ/Kg deg C
Iron	65.0	274.2	0.176	0.737
Nickel	72.1	301.9	0.157	0.657
Copper	48.9	204.7	0.118	0.494
Aluminium	92.7	388.1	0.259	1.084
Magnesium	85.6	358.4	0.333	1.394
Zinc	26.3	110.1	0.115	0.481
Lead	5.7	23.9	0 036	0.151
Tin	14.2	59.5	0.062	0.250

1 Cal/g=4.1868 KJ/kg.

The atoms in solid crystals are associated with lower free energy compared to the same atoms in a liquid state At the freezing point both solid and liquid atoms can interchange their position and giving an equilibrium at which change in free energy is zero. Therefore, in order to solidify the liquid, the free energy or temperature should be lowered. Supercooling provides driving force for solidification. In order to form new surfaces of crystals supercooling must be greater than the surface tension.

Crystallisation from a liquid metal is accompanied by the successive stages of nucleation and growth. The point of starting and relative rates of these two phenomena within the liquid metal decide the final structure of the solidified metal and establish the extent of solidification in directional or discrete manner throughout the melt.

6.1.1 Nucleation

Nucleation is the appearance of solid points (centres) in the liquid upon which further atoms can be deposited for the growth of solid crystals. Such nuclei can be formed in two ways, i.e. (a) homogeneous nucleation and (b) heterogeneous nucleation.

Homogeneous nucleation This occurs during the general movement of ordered groups of atoms within the liquid which form small zones of relatively higher 'density'. These embryonic crystals are unstable below a critical size but some of them may reach a critical size at which they become stable and start growing. Under equilibrium conditions, the nucleation depends on the relationship between the size of largest cluster and the critical size.

Once a nucleus is formed in a super cooled liquid, its rate of growth is generally quite large but limited by the rate of removal of latent heat of solidification from the growing interface. This latent heat of soliaification raises the temperature of the liquid above the original nuclei and thus restricts further nucleation. However, if the latent heat is removed rapidly, the solidification time can be determined mainly by the rate of formation of the first nuclei.

In metallic systems the actual nucleation rate is very small until the amount of supercooling reaches a critical value and then it increases extremely fast and it is impossible to measure the nucleation rate. Thus, it is easier to use the term nucleation temperature' which is a narrow range of temperature corresponding to very fast rate of nucleation. A limiting supercooling may be considered that gives homogeneous nucleation. It is found that limiting (maximum) supercooling is a linear function of melting points of metals on absolute scale, i.e.

$$\frac{\text{Amount of supercooling (maximum)}}{\text{Equilibrium temperature (melting point)}} = \text{Constant}$$

$$= 0.18 \pm 0.02$$

Nucleation can be explained in terms of the change in free energy resulting from the precipitation of a particle. This change consists of changes in both volume and interfacial energies. The formation of a solid from liquid results in negative free energy change which is directly proportional to the volume transformed. For a spherical particle, the free energy change (volume) may be written as following:

$$\Delta G_{\text{volume}} = \frac{4}{3}\pi r^3 \Delta G_v \tag{6.1}$$

where r = radius of particle, and

ΔG_v = bulk free energy change per nuit volume.

Whereas creation of new interface will require a local gain in free energy which is proportional to the surface area of particle. For spherical particle, this can be given as following:

$$\Delta G_{\text{interface}} = 4\pi r^2 \sigma \tag{6.2}$$

where σ = interfacial energy per unit area.

$$\Delta G = \Delta G_{\text{volume}} + \Delta G_{\text{.nterface}} = \frac{-4}{3}\pi r^3 \Delta G_v + 4\pi r^2 \sigma \tag{6.3}$$

The sum of volume and interfacial free energies is found to be positive for very small values of r but becomes negative at larger values of r as represented in Fig. 6.1. The maximum positive value of sum of energies (ΔG^*_{nom}) represents the critical radius above which growth is possible with a decrease in free energy. Fig. 6.1 represents an energy barrier inhibiting nucleation. However, with decreasing temperature, the critical size of nucleus diminishes and probability of homogeneous nucleation increases.

ΔG^*_{hom} is also called a critical nucleation barrier or the free energy of nucleation in homogenous systems.

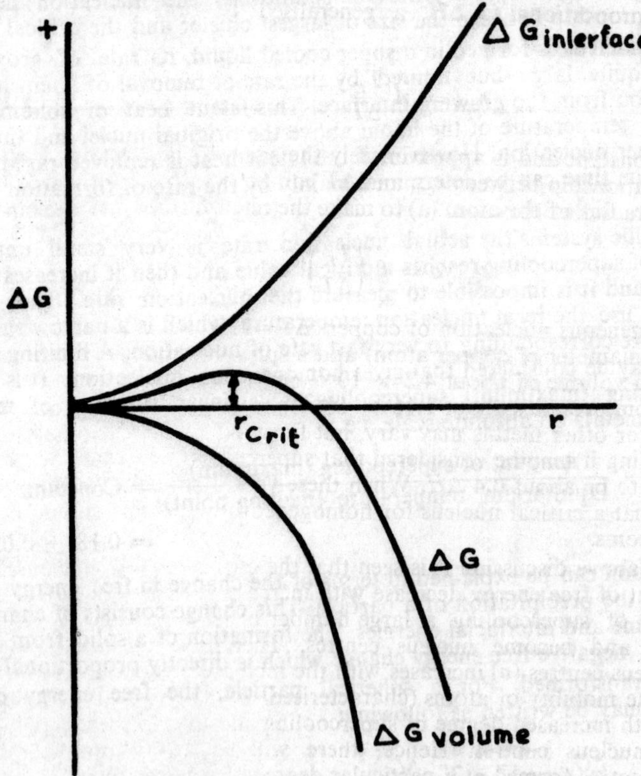

Fig. 6.1 Free energy changes in homogeneous nucleation

In the case of compressible solid the radius of critical nucleus r_{crit} may be given by the following relationship.

$$r_{crit} = \frac{2\sigma\, T_E}{\Delta T\, \Delta H} \tag{6.4}$$

where σ = interfacial energy (ergs/cm²),

T_E = Equilibrium temperature (melting point)

ΔT = Supercooling

ΔH = Latent heat of freezing and assumed to be independent of temperature.

From equation (6.4) it may be observed that the size of critical nucleus decreases inversely with increasing of super cooling. For example in case of copper, ΔH = 50 cal/gm or 1.88×10^{10} ergs/cm³, T_E = 1083 °C or 1356 °K, and $\sigma = 1.44 \times 10^2$ ergs/cm², then

$$r_{crit} = \frac{2 \times 1.44 \times 10^2 \times 1356}{\Delta T \times 1.88 \times 10^{10}}$$

$$= \frac{2.08 \times 10^{-5}}{\Delta T} \text{ cm}$$

Since σ is proportional to ΔH, a generalised relation for r_{crit}, can be written as following

$$r_{crit} = A\,\frac{T_E}{\Delta T} \tag{6.5}$$

where A is constant and is approximately the same for all metals. Further, a general relationship between r_{crit} and ΔT can be obtained by dividing both sides by the radius of the atom (a) to make the relationship (6.4) dimensionless, i.e.

$$r_{crit}/a = \beta\left(\frac{T_E}{\Delta T}\right) \tag{6.6}$$

For homogeneous nucleation of copper ΔT=230 °C and $r_{crit} = 10^{-7}$ cm (4 times the diameter of copper atom) and a spherical nucleus of this radius would have a volume of about 4.2×10^{-21} cm³ and would have about 360 atoms (1 atom occupies about 1.16×10^{-23} cm³). However, the corresponding values for other metals may vary, but for good generalisation of homogeneous cooling it may be considered that supercooling (ΔT) may be about 0.2 T_E and σ to be about 0.4 ΔH. When these values are combined together, it is found that a critical nucleus for homogeneous nucleation should be of about 200 atoms.

From the above discussion it is seen that the critical size of nucleus and the increment of free energy decrease with increase in supercooling. During the process of supercooling a large number of solid particles reach their critical sizes and become nucleus centres. Therefore, the rate of formation of nucleus centres (n) increases with the increase ofsupercooling degree. However, the mobility of atoms (characterised by the viscosity of liquid) decreases with increased degree of supercooling and thus prevents the formation of nucleus centres. Hence, there will be an optimum value of nucleation centres formed at a particular degree of supercooling.

Homogeneous nucleation or self nucleation takes place in case of pure metals and solid solutions of binary alloys. In the latter case also nucleation occurs at opproximately the same supercooling calculated from the liquidus temperature as expected for pure metal having the same melting point.

Heterogeneous nucleation The formation of nucleus of the critical size can be catalysed by suitable surface in contact with the liquid metal. Usually foreign particles suspended in the liquid, the surface of the mould, solid films of oxides on the liquid surface, etc. act as nucleation catalysts or nucleation agents. These agents alter the liquid-solid interface energy enough to assist in nucleation and resulting into reduction of supercooling required to effect nucleation. This process is referred as *heterogeneous nucleation* and generally prevails in castings atleast to certain extent. In order that nucleation agent or second phase acts as a nucleus, it must be capable of being wetted by the liquid metal forming a low contact angle and should possess some structural affinity with the crystalline solid. Under these conditions the solidification starts around solid inclusion (surfaces already existing).

In heterogeneous nucleation one part of the nucleus surface is connected with the nucleating agent and the other part with the liquid phase. Crystallographic structure of nucleation agents also plays an important role in nucleation. If the structures or crystal lattices of the solid particle and nucleus are more indentical, there will be less difference in surface tensions and thus nucleation will be easier.

In industrial practice, nucleating agents are added to promote heterogeneous nucleation, e.g. titanium and boron to aluminium alloy nutts, ferrosilicon to cast iron (to promote the nucleation of graphite) and zirconium to magnesium alloys. The lattice parameter of zirconium (HCP : a= 3.23 A°) is close to that of magnesium (HCP : a=3.21 A°), suggesting a small contact angle during heterogeneous nucleation, when close packed planes have one to one correspondence of atoms at interfare. This results in lowering of nucleation barrier, increase in nucleation rate and production of fine grains in casting. In addition to lowering of contact angle due to good lattice matching, a successful nucleating agent should be stable in the molten metal and possess a large surface area and roughness. The barrier for heterogeneous nucleation ΔG^*_{het} is given by the following relationship.

$$\Delta G^*_{het} = \frac{1}{4} \Delta G^*_{hom} (2 - 3\cos\theta + \cos^3\theta)$$

where θ = contact angle.

Nucleation in metals and alloys Both nucleation and growth of crystals occur below the equilibrium freezing temperature. Since in the beginning, conditions are unfavourable to the survival of small nuclei, enough supercooling will be required to cause homogeneous nucleation. However, in practice there is no clean metal and some foreign nuclei are always present in the form of impurities (may be present in the metal or may come from mould walls and/or atmosphere) and thus heterogenous nucleation forms the basis of solidification. The foreign nuclei may also be added deliberately to obtain a particular mode of crystallisation.

In foundry practice, generally a complex type of nucleation comprising of both homogeneous and hetergeneonous components takes place. The relation between the homogeneous and heterogeneous components depends on (i) properties of metals and alloys, (ii) degree of supercooling, and (iii) properties of foreign nucleation agents such as activity, size, and quantity. Therefore, only little undercooling a is required in solidification under foundry conditions.

After the establishment of critial nuclei, further crystallisation may take place in two ways, i.e. more solid may be deposited upon the initial nuclei, or fresh nucleation of the same or different phase may take place in the liquid metal. However, the most effective site for crystallisation is of the same phase. In case of metals the main barrier to growth is the evolution of the latent heat of crystallisation, whereas in alloys the change in composition of the adjacent liquid due to differential freezing is the main barrier.

Practically in all metals and alloys, traces of radioactive matter is present which gives off a density of energetic α-particles which may serve as nucleating agent. Cosmic rays may also act in the identical manner. Further, application of dynamic simulations such as ultrasonic vibrations may lead the formation of large number of nuclei instantly in liquids having no solid nuclei centres. The nucleation obtained by dynamic simulations is known as dynamic nucleation.

6.1.2 Growth

In both the cases of nucleation, i.e. homogenous and heterogenious, crystal growth may be accompanied by following stages:

(i) The formation of initial nucleus which does not have the equilibrium shape.

(ii) The transistion of initial nucleus into more stable shapes, i.e. spheroidal or ellipsoidal.

(iii) After attaining the critical radius, beginning of nuclei formation with equilibrium shape.

(iv) Further growth of the critical nuclei with signficance decrease in surface energy which determines the shape of the crystal.

The growth processes following the nucleation determine the final crystallographic structure of the solidified metal. The mode of growth of individual grains as well as of the general solid mass depends mainly on two factors i.e. (i) thermal conditions in the freezing zone, and (ii) constitution of the alloy.

Once a stable nucleus is formed, its growth takes place by acquiring atoms from the melt. The rate of growth depends upon the degree of undercooling and increases with the increase of amount of undercooling until it reaches a maximum and then again decreases. Therefore, the rate of nucleation as well as growth follow the same general trend with increasing degree of supercooling. However, their relative rates differe to the extent that nucleation is predominant in the early stages of solidification which results in the first layer of solid metal at the metal-mould interface, consisting of fine equi-axed grains. If it were possible to lower the temperature in a completely uniform manner, random nucleation might be expected throughout the whole liquid.

6.1.3 Freezing of pure metals

Under practical conditions of heat flow in a casting, temperature gradients produce initial nucleation at the relatively cool mould surface and the growth proceeds to form a skin of solid metal. During the growth to form solid metal skin, the latent heat of solidification is released resulting in rapid loss of undercooling. This change restricts further nucleation. However, the growth continues on some of the grains already foimed. The rate of heat transfer from the casting establishes a temperature gradient toward the casting surface and thus resulting the growth in the direction opposite to heat flow as illustrated in Fig. 6.2. Further, such growth tends to be associated with preferred crystallographic direction. The crystals which are favourably oriented grow more rapidly than their neighbours being less favourably oriented. The net result will be the formation of a columnar zone next to the outer layer of fine grains.

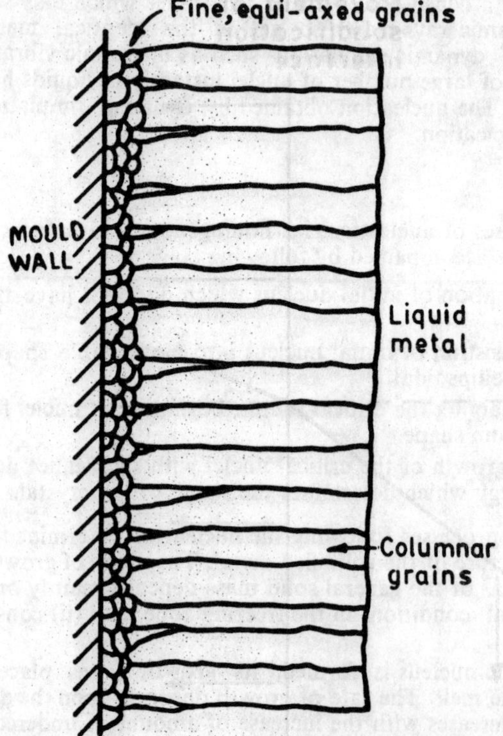

Fig. 6.2 Development of columner grain structure

At certain stage the lateral growth becomes rectricted by the mutual constraint of neighbouring crystals which depends on the number of effective nuclei and the initial rate of growth. It is followed by the advancement of an approximately plane interface by progressive deposition of atoms. It result in joining the lattice of one or other of the existing crystals. This growth takes place by the edgewise extension of close packed atomic planes and thus produces a minute terraced structure on the growth interface. This inter face is macrospically smooth and noncrystallographic in most of the metals. This typical sequence results in the production of the characteristic columnar structure which is generally observed in ingots and castings having the individual grains elongated in the direction of heat flow.

There may be two sets of thermal conditions under which solidification takes place. The first case is shown in Fig. 6.3 in which it is assumed that there exists positive temperature gradient at the solid-liquid interface and the interface advances progressively as the temperature of growth (T_G) is reached at points deeper in the casting. This mode of freezing in possible when the latent heat of solidification is inadequate to reverse the direction of the heat flow in the liquid adjacent to the interface. The advancing interface approximates to a flat plane under these conditions, i.e. slow cooling and a steep temperature gradient.

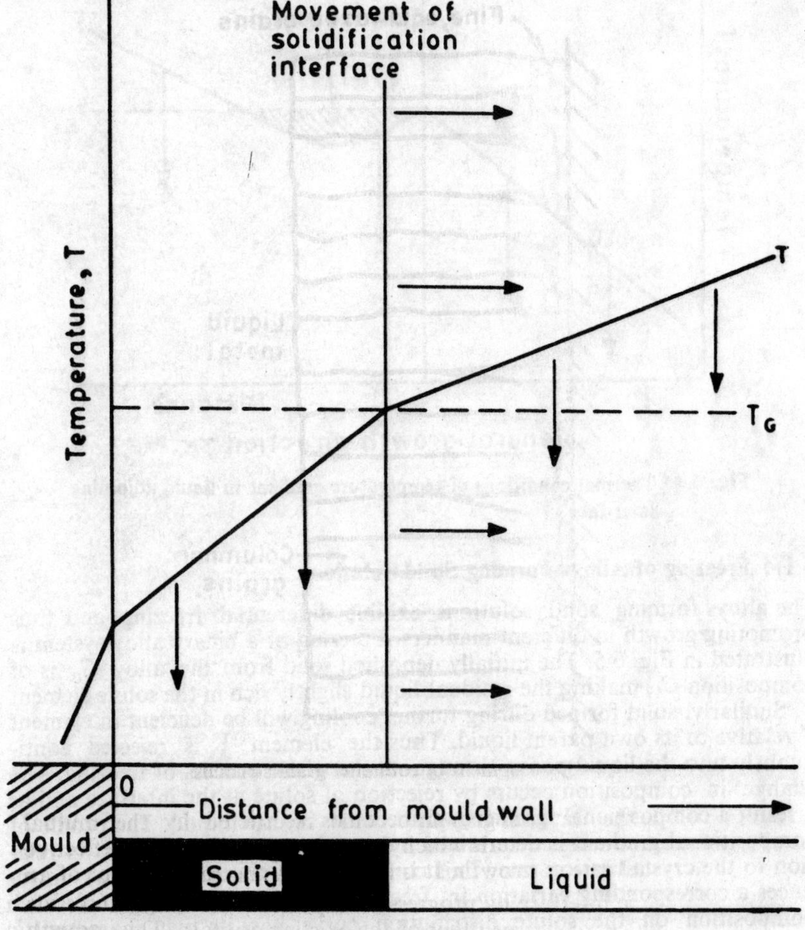

Fig. 6.3 Schematic representation of temperature distribution in plane front solidification

The second case of thermal conditions is shown in Fig. 6.4. In this case although the heat is transferred from the casting to the cooler mould and produces a positive gradient, but local evolution of latent heat is considerable to reverse the temperature gradient at the interface. Under these conditions minimnm temperature in the liquid is no longer adjacent to the interface and thus, the solidification will not occur by the advancement of smooth solidification front. There will be other modes of growth in which deposition can take place in the regions of greater undercooling. Therefore, in the formation of cast structures, microscopic heat flow may be an important factor.

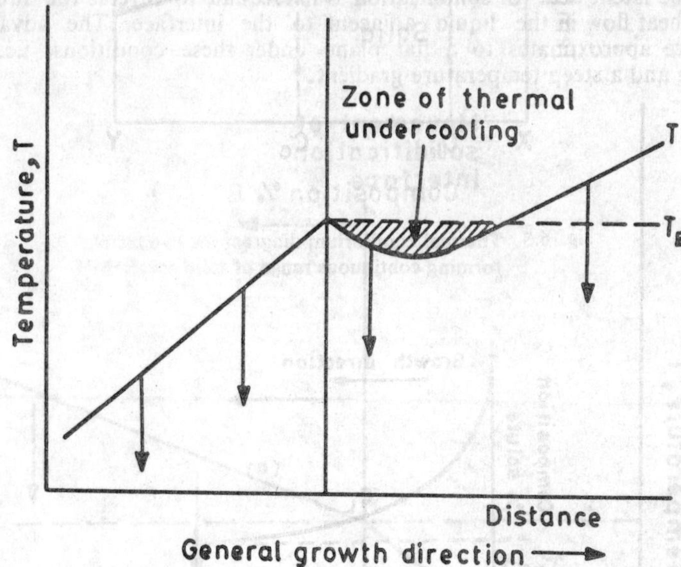

Fig. 6.4 Thermal conditions of temperature gradient in liquid adjoining inter-face

6.1.4 Freezing of Alloys Forming Solid Solutions

The alloys forming solid solutions exhibit differential freezing and thus promoting growth in different manners. Freezing of a binary alloy system is illustrated in Fig. 6.5. The initially deposited solid from the alloy C_0 is of composition C_1, making the residual liquid slightly rich in the solute element *Y*. Similarly, solid formed during further cooling will be deficient in element *Y* relative to its own parent liquid. Thus the element *Y* is rejected continuously into the liquid phase throughout the whole process of freezing. The change in composition occurs by rejection of solute at the interface and as a result a compositional gradient must exist in the liquid. The profile of compositional gradient is determined by the rate of solute transport in relation to the crystallisation rate. This compositional change (Fig. 6.6a) introduces a corresponding variation in freezing temperature (Fig. 6.6b). Each composition on the solute distribution curve has its own characteristic equilibrium freezing point below which a specific amount of undercooling will produce further crystallisation.

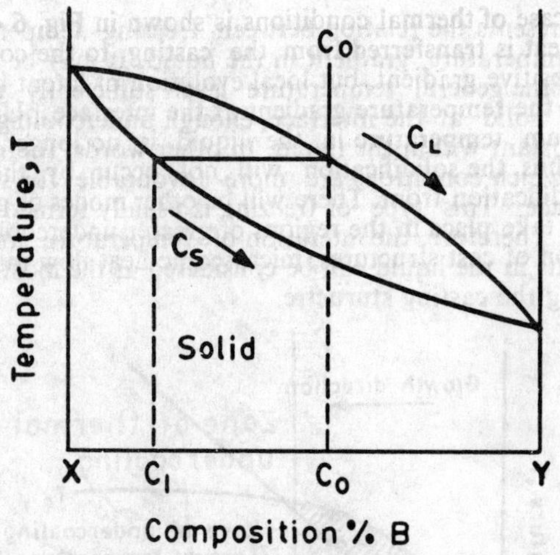

Fig. 6.5 Thermal equilibrium diagram for two metals forming continuous range of solid solution

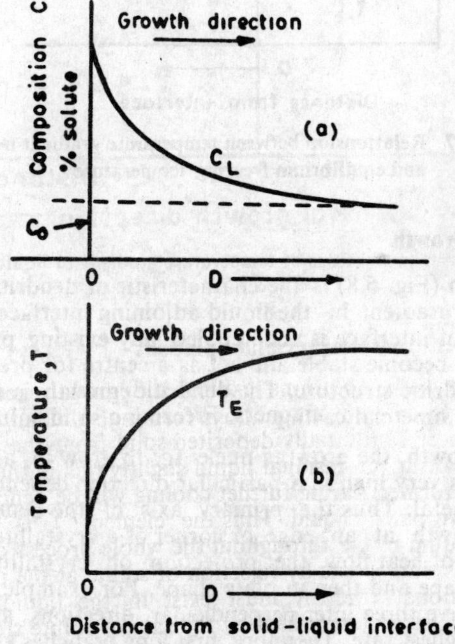

Fig. 6.6 Profile of compositional gradient of solute in liquid:
(a) solute distribution C_L
(b) corresponding change in freezing temperature

Fig. 6.7 represents the relation between freezing temperature curve and the actual temperature gradient in the liquid. It may be observed that before falling of the general temperature level sufficiently to produce fresh deposition of solid at the interface, enough undercooling may exist at the points quite distant within the liquid. In other words, there may be a some liquid within which conditions are more favourable for solidification than at the interface. This type of freezing is usually termed as *constitutional undercooling*. Therefore, the interation of temperature and the compositional gradients in the liquid can be considered as the most important single factor effecting the casting stuructre.

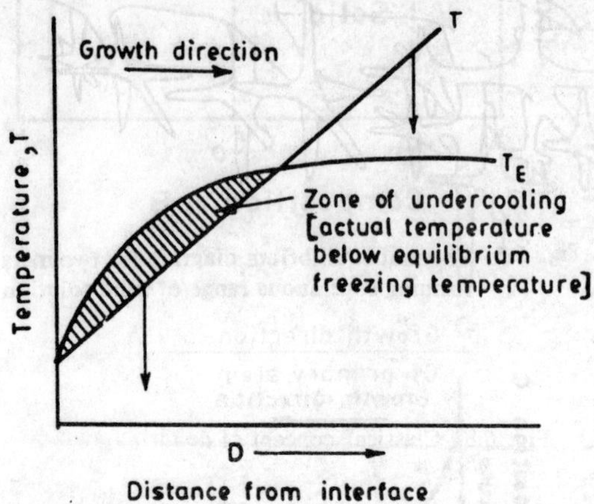

Fig. 6.7 Relationship between temperature gradient in liquid and equilibrium freezing temperature

6.1.5 Dendritic Growth

The tree like form (Fig. 6.8) is the characteristic of dendritic growth. When the temperature gradient in the liquid adjoining interface is negative, i e., the liquid ahead of interface is undercooled, any existing projections on the solid face tend to become stable and act as a centre for preferential growth giving rise to dendritic structure. The dendritic growth is most common in the freezing of commercial casting alloys forming solid solutions.

In dendritic growth, the growing nuclei try to grow in all directions but the growth rate is very high in a particular direction depending on the crystal structure of metal. Thus the primary axis of the dendrite is the result of preferred growth at an edge or corner of a crystallite. Following the general direction of heat flow the projection on crystallite develops first into a needle shape and then in plate shape. For example, the metals with cubic structure have three inter-perpendicular directions along which crystals grow with highest rate. Therefore, first long branches known as primary needles or plates grow in definite directions. The lateral growth of these primary needles is restricted by the latent heat or accumulation of solute.

The secondary and tertiary branches can also develop by an identical mechanism leading to the growth of primary needles. The secondary

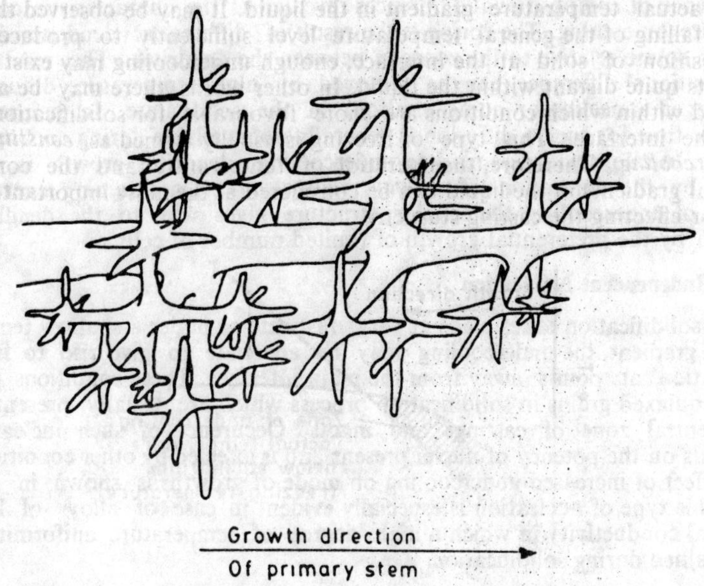

Fig. 6.8 Classical concept of dendritic growth

branches occur perpendicular starting from the projections in the primary axes. Tertiary and higher order branches are also formed similar to the secondary branches. The result is the tree like structure (Fig. 6.8) which may be considered as the skeleton of a dendritic grain. The trapped liquid metal between the dendrite branches solidifies by gradual deposition of the metal in layers on the dendrite branches. As a result, branched structure is not visible after complete solidification. However, the dendritic structure can be seen on microstructures using special etching methods. Defects in crystal lattice introduce small misorientation which gives rise to grain mosaic blocks. The dendritic growth is influenced by the following factors:

(i) *Degree of supercooling* Higher degree of supercooling gives rise to thinner and more formed dendrite crystals, increased misorientation of crystals, and smaller size of the mosaic growth.

(ii) *Anisotropy of the properties of crystals* This gives the difference in speed of crystal growth in different directions.

(iii) *Latent heat of solidification* It has direct influence on the speed of crystal growth.

(iv) *Convection current* Convection current present in the liquid meta effects the heat transfer and solidification.

(v) *Solid inclusions* Solid inclusions may absorb nuclei and thus may affect the solidification pattern.

In pure metals dendritic growth can be be observed only by interrupted freezing and decantation, whereas in alloys, it can be observed through the persistence of compositional differences revealed as the characteristic cored structure on etching. The coring is the result of differential dendrites deficient in solute relative to the zones in between the dendrites. The resulting compositional difference can be removed by high temperature diffusion (prolonged annealing or homgenisation) in the solid state. In indnstrial practice time for such a diffusion is generally not available during cooling of casting. In alloys with less marked undercooling, dendritic growth is prceded by the formation of *cellular substructure*. On attaining a certain degree of undercooling, the cellular structure gives way to the dendritic growth by the preferential growth of limited number of cells.

6.1.6 Independent Nucleation

When solidification takes place at very fast rate or under a shallow temperature gradient, the undercooling may be adequate to give rise to fresh nucleation at points away from the main interface. These conditions preduce equiaxed grains in solidification process which are usually present in the central zone of castings and ingots. Occurence of such nucleation depends on the potency of nuclei present and is affected by other conditions. The effect of increased undercooling on mode of growth is shown in Fig. 6.9. This type of nucleation is especially evident in case of alloys of high thermal conductivity in which a high degree of temperature uniformity is maintained during solidification.

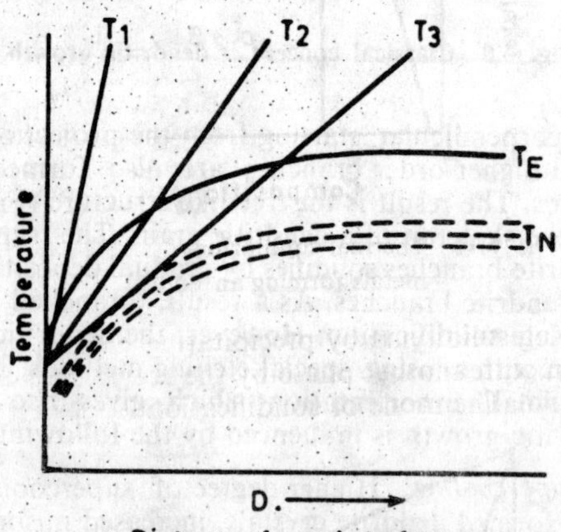

Fig. 6.9 Influence of temperature gradient on cystallisation: T_E = equilibrium freezing temp., T_N = nucleatin temp. T_1, T_2 and T_3 = temp. gradients

6.2 SOLIDIFICATION OF ALLOYS

The presence of second component in solution in the liquid effects the growth of crystals considerably even when the second component is present in small amounts. The most important effect is to produce differential composition in the solidified mass. Basically for solidification purpose, the alloys can be classified as following:

(i) Alloys solidifying by precipitating an essentially pure component but over a temperature range (Fig. 6.10 alloy A) and a very small amount of eutectic precipitation completes the solidification process.

(ii) Alloys solidifying at constant temperature by precipitating simultaniously two phases, known as eutectic alloys (Fig. 6.10 alloy B).

(iii) Alloys starting solidifying by precipitating an essentially pure component and completing the solidification by eututic type freezing (Fig. 6.10 alloy C)

(iv) Alloys starting solidifying as solutions and completing the solidification as eutectic type (Fig. 6.10 alloy D).

(v) Alloys starting solidifying as solutions and completing the solidification as solutions (Fig. 6.10 alloy E).

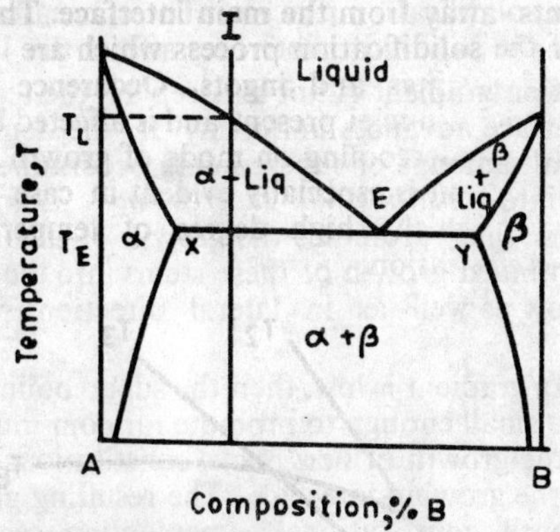

Fig. 6.10 Thermal equiiibrium diagram for two metals forming an eutectic

(vi) Alloys starting solidifying by precipitating one phase and completing solidification with another phase by the reaction of starting phase and residual liquid. The mode of solidification is termed as peritectic (Fig. 6.11).

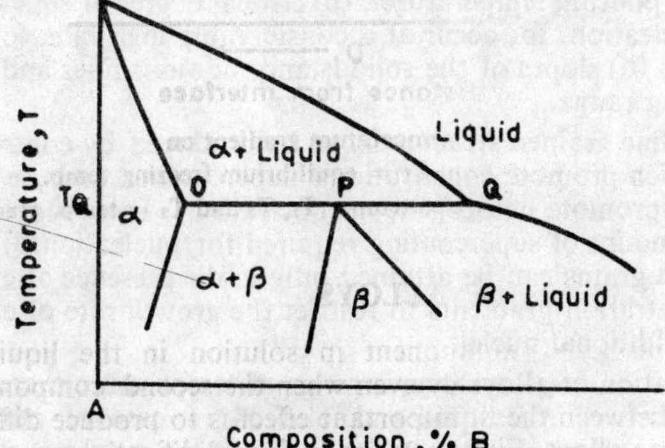

Fig. 6.11 Thermal equilibrium diagram invoiving a peritectic raction

For discussion purpose the alloy systems can be divided into three groups, i.e. (i) alloys solidifying over a temperature range (ii) alloys solidifying at a constant temperature, and (iii) alloys solidifying with a peritectic reaction.

6.2.1 Alloys Solidifying over a Temperature Range

In Fig. 6.10 these alloys are represented by alloys A,C,D, and E. In this system solidification takes place over a temperature range, the extent of which depends on the system and composition of the alloy. This solidification range increases from minimum to maximum from specific points. For example, in iron-carbon system, the solidification range is small at low carbon contents and reaches a maximum at about 1.7 pct carbon and again decreases with increasing carbon content.

In such system of solidification, the precipitating solid differs from liquid in composition, and as a result concentration gradients are set in the liquid which influence the solidification process. The concentration gradient developed just ahead of the solidifying interface affects the liquids temperature.

If the concentration gradient is not as steep as the temperature gradient, the solidifying interface advances uniformly. If the temperatre gradient is intermediate, a small amount of supercooling occurs ahead of the advancing interface resulting into *constitutional supercooling*. This condition creates the instability and thus promotcs the growth of stems which extend into the liquid. The continued growth of these stems into the liquid in a direction opposite to heat flow as well as in lateral direction results in a typical dendritic structure.

If the temperature gradient is low, then the supercooling temperature differential (ΔT) may be small enough to promote random nucleation and equi-axed grains With the growth of new nuclei constitutional supercooling develops in advance of the growing interface. The resulting grains also grow in a dendritic manner until constitutional supercooling ceases on account of thermal and compositional adjustments. The castings of commercial alloys of this group when cooled in the normal manner, they exhibit columnar dendritic grains near the surface, whereas near the center equi-axed dendri tic grains will be formed. The ratio of equi-axed to columnar grains is inversely proportional to (a) effective superheat (percentage of columnar grain structure and the size of the equi-axed grains in the centre both increase with pouring temperature), (b) critical degree of supercooling necessary for nucleation to occur at a considerably high rate, (c) thermal conductivity, and (d) slopes of the solidus and liquidus lines and proportional to the freezing range.

Relatively fine grained structures can be produced by control of alloying elements (which promote constitutional supercooling) and use of inoculating agents (these promote hetorogeneous nucleation). These inocculating agents reduce the amount of supercooling required for nucleation. However, very fine equi-axed grains can be attained only in the presence of elements giving rise to concentration gradients to restrict the growth rate of each grain, and to produce additional nuclei.

In solidification of alloys over a range of temperature there is no line of demarcation between the liquid and solid as in case of pure metals and purely eutectric alloys. First metal solidifies at the mould surface and solidification progresses inward but the dendrites formed leave behind islands of liquid metal which do not solidify until further heat is extracted. However, later on these islands of liquid also solidify and their freezing also progresses inward. Therefore, at a particular time a region consisting of both liquid and solid metal can be found in the casting. The metal of such region has a mushy consistency when worked and thus this stage is referred as *mushy stage* of solidification. The time gap between the start and end of a solidification wave is dependent on external variables and the characteristics of the alloy. Fig. 6.12 represents schematically the relationship between the conditions of solidification and the cooling curves at various points in the casting. Fig. 6.12 shows that there is no single temperature of solidification as in the case of pure metals or eutectic alloys. However, the remaining liquid remains at constant temperature after the start of solidification which is mainly due to the effect of latent heat of fusion. Only the metal near the surface (point X) is cooled fast enough to avoid the extreme thermal lag in the interior.

The width of the range over which alloys solidify (start and end of freezing) is affected by a number of variables such as (a) solidification range of alloy, (b) thermal characteristics of mould, (c) conductivity of alloy (d) temperature level of solidification, etc.

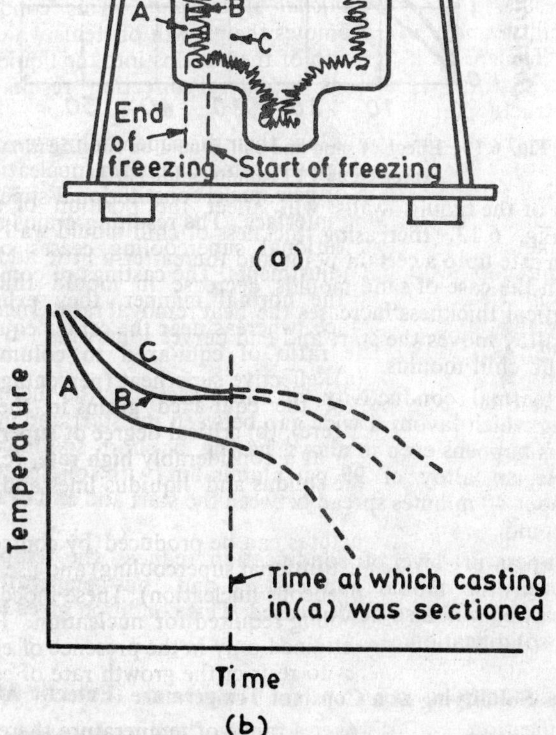

Fig. 6.12 Relationship between the conditions of solidification and the cooling curves at various points in the casting: (a) cross section through a casting, (b) cooling curves

For a given mould material, increased solidification range increases the separation between start and end of solidification. In chill mould (metallic mould) the solidification process consists mainly of the movement of the solidification band from the mould wall into isothermal liquid held essentially at the constant liquidus temperature. Whereas, the solidification in sand moulds is progressive and the solidification band is much wider or never develops. Therefore, solidification in sand moulds is more general throughout the whole casting. The effect of employing sand and chill mould is illustrated in Fig 6.13.

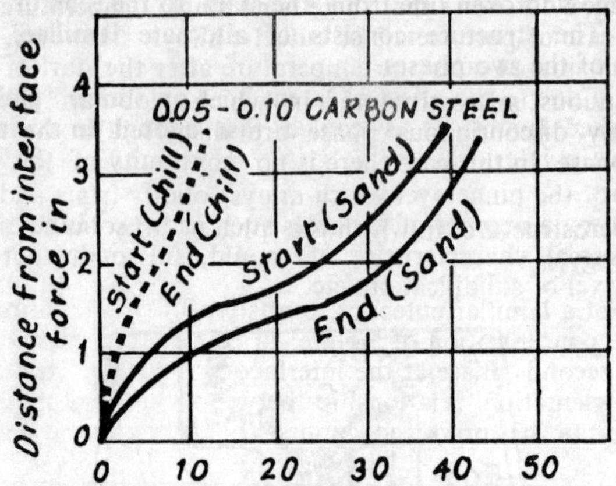

Fig. 6.13 Effect of sand and chill mould on solidification expressed as TTT solidification curves

Thickness of the mould walls will affect the position of solidification curves of Fig. 6.13. Increasing thickness of chill mould wall will increase solidification rate upto a certain point and thereafter a little additional effect is realised. In the case of sand moulds, decrease in mould thickness below a certain critical thickness increases the heat removal rates. Increased superheat of the alloy moves the start and end curves somewhat to the right in both sand and chill moulds.

The high thermal conductivity of alloys decreases the thermal gradients in the casling which favour a wide gap between the start and end of solidification. This happens even in alloys having a short solidification range. For example, an alloy of 99 pct aluminium (with solidification range of 5.5°) has about 40 minutes spread between the start and end of solidification in a sand mould.

A low temperature level of solidification reduces the temperature gradients between the mould and casting and thus temperature gradients within ths casting will also be decreased. This condition gives low degree of progressive solidification.

6.2.2 Alloys Solidifying at a Constant Temperature (Eutectic Alloys)

The eutectic solidification is characterised by the simultaneous solidification of two solid phases from a single liquid solution. Many casting alloys are either of approximately eutectic composition or consist of appreciable amounts of eutectic constituent. Fig. 6.10 represents eutectic crystallisation in which an alloy of composition B solidifies at a single temperature to gige a mixture of two solid phases, α and β of composition X and Y respectively.

Unlike the dendritic growth in case of solid solutions or pure metals, each eutectic grain is formed by the simultaneous growth of two or more separate phases close to each other. The resulting structure may be classified as following:

(a) Continuous microstructure in which both phases can be observed along some unbroken line from the start to the end of solidification process. The structure consists of alternate lamillae, or rod like elements of the two phases.

(b) Discontinuous microstructure in which globular elements of an opparently discontinuous phase are dispersed in the matrix of the second phase. In this case there is no continuity of the phase in any direction.

(c) Spiral microstructure in which the phases grow with spiral growth morphology.

The growth of a lamillar eutectic results with the nucleation of one of its phases. The concentration of solute in the liquid increases until the appearance of second phase at the interface after which both the phases grow with an orientation relationship between the two phases. The additional lamellae can be produced by first nucleation or by a process of bridgins.

The eutectic grains grow by the simultaneous edgewise extension of the plates of both phases. The each eutectic grain in the final structure may be considered as the product of associated growth from a single centre to form a microscopically distinct unit. Such units may be equiaxed similar to the equi-axed grains in single phase systems. However, it is also possible to grow columnar eutectic structures by controlled unidirectional freezing. The important differences in the morphology of eutectics are related to the concept of coupled zone' below the eutectic temperature. Within the coupled zone two phases can grow at equal velocities at a common interface resulting into regular structures like lamilar eutectics. Whereas, growth outside the coupled zone takes place at different rates producing irregular structures.

When the volume fractions of two phases are similar and the eutectic composition is roughly between the compositions of constituent phases, the structures obtained are usually lamillar or rod—like eutectic. Whereas, with the conditions opposite to above, regular structure is obtained. In the latter case the sccond phase is the subject to heterogeneous and no consistant orientation relationship exists between the two phases.

Many eutectic phases earlier considered to be discontinuous, are now known as continuous throughout a eutectic grain. The examples are flake graphite and the silicon phese in the aluminium—silicon eutectic. Therefore, the final sizes of such constituents are determined by the growth phenomena of branching and growing out rather than repeated nucleation. The differences in mode of growth and structure are mainly due to the different entropies of fusion of the constituent phases. The materials having high entropics of fusion (mostly non-metals) solidify with crystallographically faceted solid-liquid interfaces. On the other hand, most metal have low entropies of fusion and grow with non-crystallographic interfaces of planer, cellular, or dendritic form depending on the local thermal conditions as discussed earlier.

Lamillar or rod-like eutectics occur mainly in systems in which both phases possess low entropies of fusion, the example is lead-tin entectic. Whereas, irregular or complex-regular structures as in the case of Al-Si and Mg-Zn systems are produced when one phase possesses a high and the other a low entropy of fusion. The structures are always irregular with independent growth of the two sets of crystals in systems having high entropies of fusion of both the phases. Impure eutectic grevins are a also distinguished by a colony or cellular substructure similar to that obtained in solid solutions, which develop by trensition from a planer to cellular growth interface under conditions of constitutional undercooling. A typical cellular structure of Al-Zn eutectic is shown in Fig. 6.14.

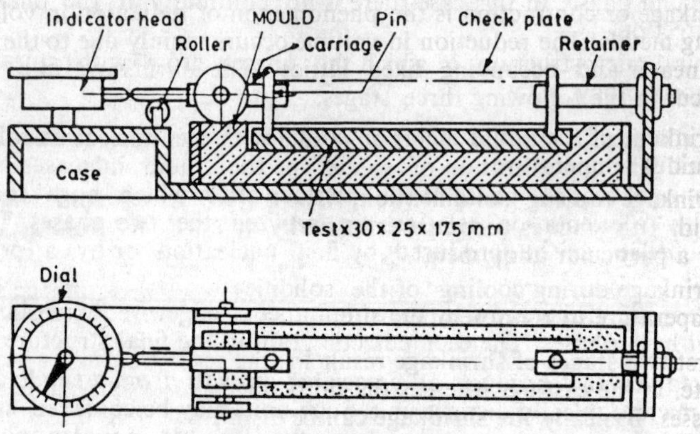

Fig. 6.14 Linear shrinkage tester

The morphology of the eutectics and the various mechanisms of eutectic crystallisation are the subject of great detail and are beyond the scope of this book.

6.2.3 Alloys Solidifying with Peritectic Reactions

Solidification involving pentectic reactions is represented in Fig. 6.11. The equilibrium solidification of compositions between O and Q involves the precipitation of primary crystals of solid solution α and their subsequent reaction with residual liquid at temperature T_Q to form a second solid phase β of composition P. For alloys having the compositions corresponding to the right of P the whole of α-phase of composition O reacts with a proportion of the liquid of composition Q to form β. The solidification of the residual liquid is completed at lower temperatures by the precipitation of additional β. In alloys having the compositions corresponding to left of point P, the whole of the liquid is consumed in reaction with a fraction of the α phase and thus on completion of solidification the structure consists of two solid phases, i.e. α and β.

However, pentectic reactions do not proceed readily with equilibrium solidlfication as the initial reaction product forms a layer surrounding the primary solid phase and the subsequent rate of reaction is controlled by diffusion through this layer. As a result, there persists a compositional gradient in the final solid. This situation results in tendency of non-equilibrium residues of primary α to be retained even in alloys having the composition to the right of point P. In case of Ag-Zn alloys, the presence of primary α-dendrites along with the β-phase has been observed in all alloys between the extreme points of the peritectic herizontal. Under these conditions the two phases precipitate over separate and successive ranges of temperature.

6.3 SHRINKAGE IN SOLIDIFICATION

The shrinkage or contraction is the phenomenon of reduction in volume of solidifying metals. The reduction in volume occurs mainly due to the atoms drawing nearer and occupying fixed lattice. The shrinkage in a casting many occur in the following three stages:

(i) Shrinkage of the liquid metal on cooling from pouring temperature to liquidus temperature.

(ii) Shrinkage during solidification, i.e. transforming from liquid to solid. This stage is accompanied usually by a specific volume charge for a particular alloy/metal.

(iii) Shrinkage during cooling of the solidified casting from the solidus temperature to room temperature.

The first two stages of shrinkage result in the severe shrinkage cavities, piping, etc. in the casting, whereas the third stage causes thermal and residual stresses. Basically the shrinkage can be distinguished as linear (E_l) and volume (E_v) which are expressed in relative values.

6.3.1 Linear shrinkage in Castings

The linear shrinkage refers to the change in linear dimension. The linear shrinkage in casting can be represented as

$$E_l = \frac{l_r - l}{l_p} \times 100 \text{ pct} \qquad (6.7)$$

where E_l = coefficient of linear shrinkage in pct

l_p = linear dimension of the pattern

l_c = length of the casting at shop temperature after solidification

Theoretically. the linear shrinkage can be calculated according the following equation:

$$E_{th} = \alpha_{av} (t_{bsh} - t_o) \times 100 \qquad (6.8)$$

where E_{th} = Theoretical coefficient of shrinkage,

α_{av} = average coefficient of thermal expansion in the temperature range t_{bsh} (at which shrinkage begins) and t_o (room temperature).

In equation (6.7) the term t_{bsh} is not known and thus determination of E_{th} is difficult. The temperature t_{bsh} depends on the alloy composition. Pure metals posses higher t_{bsh} whereas alloys containing more than 75 pct eutectics possess minimum t_{bsh}. Therefore, higher linear shrinkage is caused in pure metals whereas least shrinkage in eutectic alloys. The actual linear shrinkage may be estimated by taking into account the following considerations:

(i) Theoretical value of linear shrinkage.
(ii) Cooling rate of castings. A higher cooling rate results greater shrinkage.
(iii) Design of the mould complex and less fielding mould shapes and cores give the restrictive action of mould and thus result in less shrinkage.
(iv) Design of casting: Intricately shaped castings become the subject to the joint effect of mechanical and thermal retardation of shrinkage.

An approximate expression to estimate the linear shrinkage of the casting can be used as following:

$$E_{l,\,\text{casting}} = K \cdot \alpha_{av} (t_{bsh} - t_o) \times 100 \qquad (6.9)$$

where $K =$ empirical coefficient in which account of various factors discussed above has been taken.

The value of K for green sand moulds is less than 1 and for continuous casting $K = 1.3 - 1.5$. The values of linear shrinkage for some commercial metals and alloys are given in Table 6.2.

TABLE 6.2. Linear shrinkage of various metals and alloys

Metal/alloy	*Contraction*
Aluminium alloys	0.9 — 1.3
Magnesium alloys	1.1 – 1.4
Bronze	1.3 – 1.6
White cast iron	1.5 – 2.0
Gray cast iron	0.5 – 1.0
Alloyed steel	2.2 – 2.5

Linear shrinkage of casting starts somewhat earlier than the time at which casting solidifies completely. The shrinkage begins as soon as the neighbouring crystals combine to form a crystal skeleton strong enough to withstand the pressure of liquid metal. Therefore, the shrinkage in linear dimensions starts at the temperature lying somewhere between the liquidus and solidus. In case of pure metals, onset of linear contraction occurs obly at the critical temperature.

The linear shrinkage depends on (i) degree of superheat increased superheat increases shrinkage, (ii) rate of heat transfer, the increased rate of heat transfer increases the linear shrinkage, and (iii) alloying elements, give different effects.

During solidification some metals and alloys undergo phase transformations such as graphitization and evolution of gas in steel and irons. resulting into expansion of the casting. This is known as preshrinkage expansion which has considerable influence on centraction of high carbon alloys (when they undergo graphitization), alloy steels, and high carbon, and medium carbon steels. The alloys associated with preshrinkage expansion are ters susceptable to cracking and shrinkage cavity.

Linear shrinkage is taken into account in making patterns and corebox. The linear shrinkage can be determined by a special tester shown in Fig. 6.14.

6.3.2 Volume of Shrinkage

The type of shrinkage refer to total change in volume of metal due to liquid shrinkage, $(E_{v,l})$, freezing shrinkage $(E_{v,f})$, and solid shrinkage $(E_{v,s})$. Therefore, the total change in volume (E_v). due to shrinkage is given by

$$E_{v,t} = E_{v,t} + E_{v,f} + E_{v,s} \tag{6.10}$$

The various shrinkage values can also be written as

$$E_{v,l} = \alpha_l (t_p - t_l)$$

$$E_{v,f} = \alpha_f (t_l - t_s)$$

$$E_{v,s} = \alpha_s (t_s - t_o)$$

where α_l, α_f, and α_s are coefficients of shrinkage of solid, freezing, and solid, and t_p, t_l, t_s, and t_o are the pouring, liquidus, solidus and room temperatures respectively.

The solidification of most alloys is accompanied by appreciable contraction in volume and the amount of this differs for various 8 metals and alloys. The magnitude of volume shrinkage for common metals forming various alloys is given in Table 6.3.

TABLE 6.3 Values of volume shrinkage for metals

Metal	*Contraction freezing pct*	*Metal*	*Contraction as freezning*
Iron	3.5	Magnesium	4.1
Nickel	4.5	Zinc	4.7
Copper	4.2	Lead	3.5
Aluminium	6.5	Tin	2.3

6.3.3 Shrinkage Cavities

Shrinkage cavities in castings are voids usually of conical or spindle-like shape resulting from shrinkage of alloys during solidification. Fig. 6.15 represents shematic development of shrinkage cavity in a shaped casting (Fig. 6.15 a) shows the casting just after pouring. Even during pouring, thin walls are all to solidify and ontract. However, sufficiently strong solid skin does not form soon after pouring due to inflow of fresh metal from gates and risers. The liquid metal then starts to contract inside the skin

(Fig. 6.15 b) as it cools down to the temperature of onset of freezing or the liquidus temperature (Fig. 6.15 c). During passing of metal from the liquid to the solid state, the rate of volume shrinkage exceeds the rate of skin contraction and thus at certain stage the metal breaks away under gravity from the upper solidified skin and sags to some extent. It is followed by crystallisation of liquid metal inside the solid skin and during this state the metal shrinks so greatly that it results into a cavity (Fig. 6.15 d). If the cavity does not contain gases, the reduced pressure in the cavity may cause the thin skin to sag down into the void (Fig. 6.15 e).

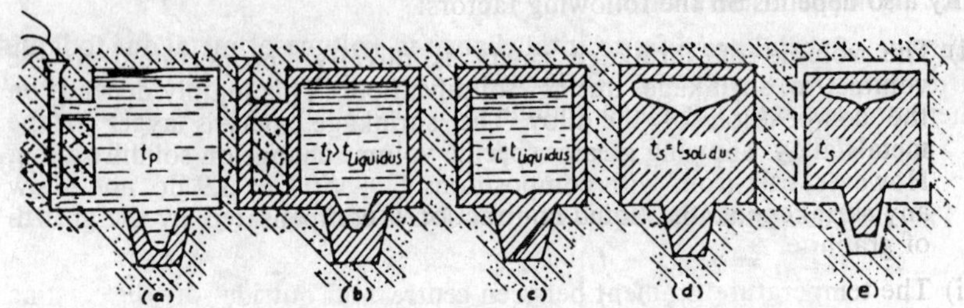

Fig. 6.15 Formation of shrinkage

The risers set up on castings are the effective means to prevent shrinkage voids. The molten metal of the riser flows under gravity into a solidifying casting and compensates for contraction. It results in the formation of shrinkage cavity only in the riser which can be subsequently cut-off from the casting. The shrinkage cavity in the form of pipe is inherent in castings from pure metals, alloys with narrow solidification range, and eutectic alloys.

The volume of shrinkage cavity differs with alloys. For example, in steel castings this is 3.10 pct of the initial volume of steel, for gray cast iron, 1.5 – 2.5 pct., for inoculated iron 2 – 3.5 pct., for white iron 2.5 – 6 pct. and for nodular iron 2.5 – 6 pct.

The relative volume of shrinkage cavity (ΔV) may be calculated by the following two formalae

$$\Delta V = E_{ve} + E_{vs} - (E_{vc} - E_{vs}) \qquad (6.11)$$

where, E_{vc} is shrinkage in the casting.

or $$\Delta V = \alpha_l (t_{lm} - t_{sol}) + \Delta V_{sol} - 1.5\, \alpha_s (t_{sol} - t_{sm})$$

(proposed by Neukhesidzi and Girshovich) (6.12)

where α_l = coefficient of linear shrinkage of liquid metal at a corresponding temperature,

t_{lm} = mean temperature of liquidmetal inside the casting at the beginning of freezing,

t_{sol} = solidification temperature

ΔV_{sol} = Volumetric shrinkage in a state of solidification

α_s = coefficient of linear shrinkage of solid metal at a corresponding temperature,

and t_{sm} = mean temperature of solid metal at the end of solidification of the entire casting.

The volume of the shrinkage cavity is usually 6-8 pct of the casting volume. The equation (6.11) shows that the volume of shrinkage cavity is mainly dependent on the shrinkage of the liquid metal inside a solidifying casting and the contraction during solidification. The volume of shrinkage cavity also depends on the following factors:

(i) The volume of shrinkage cavity is directly proportional to the volumetric shrinkage during solidification. This mainly depends on the composition of the alloy. The shrinkage cavity is layger for the metals/alloys showing greater specific volume change on solidifi c ation. However, cast irons with compositions $C + Si/7 = 3.9$ do not show any shrinkage as the liquid contraction is compensated by the growth of graphite.

(ii) The temperature gradient between centre and outside of the casting also affects the volume of shrinkage cavity. Less amount of superheat reduces the volume of shrinkage cavity.

(iii) Conductivity of metals also affect the shrinkage cavity. When high conductivity metals/alloys are cast in metallic or cold moulds, a solid crust is formed immediately which causes larger shrinkage cavities. This problem can be minimised by using preheated moulds and slow pouring.

6.3.4 Location, Shape, and Size of Shrinkage Cavities

For a given metal or alloy, the location, shape and size of shrinkage cavities largely depend on the cooling conditions. The typical cases of cooling conditions are shown in Fig. 6.16. Fig. 16 (a) represents the case of bottom cooling. In this case no shrinkage cavity is formed and there is only reduction in metal volume. Continuous casting and very slow pouring represent this type of cooling. Fig. 6.16 (b) shows the case where cooling is affected only from sides. This results in solidification from the outside and thus forming a conical piping having the height equal to that of casting.

Fig. 6.16 (c) represents the case of cooling from sides as well as bottom. In this case a funnel shaped cavity is formed only in the upper portion of the casting. If the height/diameter ratio of casting is less and the rate of bottom cooling is high, the shrinkage cavity will be less in size and volume. Fig. 6.16 (d) represents the normal case of cooling in which cooling is accompanied in a closed mould. In this case the upper layer of the metal gets solidified very fast due to rapid loss of heat caused by radiation and convection with air. This results in the formation of solid skin on the top and proceeding of solidification inward. The formation of cavity is relatively near the top having the identical shape as in the third case (Fig. 6.16 (c)).

Fig. 6.16 (e) represents the case of complex shape. This also results in the similar condition as in Fig. 6.16 (d). In this case the metal draws away and internal shrinkage cavity will be formed in the regions having largest liquid metal. Such cavities can be avoided by the use of chills and risers (discussed in chapter—7).

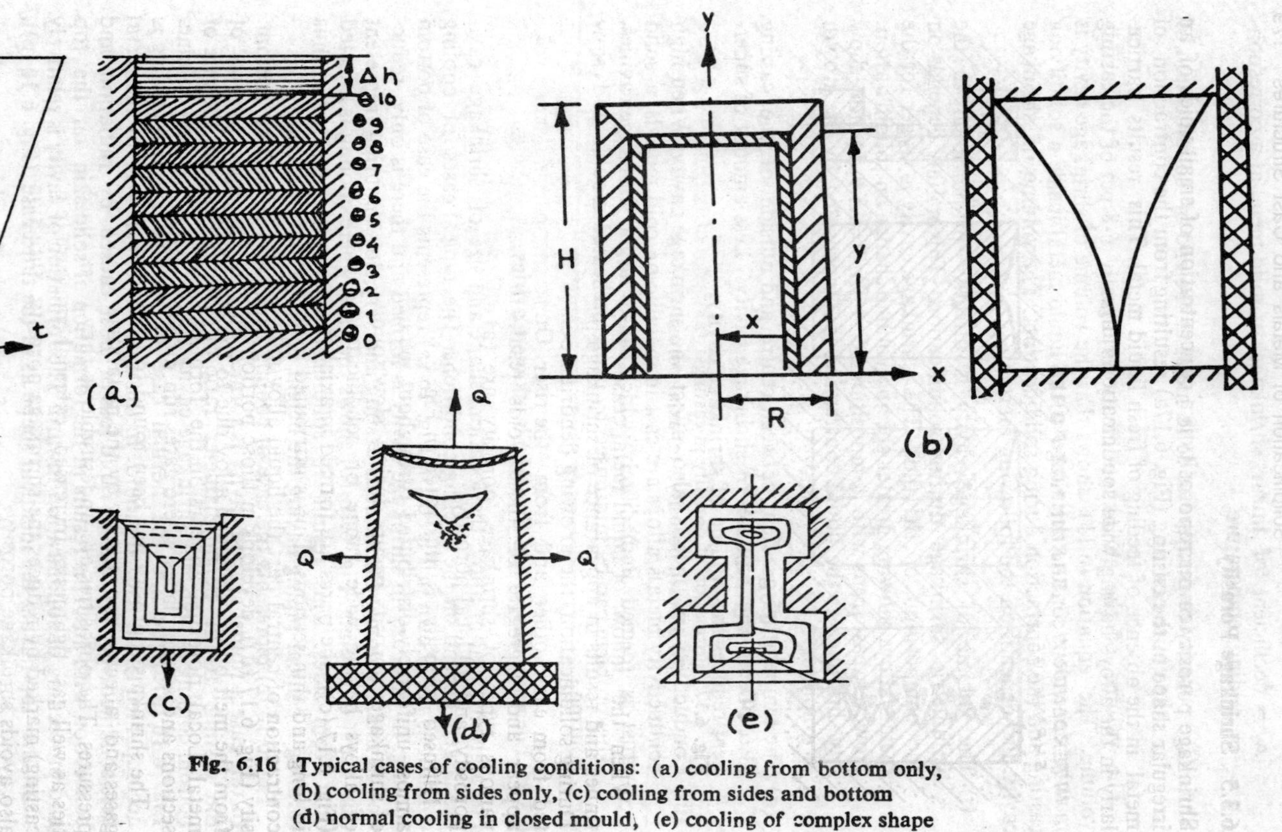

Fig. 6.16 Typical cases of cooling conditions: (a) cooling from bottom only, (b) cooling from sides only, (c) cooling from sides and bottom (d) normal cooling in closed mould, (e) cooling of complex shape

6.3.5 Shrinkage Porosity

Shrinkage porosity or microporosity is a concentration of small voids of an irregular shape in the casting (Fig. 6,17a) resulting from the contraction of metal in the absence of feeding of fresh liquid metal. This results particularly in the alloys having wide solidification range.

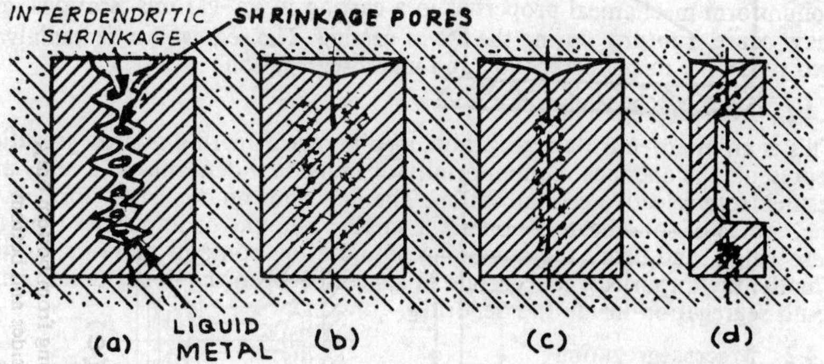

Fig. 6.17 Formation of porosity in castings: (a) voides of irregular shape, (b) diffused porosity, (c) axial porosity (d) local porosity,

When the inflow of liquid metal ceases, the volume shrinkage still continues and results in appearance of shrinkage pores in interdendritic spaces. During solidification, the gr owing dendrites combine to produce cells isolated from each other and from the riser. On further cooling, small voids appear and aggregate of these voids results in shrinkage porosity. The various types of shrinkage porosity are (a) diffused porosity, (b) axial porosity, and (c) local porosity.

Diffused porosity is illustrated in Fig. 6.17 (b) in which small voids are almost uniformely distributed throughout the bulk of the casting. This type of shrinkage porosity occurs during slow solidification of massive castings of alloys having wide range of solidification. In case of axial porosity (Fig. 6.17 (c)) the pores are formed in axial portions of castings and also in long and thin sections. This is attributed to the fact that at the time of contraction of central part the liquid alloy/metal flows poorly. Local porosity (Fig. 6.17 (d)), develops in those portions of castings which are isolated from the melt by the solidified alloy due to the obstruction in flow of liquid metal. Local porosity appears in the form of large sized voids in massive sections and at the places where gates run into the casting.

The shrinkage porosity has been found to be promoted by the dissolved gases and affected appreciably by the metallostatic and externally applied pressures. The porosities are undesirable as they reduce mechanical properties as well gas tightness. Increased external pressure (e.g. in pressure die casting) marked by reduces the shrinkage porosity. Directional solidification also avoids shrinkage porosity.

6.4 SEGREGATION

Segregation is referred to the nonuniform distribution of alloying elements, inclusions and impurities in castings or ingots resulting during solidification. The reason for this phenomenon is mainly the difference between the solubilities of individual components of an alloy in its solid and liquid phase. Larger difference in solubilities causes more non-uniform distribution of an impurity over the cross section of casting and thus results in higher degree of segregation. For example, in case of iron-carbon alloys, sulphur, phosphorus and carbon segregate most heavily. Segregation is the cause of nonuniform mechanical properties in a casting in its various sections and may cause rupture during the use of casting. The segregation is mainly of two types, i.e. microsegregation and macrosegregation.

6.4.1 Microsegregation

This is an interdendritic segregation, also known as coring microsegregation represents the chemical nonuniformity of individual dendrites in a alloy and results from selective crystallisation. The first arms on dendrites formed in the course of solidification contain a smaller amount of impurities than the base alloy and hence the remaining liquid alloy contains more impurities. Finally, the impurities concentrate in spaces between the dendrite arms and result segregation inside the dendrites.

6.4.2 Macrosegregation

This refers to zonal segregation of chemical constituents of an alloy in different parts of the castings during solidification. In addition to the selective crystallisation, the progress of macrosegregation is greately affected by the processes causing segregating elements to move from one portion of the casting to another. Thus the processes leading to zonal segregation may include (i) diffusion of impurities from the two phase zone of the solidifying casting within the still liquid alloy, (ii) convection of alloy streams in the liquid portion of the casting, (iii) floating of low density impurity laden portions and (iv) action of centrifugal forces.

One type of macrosegregation may be referred as *gravity segregation* in which alloying components differing in density separate out mechanically one from the other. This is the case in lead bronze in which slow cooling causes the sinking of lead into lower portions of the casting. Macrosegregation may also appear as *negative* or *inverse segregation*.

In the common zonal segregation the axial portion of casting becomes richer in impurities which lower the melting point of the alloy. The amount the impurities in the surface portion of the castings is less than the mean content of impurities in the casting. On the other hand, in inverse or negative segregation the axial zone of casting contains a smaller amount of low melting constituents than in the outer parts.

The development of any type of macrosegregation depends mainly on the character of alloy solidification and the conditions prevailing during solidification. Alloys subject to progressive solidification show common zonal segregation, whereas the alloys solidifying almost over the entire volume show negative segregation.

The effect of microsegregation can be overcome by annealing the castings. However, heat treatment can not correct the effect of macrosegregaion. Macrosegregation can be minimised by (a) making the casting sections more uniform, (b) avoiding sections of extra mass at junctions, (c) feeding metal to the mould at several places, (d) increasing the rate of metal cooling and (e) pouring the alloys at a minimum overheat above the liquidus temperature.

6.5 DIRECTIONAL AND PROGRESSIVE SOLIDIFICATION

Directional solidification refers to the freezing stertingin those parts of the mould furthest from the feeder heads and then solidification progresses through the casting towards the feeder heads. The feeder heads are designed to solidify last in order to supply liquid metal throughout freezing. The principle of directional solidification is invariably used in feeding of casting. The cooling of casting may be controlled by proper designing of mould and casting. Directional solidification controls the soundness of castings.

The condition of having a partially solid partially liquid zone growing from the outer side inward is called *progressive solidification* which is the result of freezing mechanism and thus can not be avoided. However, the degree of progressive solidification can be controlled. A rapidly cooled casting resulting in a short distance between the start and end of freezing is referred to have a high degree of solidification. On the other hand, a slowly cooled casting would have a low degree of progressive solidification.

Directional solidification is the result of various factors such as casting design, location of gates, and risers, use of chills, and other means for controlling the solidification process. Is a casting is so designed with respect to feeding system that the sections most distant from the available liquid metal solidify first, then there will be successive feeding of the contracting metal by still liquid metal until the heaviest and last to solidify section is reached. The heaviest section in turn can be fed by extra reservoirs of metal known as *risers* or *heads*. These are attached to the castings at the suitable positions to supply continuously the hot liquid metal to the shrinking casting until it is completely solidified.

Directional solidification is achieved for the development of special properties. The development of columnar or of mixed microstructures in many castings is obtained as a result of casting geometry and foundry conditions such as moulding materials and casting methods. However, the methods are also available to produce aligned structures of preferred orientation by the growth of columnar grain s in the direction of heat flow which is accompalished by controlled directional solidification. The primary requirement of direc tional solidification is the maintenance of steep temperature gradients throughout the whole solidification process. The steep temperature gratdient can be obtained by differential cooling, using chills in combination wih insulation, preheating the mould, using exothermic mould materials or other means of local heating.

The main advantage of directional solidification has been the improvement of high temperature properties by the elimination of grain boundaries in the direction transverse to the applied stress.

One of the important possibilities resulting from controlled undirectional solidification is the use of multiphase alloys to produce aligned dispersions, for example, aligned eutectic structure in aluminium-nickel system. If an eutectic alloy is chosen for the contrasting properties of its separate phases, a fibrous composite materials can be produced by casting through unidirectional solidification. The principle of directional solidification has been adopted for the production of cast magnets in which an isotropic properties could be used with advantage. Precision cast gas turbine blades have also been produced both with columnar and with single crystal structures of controlled orientation.

6.6 STRUCTURE OF CASTINGS

The structure of solidified castings can be seen as an out come of the following three main influences:

(a) *Alloy composition.* This governs the basic mode of crystallisation and determines whether structure will consist of single phase or eutectic grains or both. Since the alloy composition is also characterised by specific distribution and diffusion coefficients for solute in liquid and solid phases, it determines the relative tendency of constitutional undercooling

(b) *Thermal conditions.* The temperature distribution and cooling rate in a casting influence the structure of castings, which in turn depend on the initial temperature conditions and the thermal properties of metal/alloy and mould.

(c) *Nucleation and growth conditions.* The presence of foreign particles or solute in the liquid either as trace elements or added deliberately, influences the nucleation and growth conditions.

Since thermal conditions may vary widely at various stages during cooling of a single casting, the overall structure of casting may consist of separate zones.

6.6.1 Interaction of Temperature and Compositional Gradients in Determining Structure

In case of alloys forming a continuous range of solid solutions the mode of crystallisation is governed by the interaction of temperature and compositional gradients in the liquid. Columnar growth from the mould wall towards the centre of the casting is favoured by the presence of a s eep temperature gradient, which increases the probability of renewed growth at the interface before sufficient undercooling to bring about further nucleation elsewhere. Slow cooling also promotes columnar growth and allows more time for solute transport in the liquid and thus diminishes the concentratiore gradient at the interface. Slow cooling is further associated with an inhen rently low rate of nucleation relative to the growth rate in all crystallisation-processes.

The effect of foundry variables on structure of castings should be determined in terms of their effect on the temperature gradient (G) and the freezing rate (R). The ratio G/R is an important parameter for the mode of growth and final grain structure in solid solution alloys. Progressive change in the parameter G/R from a high to a low value is accompanied by successive transitions in the mode of freezing as the undercooling effect becomes more pronounced.

Fig. 6.18 represents the successive stages of solidification morphology of a given alloy and Fig. 6.19 represents the associated thermal conditions for various stages. A very high G/R ratio results in columnar growth with the advance of plane interface which gives way to the cellular growth. Decreasing values of G/R give about the cell-dendritic transition with columnar growth occurring on probes of solid some distance ahead of the main interface. Further lower values of G/R give rise to heterogeneous nucleation in the undercooled zone and this results in the growth of ne w grains in positions which are remote from the existing interface. These grains may undergo dendritic growth or radial growth with spherical morphology analogous to the plane interface condition in unidirectional solidification.

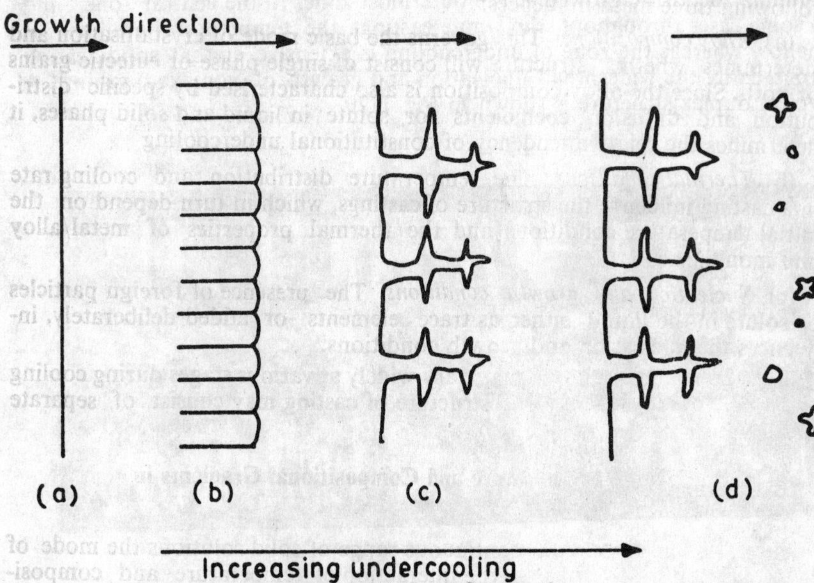

Fig. 6.18 Successive stages of solidification morphology: (a) planer interface, (b) cellulor interface, (c) dendritic growth, (d) independent nucleation

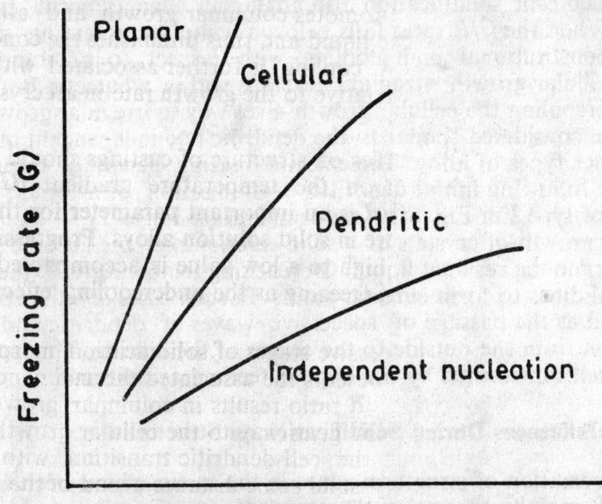

Fig. 6.19 Influence of temperature gradient and freezing rate on solidification morphology

Since the thermal conditions change during solidification, the separate structural zones formed in castings may be explained on the basis of critical changes in the G/R ratio. Initial solidification takes place under a considerably high temperature gradient which is usually sufficient to bring about columnar dendritic growth in the outermost zone. In the central zone, and in some cases throughout the whole casting, the temperature gradient is shallow whereas the zone of undercooling is extensive, so that solidification proceeds by the widespread nucleation and growth of equiaxed grains. A typical duplex structure is shown in Fig. 6 20.

Fig. 6.20 Typical duplex microstructure in cast metals (columnar and equiaxed grains)

In case of eutectic solidification also analogous changes occur as discussed above. When the G/R ratio falls below a value representing a critical degree of constitutional undercooling with respect to an impurity, the colony or cellular growth structure develops within a eutectic grain. With further undercooling the cellular growth gives way to irregular growth forms which can be considered similar to the dendritic and independent nucleation stages in other types of alloy. The eutectic freezing generally occurs as a second stage following initial dendritic crystallisation of a primary phase. In an alloy of type I in Fig. 6.10, the temperature range $T_L - T_E$ is marked by dendritic growth of crystals of α solid solution and the accumulation of composition β in the residual liquid. On reaching temperature T_E the remaining liquid solidifies to form eutectic grains. In such cases solidification may be considered as the passage of successive waves of dendritic and eutectic crystallisation from the outside to the centre of the casting. This pattern of freezing is well represented by the hypo-eutectic gray cast iron.

6.6.2 Other Influences During Solidification on Structure

So far the formation of structural zones in the castings have been explained in terms of mainly thermal and compositional effects on nucleation and growth conditions. However, considerable effects also arise from crystal multiplication and from the movement of individual crystals or grain blocks within the liquid but away from the sites of their original nucleation. Such movement may result due to various reasons such as (a) turbulence originating during pouring, (b) mass feeding, (c) thermal convection, or (d) gravitational separation due to the difference in density between liquid and solid phases.

Dendritic fragmentation is an important mechanism in the formation of equi-axed zone in castings. Dendrite arms in the columnar zone get detached by local disturbance caused by thermal fluctuations and changes in growth rate. The detached dendrite arms are carried by convective stirring and turbulence into the central region where they grow independently in undercooled liquid. In addition to the fragmentation as the source of centres for equi-axed growth, the contact with the cool mould surface on pouring also initiates the nucleation of many crystallites which get distributed widely by turbulence caused due to pouring. In castings poured with little sperheat these crystallites continue to grow rapidly to form completely an equi-axed structure. This type of nucleation is known as big-band' nucleation.

In castings poured with higher superheat, the initial crystals may result completely or may remain in considerable numbers to participate in the formation of equi-axed zone. Crystallite may also get detached from the uper metal-mould interface or from the exposed and chilled free surface of the liquid and fall through the melt. This phenomenon usually occurs in ingot soldification.

In a brief it may be said that formation of equiaxed region is formed due to (a) heterogeneous nucleation *in situ*, (b) crystal multiplication and (c) transport of crystallites by mass movement of liquid or by gravity.

6.6.3 Importance of Cast Structure

The final metallographic structure of castings is governed by the following factors:

(a) Thermal properties and composition of alloy.
(b) Design and dimensions of castings.
(c) Superheat and final pouring temperature.
(d) Thermal properties of the mould.
(e) Conditions for heterogeneous nucleation.
(f) Conditions influencing the movement of crystallites during solidification.
(g) Subsequent heat treatment.

Though the structure of casting is mainly a function of alloy composition and geometry of casting, it is also affected by various measures in casting steps such as treatment of liquid metal and variation of cooling rate within the mould. In general, the manipulation of the solidification process in castings is directed to overcome feeding problems rather than structure.

The importance of cast structure lies mainly in the structure sensitive properties having bearing in engineering applications. The wrought metals/alloys can be further worked for changing both structure and dislocation density, but in case of castings the initial microstructure is the main vehicle for the control of properties except the possibility of heat treatment in some cast alloys. Following three types of metallographic feature are responsible for the critical characteristics of the cast structures.

(a) *Grain Structure (Size shopc and orientation)*. The boundaries of primary grains formed during solidification are generally observed by microscopic or macroscopic examination. However, their detection may become complicated by solid state transformations taking place on further cooling from solidification temperature to shop temperature. Boundaries may be still more difficult to distinguish in case of eutectic grains.

Grain shape and orientation are important in relation to properties of the cast metal and to the problems of feeding. In an equi-axed region, whilest each grain possesses its own crystallographic orientation, the statistical effect of a mase of grains gives isotropy of properties. Whereas, in an extensive columnar zone the boundary alignment and common orientation can cause anisotropy. The influence of directionality on mechanical properties of single crystals and wrought materials is well known.

In many commercial casting alloys directionality is much less marked which may be due to (i) control of properties by dentrite or eutectic substructure or (ii) modifying the influence of as cast structure by solid state transformations. Therefore, the degree of anisotropy depends both on alloy type and on structural characteristics.

The primary grain size is one of the most important features emerging during solidification. In addition to the interinsic effect of grain size, it influences the distribution of solute and of second phase constituents. Direct effects of grain size are expressed in well known relationships arising from the function of grain boundaries as barriers to the movement of dislocations. This influence is also present in cast alloys but the most important effects of grain size in castings are mainly due to the influence of impurities and alloy phases in grain boundaries.

(b) *Substructure*. Important features of substructure include the distribution and state of division of alloy constituents and microconstituents (including solutes, second phase networks and separate phases within the eutectic grain). This also includes dendrite substructure. The normal cast structure can also be complicated by the presence of shrinkage cavities or macroscopic segregation of alloying elements.

(c) *Sub-grains*. Within the individual grains, both dendrite and eutectic substructure may be significant which may outweight the importance of grain size. In dendritic structures, dendrite cell size and secondary arm spacing usually determine the form and distribution of second phase constituents and the range of compositional fluctuations, arising due to microsegregation. Therefore, it exercises the same function as the primary grain size in other cases. The importance of the morphology of the phases within the eutectic grain can be illustrated by the effects of graphite dispersion in cast iron. When the graphite changes to spheroidal form, the limitation on mechanical properties is shifted from the graphite to the matrix and thus giving both ductility and increased strength. However, these considerations do not apply to mechanical properties of grey cast iron in which damping capacity increases with cooling rate and with carbon equivalent of iron.

6.6.4 Control of Cast Structure

Though the structure control is mainly of interest for its effect on the mechanical properties of the castings, it is equally important in producing a sound casting. Therefore, refinement of cast structure may not be necessarily a step towards superior mechanical or physical properties, but it may be for the suppression of hot tears or for benefits to be achieved infeeding. Control of cast structure has been obtained by working in two directions. Most widely established measures are directed towards the refinement of grains and constituent size. Special treatments for this purpose are well established features in the production of light alloys and cast iron castings. The second direction consists of grain size control in cast steels and other alloys. However, the second form of structure control has received limited attention, and it involves the manipulation of grain shape and crystallographic orientation to produce anisotropy of structure and properties.

Metallographic structure is determined by the three significant factors, i.e. nucleation, growth behaviour and crystal multiplication. The main aim in control of structure is to obtain the refinement through one or more treatments. The effect of most refining treatment is the reduction or elimination of the columnar region followed by progressive increase in the number of equiaxed grains. Refining treatment may also influence growth of substructures due to the effect on freezing rate. Grain refining treatments are quite at advanced level in case of aluminium and magnesium alloy castings resulting into marked improvements in feeding, tear resistanecand mechanical properties. In grey cast iron, inoculation is invariably used to improve the graphite dispersion by increasing the nucleation of eutectic grains. More radical refining treatment can be illustrated in the production of spheroidal grey cast iron. Various practical treatments for grain size control and modification of structure include (i) variation of cooling rate, (ii) chemical treatmeet of the liquid metal, (iii) manipulation of superheat of liquid mtetal, and (iv) agitation during solidicfiation.

(i) *Variation of cooling rate.* Rapid cooling results into fine grain structure which is due to the influence of undercooling on the relative rates of nucleation and growth. The undercooling of a liquid metal to a lower temperature increases the number of effective nuclei in comparison to the growth rate since the growth rate is restricted by the rae of latent heat dissipation. On the other hand, slow cooling favours gtrowth from few nucilei and results in coarse grain structures. The refining effect of rapid cooling rate is applicable to primary grain size as well as to substructure. However, in case of substructure, the effect is upto the growth process rather than upon nucleation.

In cast irons increased cooling rate results in increased undercooling as a result of which large number of eutectic grains will be formed along with finer graphite within grains (with a ferrite matrix). This type of structure is associated with inferior mechanical properties and wear resistance. Further increased cooling rate results in the formation of the metastable Fe_3C constituent which depends on the balance of composition and thus wholly white cast iron structure is produced.

In castings the cooling rate is mainly governed by the design and thermal properties of the casting, but is also affected by the mould material. In practice the highest heat diffusivity is obtained with the metal moulds used in gravity die casting. This increased cooling rate results in superior properties usually associated with chill castings, thin sections, and suface zones. Therefore, extensive use of chills is an important featur in the production of high quality castings. The use of chills refines the microstructure and assists feeding by steeping the temperature gradient and thus the castings obtained possess high and guaranteed level of mechanical properties.

The use of very high cooling rates can produce fine structures and thus offers a possibility for the production of high strength alloys by splat cooling. The splate cooling refers to projection of successive molten drops against a chilled surface to produce solid mass with an ultra fine cast structure. However, the technique of splat cooling is suitable only for ingots and certain simple shapes.

The pouring temperature bears the maximum influence on the cooling rate. An increase in the pouring temperature diminishes the freezing rate by preheating the mould and thus reduces the rate of heat transfer during solidification. Thus, to obtain maximum undercooling for grain refinements a low pouring temperature is desirable. However, there is little scope to control cooling rate through pouring temperature and mould properties due to the other limitations to other foundry requirements such as fluidity and feeding.

(ii) *Chemical treatment of the liquid metal* Increasing of normal alloy content results only a small refinement in grain size, but inoculation gives highly effective grain refinement. Inoculation refers to the addition of small amount of such substances to the liquid metal which promote nucleation. In certain cases the function of such addition may be to modify the growth rather than the nucleation process. The addition of these small substances results in a disproportionately large refining effect. The basic requirement of heterogenous nucleation is the ability of liquid metal to wet the foreign particles. The nucleant should be capable of survival in superheated liquid metal and should be in a sufficiently fine state to be present as a widely dispersed suspension.

Various substances ore effective as nucleants in different non ferrous alloys. In aluminium alloys effective grain refinement can be obtained by small additions of titanium and titanium-boron alloys, where as for magnesium alloys carbon and zirconium are effective agents. Iron is an effective nucleant for copper. In case of copper and aluminium there oxides also exercise considerable refining effect and thus the atmosphere has a potential effect on the structure of metals.

Various explanations have been given for the refining function of alloy additions. The most direct explanation is the formation of stable particles as nuclei in the melt for which generylly small additions are adequate. Another important effect is considered to be the growth restriction due to solute concentration gradients, and constitutional undercooling in the liquid adjoining the crystal.

The carbide is an active nucleant for titanium. The occurence of stable carbides and nitrides of Zr, Nb, V, W, M, Ta and B has been observed to be a common feature in promoting refinement in aluminium alloys. In magnesium-aluminium alloys, Al_4C_3 is formed on inoculation of the liquid alloy with carbon. According to peritric hypothesis, refinement by most inoculants in aluminium is attributed to peritectic reactions which are common to their respective systems with the parent metal. For example, such a peritectic reaction in the magnesium-zirconium system is responsible for very fine cast structures. In certain circumstances both the mechanisms may be operative.

The grain refinement of cast structure of carbon steels can be effected by the addition of Ti, Nb and TiC which decrease the dendrite arm spacing considerably. It has been observed that nitrogen plays an important role in grain refinement of sustenitic steels.

The purpose of inoculation in cast irons is to promote increased nucleation of eutectic grains for a given rate of cooling. This in fact decreases the degree of undercooling with slower growth rate and reduced branching of graphite flakes. However, the overall effect is to promote a uniform distribution of medium flake graphites. In cast irons this is achieved by the addition of ferrosilicon or calcium silicide at the pouring stage. The graphite structure is, however, effected by many other elements, e.g. sulphur gives a powerful effect in nucleating more eutectic grains and coarsening the graphite flakes. The presence of titanium neutralises the effect of sulphur and thus titanium acts as a scavenging element and a graphite flake refiner. The spheroidal graphite structure is promoted by the addition of magnesium and cerium which suppress the nucleation of flake graphite eutectic grains and bring about a condition of marked undercooling. This results in heterogeneous nucleation of graphite on a large scale with the production of irregular discontinuous structure which is responsible for the ductility of spheroidal cast irons.

Another important example of chemical inoculation is the modification of the aluminium-silicon eutectic by small additions of sodium or other elements.

Most chemical treatments involve additions to the liquid metal either within the furnace or during pouring. The mould washes can also be used as medium for the introduction of nucleants, particularly to refine small investment castings.

(iii) *Manipulation of super heat of liquid metal.* Superheating of metal to a higher temperature than that used for pouring also exercises a considerable influencc upon nucleation. Superheating can be embloyed either in the course of normal production or as a deliberate preliminary treatment. The effect of supreheat gives added significance to the pouring temperature itself. Increased pouring temperature and holding time result in decrease in the number of eutectic grains in grey cast iron. Excessive superheat may even. suppress nucleation of graphite completely and result in a chilled structure However, in certain cases the superheat produces the opposite effect of grain refinement, e.g. in case of magnesium aluminium alloys.

The mechanism of refining by superheat has been explained as the solution of relatively large particles of nucleating compounds and their precipitation in more finely divided and crystallographically favourable forms on cooling towards the liquids temperature.

(iv) *Agitation during solidification* Nucleation can be brought about by physical disturbance of undercooled liquid, e.g. stirring or gas evolution. This may be characteristic to the widespread distribution of nuclei initially produced at the surface, or to the fragmentation of already growing crystals. It has been reported that vibration can be used effectively in inducing grain refinement and it offers possibilities of practical treatment particularly in case of small castings. The refining effect may be explained as being the result of increased pressure wave accompanying the vibration cycle which reactes nuclei. The nuclei then grow at sufficiently fast rate to survive the following reduced pressure wave. In cast irons also vibration reduces the size of graphite flakes.

Electromagnetic stirring with the aid of induction coil is another possibility of practical treatment. However, it is not always possible to produce crystal multiplication by stirring. In some cases the predominent effect of vibration or stirring is the elimination of solute barrier and the remelting of equi-axed grains which favour the columnar growth.

The structures in large sand castings is relatively coarse in the absence of the additional refining effects. Thus the efforts should be made to achieve close control of grain structure by suitable means.

7

Flow of Metals and Gating System

Since the behaviour of liquid metal and its subsequent solidification and cooling determine soundness of casting, the pouring of molten metal and its flow into the mould are critically important. Pouring a casting looks to be a simple procedure but in practice it is quite complex due to the following reasons:—

(i) Metals are not inert chemically.
(ii) The liquid metals absorb gases.
(iii) The metal exhibit shrinkage on cooling.
(iv) The liquid metals are erosive to the mould and
(v) Different metals posses different specific gravities.

The success of the pouring operation depends partly upon certain properties of the metal/alloy itself (for example, the composition and temperature which influence the flow of liquid metal) and partly upon properties and design of the mould (such as nature of the moulding material and gating system used to introduce the metal into the mould cavity). In addition to these factors, due consideration is given to (i) forces acting up on the mould and (ii) volume contraction during solidification of liquid metal. All these aspects are dealt in this chapter.

7.1 FLOW OF METALS

The flow of liquid metals through channels and mould is associated with some specific characteristics which are not present in the flow of ordinary fluids.

These specific characteristics are as follows:

(i) Molten metals are not necessarily homogeneous.
(ii) During the flow of liquid metal the temperature decreases with time and hence the process of metal flow is irregular and unstable.
(iii) The hydraulics deals with the fluids flowing in open channels or gas impermeable tubes, whereas motion of liquid metal in mould is with

porous and gas permeable walls. Thus, it can be considered as the case of pressure head of liquid flow in permeable and perforated tubes.

(iv) Metal flows along with the process of phase change and volume change due to solidification process.

In the process of mould filling three types of metal flow may occur in practice, i.e. metal flow in top casting, metal flow along the channels, and metal flow in moulds.

7.1.1 Metal flow in top casting

This is the case mainly of ingot casting and rarely in the shape of casting. In this instance free fall of liquid metal is observed which results in (i) intensive oxidation of metal, (ii) air injection into the metal and (iii) heavy metallostatic impact of metal jet against the mould surface. When the free jet contacts with the mould bottom, its kinetic energy is transformed into potential energy which results in increase of pressure above the jet and the metal begins to flow along the bottom of the mould.

The jet energy (F) can be given as following:

$$F = \frac{m\,v^2}{2g} = \frac{A\rho v^3}{2g} \tag{7.1}$$

where, A = area, v = velocity, ρ = specific gravity and g = acceleration due to gravity.

The jet energy (F) will be consumed in the liquid in the formation of vortex under the jet. From hydraulics, for any continuous potential flow symmetrical to a certain axis.

$$v_n \cdot r = \text{constant} \tag{7.2}$$

where v_n = velocity component normal to the radius of rotation and r = radius from the centre of rotation.

As a result, the maximum velocity of metal flow will be in the centre of the vortex flow. On the other hand metallodynamic pressure in liquid decreases with the increase of its velocity (from Bernoulli's theorem) which hamper the air and slag liberation from liquid metal in the course of vortex formations. The initial impact of liquid metal always involves air inclusion. The subsequent air inclusion depends on many factors such as:

(i) Shape, dimensions and velocity of the jet.
(ii) Reynold's number of the jet flow, and
(iii) Stability of oxides on the metal surface.

In foundry practice, the top casting process is avoided as far as possible.

7.1.2 Metal flow along channels

Though simple laws of hydraulics can be applied to the flow of superheated liquid metals, following specific characteristics are observed.

(i) If superheat is less, solid crusts of metal may be formed on the walls of the channels resulting into decreased cross section for metal flow. This crust increases with time according to the formula

$$\partial = K\sqrt{t} \tag{7.3}$$

where ∂ = thickness of crust, t = time, and K = coefficient of solidification having different values depending on the conditions of solidification and the properties of metal. As a result, the cross sectional area of channel ($\pi D^2/4$ or πR^2) will change to as $A = \pi\,(R - K\sqrt{t})^2$.

(ii) The prorerties of liquid metal may change, particularly the viscosity of metal due to drop in temperature during the flow. However, the hydraulic characteristics are not mud changed. The viscosity is characterised by Reynold's number. Reynold's number is the ratio (dimersionless) of two kinds of forces in flow of liquid which are (i) viscous force promoting the laminar flow and (ii) inertia forces promoting turbulent flow, i.e.

$$\text{Reynold's number, } R_e = \frac{\rho v d}{\eta} = \frac{vd}{\nu} \tag{7.4}$$

where ρ = density, η = dynamic viscosity of liquid, d = diameter of the flow channel, $\nu = \eta/\rho$ = kinematic viscosity and v = velocity of metal flow.

The lower critical velocity for flow in channels corresponds to R_e = 2,000 for all fluids and the upper critical velocity generally corresponds to R_e = 30,000 – 40,000. The velocity of liquid metal flow (v) may be given by

$$v = \sqrt{\frac{2gH}{\frac{1 + \lambda L}{D} + \sum R_i}} = \mu\sqrt{2gH} \tag{7.5}$$

where λ = coefficient of friction losses, $\sum R_i$ = total local resistance, L = length of channel, D = diameter of channel, H = metallostatic head, and μ = flow coefficient.

For example, in case of pig iron for a channel diameter of 1 cm with metallostatic head of 12 cm, the velocity will be about 130 cm/sec. Kinematic viscosity of pig iron at 100 °C superhesting is about 0.011 stokes. Hence Reynold's number R_e will be equal to about 13000. Similarly for steel with 1 2 cm channel diameter, the velocity will be about 120 cm/sec. In the case R_e = 15000 (v = 0.008 stokes at 100 °C superheating).

In the similar way, R_e values for other metals can be found which are generally above 15000 for nonferrous metals and alloys. Thus, the flow of molten metals in all the cases is of turbulent nature.

(iii) The resistance due to friction and local resistances also influence the flow of liquid metal. Coefficient of friction losses (μ) depends on the roughness of channel walls and internal friction, i.e. Reylond's criterion. In case of laminar flow roughness of wall does not affect the value of friction coefficient due to the boundary layer and the losses are only due to Reynold's criterion. However, under turbulent conditions coefficient of friction depends on roughness of the wall as well as Reynold's number. In the event of excessive turbulence the boundary layer is absent and the friction coefficient does not depend on the R_e value. In such cases the flow coefficient becomes constant. Friction coefficient for turbulent flow can be found out by the White and Kolbook's formula

$$\frac{1}{\sqrt{\mu}} = -2 \log \frac{2.51}{R_e\sqrt{\mu}} + \frac{\Delta}{3.7\,d} \quad (7.6)$$

or by the approximate formula of Altshul

$$\mu = -\frac{1}{1.74 + 2 \log (\Delta/r)} \quad (7.7)$$

where $\Delta =$ average height of projectioas on the inner surface of channels, $d = 2\,r =$ actual or equivalent diameter of channel.

Local resistance coefficients can be evaluated from experiments by finding out the head or energy losses which occur at the orifices, channel entrances, bends, junctions, contractions and enlargements.

7.1.3 Metal flow in mould

In flow of molten metal in the mould space is the flow of lipuid of altering viscosity and with heat transfer taking place from the mould walls to the atmosphere. Therefore. system can be described only with the help of various complex differential equations such as (i) of flow supplemented by the viscosity alterations, (ii) of heat exchanges between the liquid metal, mould and the atmosphere, (iii) of heat conductance in the metal volume and (iv) of flow continuity.

The solution of such a set of differential equations is possible only for some simpler cases under many assumptions and simplifications. Therefore, analytical method is hardly of any use in foundry peactice. However, modelling may be useful for understanding of some aspects of the problem. The modelling is beyond the scope of this book.

7.2 FLUIDITY OF MOLTEN METALS

Fluidity is the most widely recognised term to describe certain aspects of metal flow behaviour. In general, fluidity may ba defined as that property of metal which enables its flow through mould passages and to fill all the spaces of the mould providing sharp lines and accurate reproduction of design details. Thus insufficient fluidity may cause short run castings or poorly defind surface features. However, the fluidity is not a single physical property as density and viscosity but it is acomplex charactsristic related to behaviour under specific conditions within a mould. In foundaries many other synonymous terms of fluidity such as runability, fluid life, flowing power, shrinkageability etc. are also used.

The fluidity is a function of many variables and can only be defined on the basis of emperical tests in which metal flows in a narrow channel until it solidifies and the length of flow is considered to be the measure of fluidity.

The physicists define fluidity as a reciprocal of the viscosity coefficient and consider the term of fluidity in absolute meaning. Viscosity may be defined as the force required to move a surface of unit area at unit velocity past an equivalent parallel surface at unit distance. Thus it is the measure of the capacity of a liquid to transit a dynamic stress in shear. For a liquid flowing, its viscosity will determine the extent to which the drag imposed by the passage wall is transmitted to the bulk of the liquid. Therefore, viscosity influences the rate of flow which is found to have simple relationship with the viscosity. The viscosity more directly related to the capacity of liquid to flow under its own pressure head is known as *kinematic* viscosity, i.e. absolute viscosity divided by the specific gravity of liquid.

. Whatever the viscosity of liquid, the most important characteristic of the liquid state is its ability to conform in time to the shape of its container. In case of liquid metal it happens rapidly as the viscosities of liquid metals are usually low. In castings, the failure to fill the mould cavity in fact is the result of premature solidification rather than of high viscosity. Therefore thermal conditions and mode of solidification are critical factors for the termination of flow. Thus in foundry technology, concept of fluldity should take these aspects into account.

7.2.1 Fluidity measurcmnnt

The fluidity of a liquid metal can not be assessed from individual physical properties and thus empirical tests have been developed to measure the overall characteristic. A number of methods are available to measure the fluidity, which are based on the conditions analogous tc the casting of metals in the foundry. The fluidity is measured in terms of total distance covered by liquid metal in standardised system of enclosed channels before the flow ceases. Another parameter in fluidity test is the flow time or fluid life. The various methods of fluidity measurement can be divided into two groups i.e (i) tests with constant section channel, known as rod test, and (ii) test with variable section channel, known as *V* — test. However, for the practical purpose, the various tests can be considered as (i) straight rod fluidity test (ii) spiral fluidity test, (iii) vacuum flidity meter, (iv) *U* — shaped probe test, and (v) wedge test.

(i) *Straight rod fluidity test* In this test the liquid metal is allowed to flow in a straight flow channel of constant section. This test involves excessive length of channel and sensitive to angles. This test can be used for certain nonferrous metals but it is discontinued in favour of spiral test.

(ii) *Spirol fluidity test* Numerous variations of spiral test are available. A typical spiral fluidity test is illustrated in Fig. 7.1. Variations in spiral test are mainly concerned with the problem of obtaining the identical conditions of flow as in case of castings. The moulds used for testing may be of green sand, dry sand, metal or graphite which should be of the same composition as used in actual metal casting. Various designe of reservoir systems have been attempted to regulate the pressure head and constant speed pouring devices have been devised to ensure a uniform rate of metal delivery to the system.

(iii) *Vacuum fluidity meter* A closer approach to standardisation is achieved in vacuum fluidity meter devised by Ragone, Adams and Taylor—for non ferrous metals. The apparatus used is illustrated in Fig. 7.2. The metal to be tested is allowed to flow through a smooth glass tube under suction induced by partial vocuum. In this case the pressure head is accurately known and thus it eliminates human factor in pouring. A variation

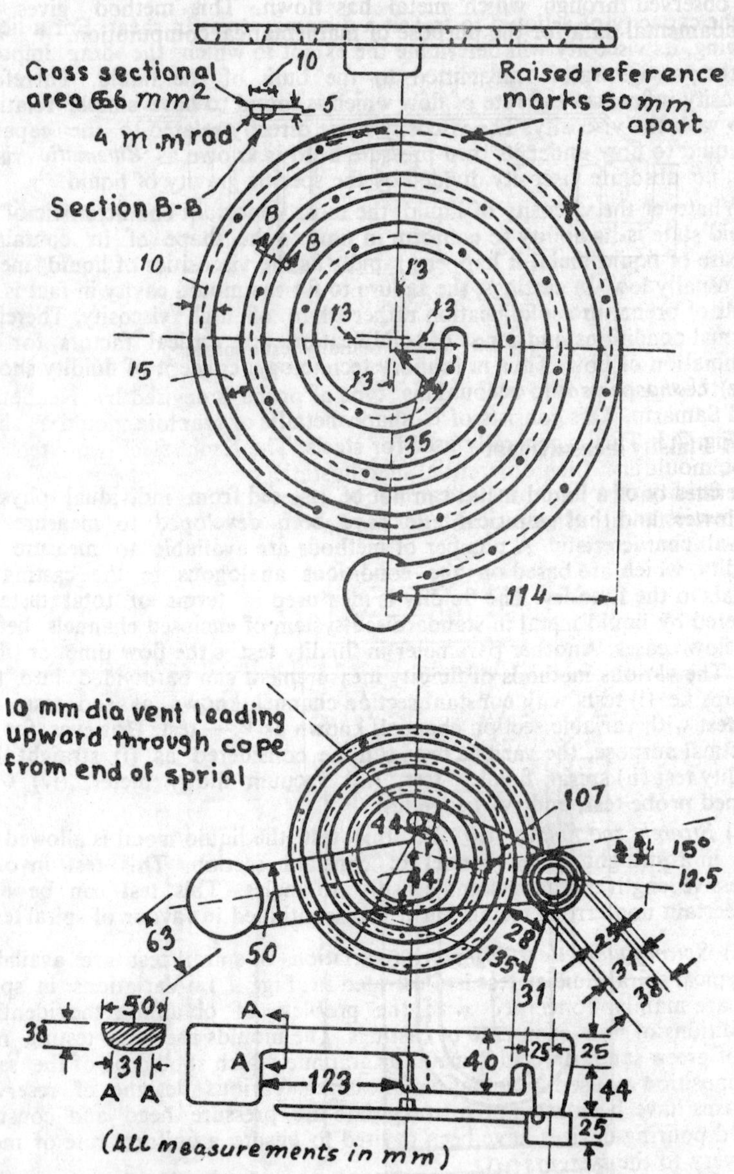

Fig. 7.1 Spiral fluidity test: (a) Standard fluidity spiral, (b) arrangement of down gate and pouring basin for standard fluidity spiral

of vacuum fluidity meter for steels has been devised by Dietert and Co. (U. S.A.) in which stainless steel tube is used in place of glass tube. Since the colour of the outside of the tube changes due to oxidation, the length can be observed through which metal has flown. This method gives more fundamental data for the purpose of mathematical computation.

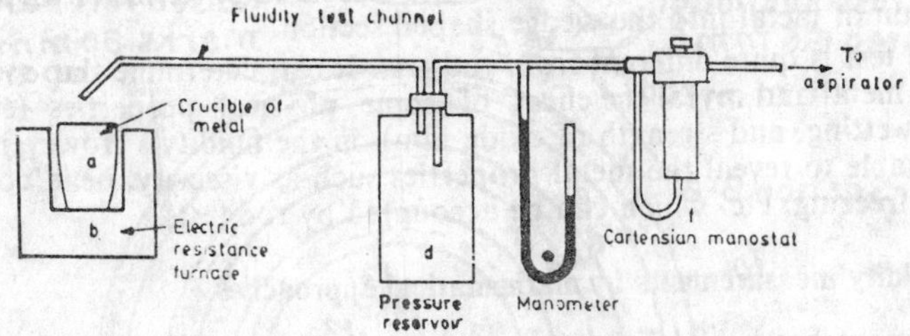

Fig. 7.2 Vacuum fluidity test appratus

(iv) *U-shaped probe test* This type of probe is devised by Nekhendzie and Samarin. This consists of *U*-shape metallic or graphite mould as shown in Fig. 7.3. This is generally used for steels. The liquid steel is poured in *U*-type mould in which differential metallostatic pressure operates. Since it operates on the variable head principle, the meter becomess more sensitive at lower fluidity. It measures only the flowing power of the metal.

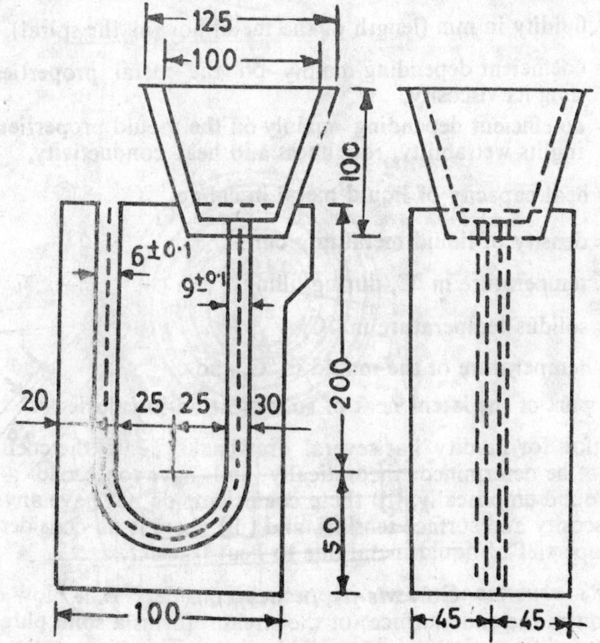

Fig. 7.3 U. shaped probe test

(v) *Wedge test* In this case the mould is made of variable section in the shape of a wedge. The fluidity is determined by filling the metal in the cavity of mould. In this system the metal does not fill the whole cavity, but it flows into the narrowing section. The oxide films and surface tension prevent the filling of mould The fluidity is measured in term of depth of penetration of metal into the wedge shaped section.

Wedge test is quite different from rod test. It can determine the mobility of liquid metal and reveal the effect of some physical properties (surface tension, wetting. and strength of oxide film) on the fluidity. However, this test is unable to reveal the metal properties such as viscosity, heat content, mode of freezing, etc. which can be accounted by rod tests.

7.2.2 Fluidity measurements by mathematical approaches.

Many approaches have been attempted to evaluate the fluidity theoretically from the physical properties of metals. The important approaches are described below.

(i) *Approach by Portevin and Bastien of France*: This was the first approach proposed for the interpretations of fluidity of liquid steel in mathematical form. This can be expressed as following:

$$\lambda = \alpha \frac{cd\,(\theta_f - \theta_s)}{\theta_f - \theta_m} + \beta \frac{dL}{\theta_f - \theta_m} \tag{7.8}$$

where λ = fluidity in mm (length of the metal flow in the spiral),

α = coefficient depending mainly on the metal properties including its viscosity,

β = co-efficient depending mainly on the mould properties including its wettability, roughness and heat conductivity,

c = heat capacity of liquid metal in cal/cm^3 °C,

d = density of liquid metal in g/cm^3,

θ_f = temperature in °C, during filling,

θ_s = solidus temperature in °C,

θ_m = temperature of the mould in °C, and

L = part of the latent heat of solidification in calories.

This equation for fluidity has several drawbacks, i.e. (i) the coefficients α and β can not be determined theoretically and have too broad a physical meaning if found empirically, (ii) these coefficients do not have any connection with viscosity and surface tension, and (iii) there is no consideration of changing properties of liquid metal due to heat transfer.

(*ii*) *Gulyev's equation* Gulyev's approach considers that flow of metal ceases as soon as the end surface of the stream forms a solid plug (due to the cooling conditions in the channel) having sufficient strength to withstand the metallostatic load. The proposed equation for liquid steel is,

$$\lambda = \sqrt{\frac{2gH}{1 + \sum x}} \times (\mathrm{Ar}^2\, l_{l1}\, \mathrm{BT} + \mathrm{CHr}) \tag{7.9}$$

where H = metallostalic head, $\sum x$ = total loss of the head, r = radious of the channel, T = superheat, and A, B and C are constants related to the thermophysical properties of the liquid metal and mould, the strength of the surface film, surface tension and individual characteristics of the liquid metal.

In this equation also the constants $A,B,$. and C have to be determined experimentally. However, this equation gives the explanation for termination of flow even when enough latent heat and viscosity are available for the flow.

(*iii*) *Nekhendzie and Samarin's equation* They have considered that alloys of solid solution type (e.g. steel and Cu-Ni alloy) stop flowing when the fluidity is zero. They assumed this stage when about 20 pct of latent heat of molten metal has been removed through heat transfer. The equation is,

$$\lambda = \frac{Ad\, C_p\, (\theta_f - \theta_s) + dL}{t_{metal} - t_{mould}} \tag{7.10}$$

where A =constant depending on the physico—chemical and other properties of the given metal and mould, d = density, C_p = specific heat, dL = part of the latent heat abstracted on which fluidity depends, and $t_{metal} - t_{mould}$ = characteristic of heat exchange between metal and mould during the time the metal is in liquid state on cooling from θ_f to θ_s.

This equation offers the advantages that (i) only one coefficient has to be determined experimentally, and (ii) relative fluidity of different metals can be determined if the heat transfer terms ($t_{metal} - t_{mould}$) are maintained constant throughout the experiments. The main drawback of this equation is that thermal conductivity of the liquid metal has not been taken into account, whereas it plays an important role in heat transfer between mould and metal.

7.2.3 Factors Influencing Fluidity

The characteristics of liquid metal as well as mould are responsible for the fludity of a metal. The metal characteristics which influence fluidity are metal composition. viscosity, heat content, surface tension and wetting, solidification characteristics, temperature of solidification (melting temperature), specific weight of liquid metal, superheat, and other metallurgical factors such as surface oxide films, absorbed gas films, suspended inclusions, and inclusions precipitating during freezing. The mould characteristics which influence fluidity are thermophysical, gas permeability and state of the surface. The influence of these factors are discussed below.

A. Metallurgical Factors Influencing Fluidity

(i) *Metal composition* This is one of the major factors influencing fluidity as the composition of an alloy changes the solidification process. The useful comparisons of the fluidities of different alloys can only be made at constant superheat. Under these conditions a marked relationship is observed between alloy constitution and fluidity. Pure metals and eutectic alloys possess high fluidity, whereas alloys forming solid solutions show poor fluidity. An inverse relationship between fluidity and solidification range has been established by many workers. Poorest fluidity is shown by the alloys having greatest solidification range. A long solidification range is indicative of mushy condition of metal having interlacing dendrites surrounded by liquid at practically its solidification temperature, and thus this condition would restrict flow of metal. On the other hand, pure metals freeze by gradual inward growth of solid metal which does not restrict the flow of interior liquid metal. The relationships between composition and fluidity for lead-tin,

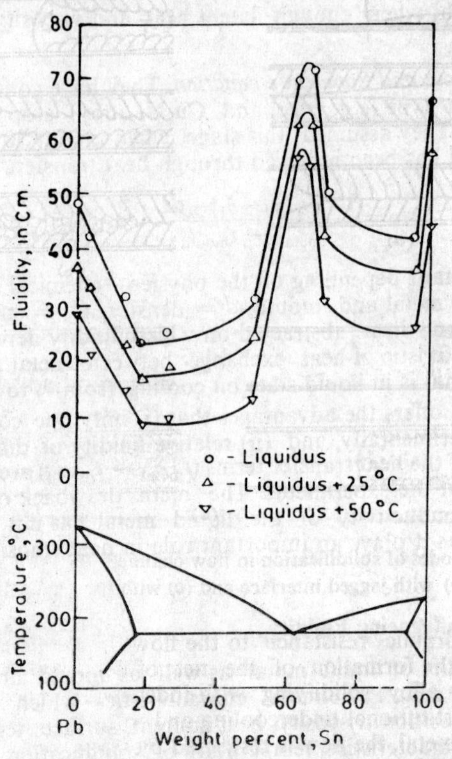

Fig. 7.4 Relationship between composition and fluidity of lead-tin alloys

(ii) *Solidification characteristics* Differences in the flow behaviour of various types of alloys can be attributed mainly to their characteristic modes of solidification. In case of alloys solidifying in narrow freezing range the solidification occurs by progressive advance of a plane interface (the layer of solid metal has relatively smooth surface) from the mould wall and the flow can continue until the channel is finally choked which is found to occur near the point of entry of metal. Pure metals solidify in this manner and show sufficient fluidity even when poured at the liquidus temperature and thus the metal continues to flow during evolution of latent heat of solidification. The alloys having long range of solidification solidify with highly developed dendrites which are spread over large part of the mould and have jagged surface. They solidify with mushy mode. Therefore, flow of metal stops much before the solidification of the alloy in a given section of the

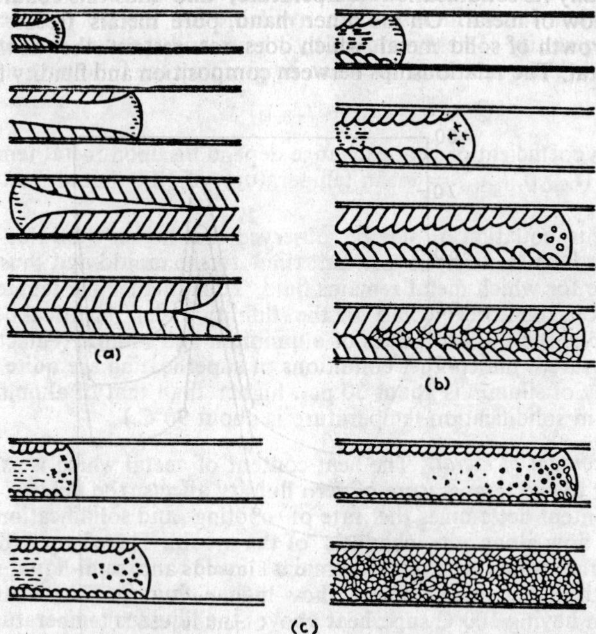

Fig. 7.5 Modes of solidification in flow channel: (a) With planer interface, (b) with jagged interface and (c) with independent crystallisation

mould. The hydraulic resistance to the flow of the liquid metal increases sharply due to the formation of the net of dendrites and the rate of flow falls sharply. In alloys solidifying with independent crystallisation (occurring due to constitutional undercooling and other phenomena) in the main mass of liquid metal, the flow is restricted by the presence of free crystals in the liquid m tal of the advancing stream. These alternative modes of solidification are illustrated in Fig 7.5.

(iii) *Viscosity of liquid metal* Higher the viscosity of metal, less is its fluidity. The coefficients of viscosity of liquid metals and alloys are not much different in values compared to water. For example, at 1555 °C steel has viscosity 0.0260 gm/cm sec., viscosity of zinc is 0.016 to 0.014, and of aluminium is 0.02, whereas of water at 20 °C it is 0.0107 gm/cm. sec. The influence of viscosity on fluidity of liquid metals is observed mainly in the process of cooling when it flows along the channels of the mould (The coefficient of viscoity increases with lowering of temperature). In some cases the introduction of other components such as silicon and sodium in aluminium changes the viscosity of the melt drastically. The flow of metals in mould channels is determined by the value of viscosity coefficient and its change during the cooling of melt.

(iv) *Temperature of alloy solidification or melting temperature* The higher the melting temperature of metal. the more intensive is the heat exchange between metal and mould and there will be more heat removal per unit time. This condition results in less time available for the metal to be in filling state. The quantity of heat removed by the mould through unit area of the surface during the time interval x can be given by the following equation.

$$Q = \alpha\,(t_{metal} - t_{mould})\,x \qquad (7.11)$$

where α = coefficient of heat exchange depending upon metal temperature. i.e. $\alpha = f\,(t_{metal})$, t_{metal} = mean temperature of the liquid metal during pouring.

From this equation it may be observed that higher the value of t_{metal}, more is the heat transferred per unit time to the mould and thus less will be the time for which metal remains fluid. Thus higher the temperature of solidification, the lower would be the fluidity under identical conditions. For example, the fluidity values of aluminium and silumin (eutectic Al—Si alloy) measured under the conditions of superheating are quite different. The fluidity of silumin is about 30 pct. higher than that of aluminium (the difference in solidification temperature is about 90°C.).

(v) *Heat content of metal* The heat content of metal which is extracted in its cooling to the temperature of zero fluidity affects the fluidity. This part of heat content determines the rate of cooling and solidification of metal during its flow along the channels of the mould. Therefore, it determines the time for which the metal remains liquid and semi-liquid state. The metals with higher heat content show higher fluidity. For example, the aluminium having 100°C superheat above the liquidus temperature posses-ses the heat content equal to $232.5 + 100 \times 0.25 = 257.5$ Cal/cm^3 and magnesium superheated to 100°C above the liquidus temperaure contains $136 + 0.3 \times 100 = 166$ Cal/cm^3. The heat content of aluminium is almost 1.5 times more than of magnesium and due to this fact the fluidity of aluminium is about two times higher than that of magnesium with equal superheating. The heat content of the alloy/metal can be rcised by increasing the superheat.

(vi) *Surface tension and wetting* Hoar and Atterton have demonstrated the importance of surface tension with respect to flow in small passage and they found a direct relationship between surface tension and the pressure required to produce penetration of liquid metal into surface voids in sand compacts. The surface tension on the demarcation surface between metal and mould influences the mould filling with metal. It has been agreed by most workers that the surface tension factor becomes significant in the channels of small size (0.5—5 mm). In practice the surface tension may have a value as high as 1500 dyn/cm (1.5 N/m) and this is further modified by the influence of surface films.

The effect of surface tension on the fluidity is the phenomenon of wetting of the mould with metal. The ability of liquid metal to wet the mould is determined by the contact angle of wetting. If the meniscus is concave (contact angle $< 90°$), wetting condition will prevail and if meniscus is convex (contact angle > 90), non-wetting condition will prevail. Under wetting condition the liquid metal is easily drawn into narrow channels of the mould enabling better filling of metal and thus a better fluidity is observed. On the other hand, under non-wetting condition the flow of liquid metal flowing through the mould channels will slow down near the mould surface due to friction and thus the fluidity of metal will be lower.

In foundry practice in most cases the metal does not wet the mould surface and hence the effect of surface tension of metals on fluidity is usually negative. Hence in order to fill narrow channels of the mould it is essential to increase the pressure of the liquid metal.

(vii) *Specific weight of liquid metal* The metals having higher specific weight possess higher rate of metal flow into the channels of the mould which can be observed from the following equation.

$$V = kA\sqrt{2gH} \tag{7.12}$$

where V = velocity of metal flow, k = coefficient of discharge, A = section of the channel, g = acceleration due to gravity, and H = metallostatic pressure height = dh where d = specific weight of liquid metal, and h = height of the liquid column.

From the above equation it is clear that higher is the specific weight of metal, higher is the metallostatic pressure H and thus higher will be the flow rate which in turn indicates that fluidity of the alloy will be higher.

(viii) *Oxide film, inclusions, absorbed gas films etc* The presence of oxide film on the metal surface, absorbed gas film, inclusions etc. influence the fluidity of liquid metals considerably. The surface tension value gets modified by the presence of surface films existing on metals in normal atmospheres. In some cases these films may be visible in case of aluminium bronxes where as in other cases they may be detected by their marked effect upon the emissivity of the metal surface. For example, liquid aluminium carries surface films of oxide which increase surface tension by a factor of three. The reduced fluidity of steels containing aluminium has also been explained by the presence of alumina film. Similarly titanium produces oxide films on stabilised stainless steels and nickel base alloys. Chromium also produces strong oxide surface films.

The presence of restrictive oxide film determines the final outline of the casting in confined corners. Alloys carrying the films are generally prone to poor definition and to the formation of surface laps and wrinkles. In such cases, special care should be taken in gating.

The viscosity coefficient of liquid metals is also influenced by the impurity of the melt with non-metallic inclusions. The alloys clogged up by hard non-metallic inclusions (such as nitrides, carbides, oxides, etc.) show low fluidity. Magnesium, aluminium, and cast iron are associated with this problem.

B. Mould Factors Influencing Fluidity

(i) *Thermal Properties* The rate of cooling to the temperature at which fluidity is zero, is mainly determined by the heal diffusivity (D) of the moulding material.

$$D = (kc\rho)^{1/2} \tag{7.13}$$

where k = thermal conductivity, c = specific heat and ρ = density of metal.

In some cases the cooling rate may be further effected by the absorption of the latent heat of vaporisation of volatile constituents of mould. In case of high conductivity of mould materials, more rapid will be the heat extraction from the liquid metal during the flow and thus less time will be available for the metal to remain in motion. Cooling is also accelerated by evaporation of water from the green sand moulds and thus the metal flow is retarded in thin sections.

The flow of metal in metal moulds is poorer than into the sand mould and similarly poor metal flow is obtained in green sand mould, cold mould, uncoated mould compared to dry sand mould, hot mould, and coated mould, respectively. Hence, for thin sections dry sand mould is used and in case of investment castings the rate of heat removal can be further reduced by the use of high mould preheat temperature. However, an excessive preheat (above 300 °C) leads to the rough structure of the casting and quick wear of the mould. The surfaces of metallic moulds should be coated with heat insulating paints such as ZnO, TiO_2, talc, chalk, asbestos, etc. Incase of green sand moulds, the filling can be improved by drying the gate channels or by making the gate system in core sands.

(ii) *Gas permeability of mould material* When the metal enters the mould, air should be displaced quickly from the mould. This can be achieved through vents, and open risers and through the permeable structure of moulding material. If these means are not sufficient, the back pressure of air will retard the flow of metal and will result an apparent loss of fluidity.

(iii) *Mould surface effect* Flow of metal into a mould is restrained by fractional forces dependent on the roughness of the mould surface. A rough surface prevents the metal flow and results in poor filling of the mould. The roughness of mould surface is related to the grain size of the moulding material. Fine grained sands are preferred for obtaining better metal flow. Apparent fluidity can be greatly increased by mould coating which reduces the friction by providing a smooth surface. Increased flow of metal can also be obtained by employing a reactive coatings which influence wetting characteristics at the metal-mould interface. For example, hexachloroethane may be used which generates active chlorine between metal and surface of the mould and thus reduces the resistance to metal flow. In aluminium alloys, hexachloroethane increases the flow distance which is most probably due to reduction in restrictive oxide films.

The flow of metal in a given mould cavity is thus determined by both metal and mould characteristics. With respect to metal characteristics, the metal flow is primarily a function of composition and superheat. However, the metal composition is governed by the alloy specification and thus can be varied only within narrow limits. Hence, casting temperature remains as the metal variable for practical control of metal flow. Such control should be exercised with many other factors in mind, such as feeding technique, metallurgical structure, and prevention of hot tears and metal penetration. The mould characteristics influencing metal flow are mainly inherent in the moulding process and its associated materials. Therefore, successful filling of mould depends mainly on the use of suitable gating system to a given combination of metal and moulding materials.

7.3 POURING LADLES

The vessels used to carry the molten metal from furnace to the mould are known as pouring ladles. Generally, the molten metal is temporarily stored in large holding ladles from which metal can be tapped off as and when required. These holding ladles are constructed of steel shell lined with refractory bricks. These ladles are generally designed to receive metal from the furnace simultaneously with the pouring off of metal from them into smaller ladles known as pouring ladles. Depending on the size, pouring ladles may be either crane operated or hand operated.

The ladles also differ in design as to facilitate the prevention of impurities such as slag or dross from entering the mould. The common types of pouring ladles are *lip pouring*, *teapot-pour*, and *bottom pour* ladles. In ordinary lip-pour ladles, slag is cleaned off before pouring and the metal is skimmed clean at pouring lip by means of metal skimming bars. In some ladles, a refractory dam is provided in front of the lip which helps in holding back the impurities. The teapot pour ladles have refractory spout extending down beneath the metal surface, so that only clean metal passes through the spout. Bottompour ladles consist of tap-hole at the bottom which is stoppered by a suitable refractory covered stopper rod manipulated from the outside.

Care should be taken to avoid metal heat loss from ladles and for this reason, the ladles are thoroughly dried and preheated. In case of large ladles, a thick layer of slag may be used to prevent metal heat losses by radiation. In case of small ladles, heating coils round them can be used to compensate the heat loss. In order to avoid cold shuts and entering of slag or dross to the mould, the pouring should be carried out in such a way as to maintain a smooth and uninterrupted flow of metal.

7.4 GATING SYSTEM OF CASTINGS

The gating system may be defined as that part of the mould cavity through which moltem metal flows to fill the casting impression. The design of gating system is the principal means for controlling the pattern of metal flow within the mould. The importance of gating system arises for the following purposes.

(i) To achive a proper rate and direction of metal flow to ensure complete filling of the mould before freezing.

(ii) To obtain the smooth and uniform metal flow with least turbulence in order to prevent entrainment of air and mould gases in the metal stream, metal oxidation, and mould erosion.

(iii) To promote the ideal temperature distribution with the completely filled mould cavity in order to obtain favourable pattern of cooling with respect to feeding and thus to minimise shrinkage, unsoundness and distortion.

(iv) The gating system should incorporate traps to prevent the passage of dross, and non-metallic inclusions (whether dislodged in the gating system or introduced with the metal) into the casting.

(v) To keep the metal velocity sufficiently low to avoid mould and core erosion.

(vi) To avoid the necessity of high pouring temperature.

(vii) To obtain maximum yield of casting.

(viii) To provide ease of pouring with available ladle and crane equipment.

A gating system should be designed to take account of the weight and shape of the casting, fluidity of liquid metal, and the relative susceptibility of metal to oxidation. The techniques of gating may vary depending on these conditions, but the basic objectives remain the same, i.e. to have minimum cost in moulding, minimum time in fettling, and minimum consumption of metal.

The basic elements of a common gating system are (i) an open feeder head situated at the top or the side of the casting, and known as *pouring cup* or *bush* or *basin* through whieh metal is poured, (ii) *sprue* or *down sprue* through which metal enters the runner, (iii) *runners* which conduct the liquid metal to the *ingates* and (iv) *ingates* through which the liquid metal finally enters the casting impression. Fig. 7.6 illustrates the basic elements of a common gating system. The point at which the metal enters the casting cavity is known as *primary choke* or *choke* which restricts or regulates the rate of pouring. The pouring cup or pouring basin minimises the splash and turbulence and allows the entry of clean metal into the downsprue. The entry of dirt or slag into the downsprue can be further prevented by providing a *skim core strainer* or a *sprue plug* in the pouring basin. A *splash core* may be placed in the bottom of the pouring basin or at the bottom of the downsprue to prevent erosion of the gating system.

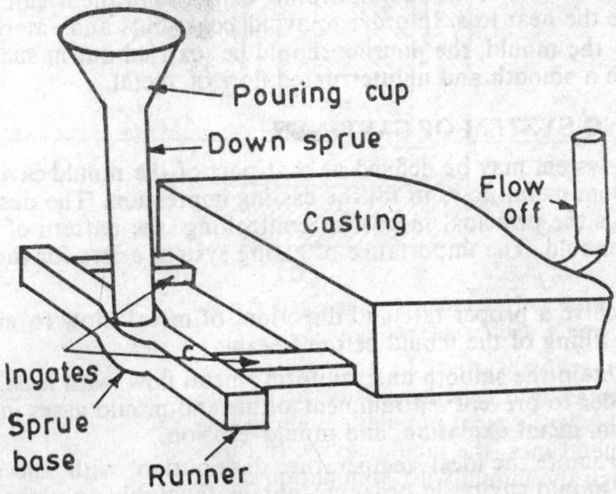

Fig. 7.6 Basic elements of gating system

The gating system may have a single passage or the metal stream may be subdivided and directed into the casting at widely dispersed points. The large castings usually require more than one sprue and the individual sprues are filled either from separate ladiles or from common launder or tundish.

7.4.1 Single Gates

The molten metal can be allowed to flow in a mould cavity in many ways. Since the design of individual gating system depends on its primary objectives, a gate may be designed for (i) case of moulding (ii) avoiding turbulence, (iii) preventing washing of sand from the mould walls, and (iv) avoiding inclusion of slag or dross with the metal. The principal variations of single gate system are bottom gating, top gating and side gating which are illustrated in Fig. (7.7).

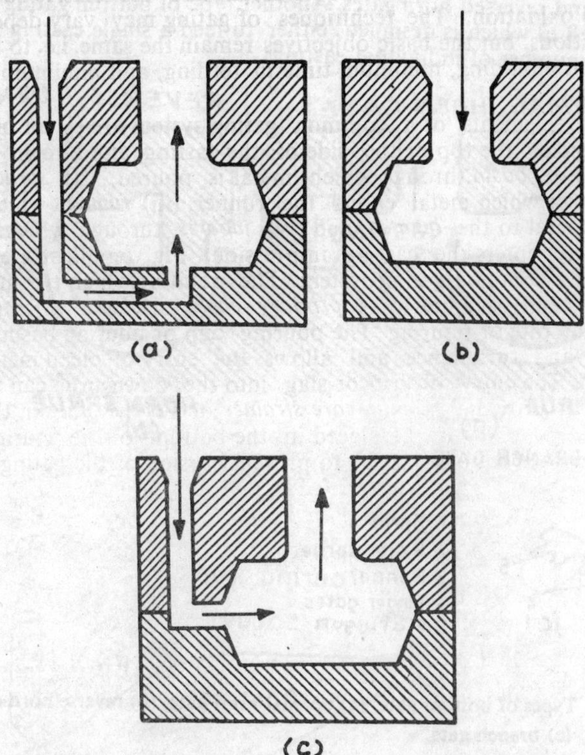

Fig. 7.7 Principal variations in single gate system: (a) bottom gatting, (b) top gating and (c) side gating

(i) *Bottom gating system* In this system the gate provides the entry of molten metal into casting cavity at the bottom of the drag half of the mould. This is the most popular system in foundries as it provides the quiet entry of metal into the mould cavity. In this method the metal rises steadily through the mould with elimination of splashing and tending to carry the dislodged moulding material to the surface.

When bottom gates are used with top feeder heads, the resulting temperature gradients are opposed to feeding, but this effect is minimised by various measures. Though the moulding of bottom gate is quite complex, the method is used for heavy castings, particularly of steel. In steel castings the bottom gating reduces erosion and gas entrapment and prevents splashing which may otherwise cause cold shots. Many variations of bottom gating find practical applications. The most widely used form of bottom gate is horn gate which is illustrated in Fig. 7.8 This may be either normal or reversed gate. The horn gate is designed, with its smooth curves and progressive change of dimensions to minimise erosion and oxidation. The normal horn gate gives jet effects at the point of metal entry,and thus reversed horn gate is generally adopted for the alloys susceptible to skin formation such as aluminium bronze. However, the openions differ in the effectiveness of normal and reversed horn gates. Another type of bottom gating is branch gate (Fig. 7.8 c) which is designed either to feed a single casting at several points or a number of individual castings.

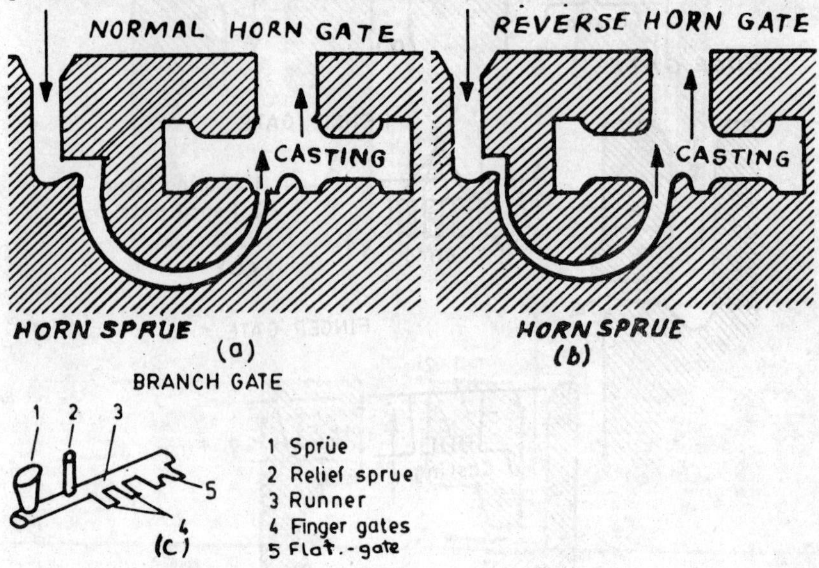

Fig. 7.8 **Types of bottom gate: (a) normal horn gate, (b) reverse horn gate and (c) branch gate**

(ii) *Top gating system* In This system the metal is poured through a top gate or directly into an open feeder head. The basic elements of this system are also essentially the same as in case of bottom gating. In this case, the molten metal impinges against the bottom of the mould cavity until a pool is formed which is kept in agitation until the mould is filled completely. This system suffers from erosive effect of the unconfined stream and oxidation due to splashing. However, the mould surface can be protected by the use of preformed refractory tiles and the erosion intensity can be reduced by the use of *pencil* gates if the fluidity of metal permits. But pencil gates and other methods dividing the metal stream are unsuitable for alloys which are sluggish or prone to rapid oxidation. The top gating system is used with success for cast iron. Some variations of top gating are known as *open pour gate, edge gate, pencil gates, fingure gate. wedge gate, and ring gate* which are illustrated in Fig. (7.9).

The top gating possesses various advantages such as (a) simplicity for moulding, (b) low consumption of additional metal, and (c) favourable temperature gradients to feeding from top heads (due to proportionately rapid cooling of the first metal poured).

(iii) *Side gating system or parting gate* This system consists of gates entering the mould cavity along the parting line. This system simplifies the mould-

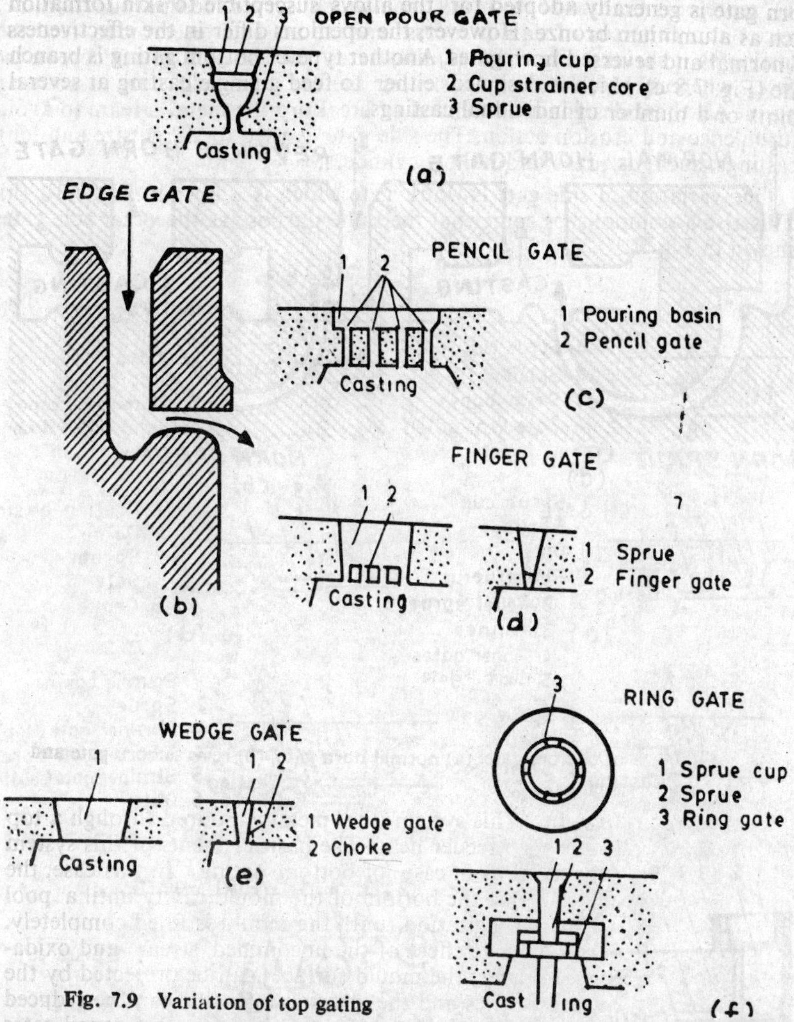

Fig. 7.9 Variation of top gating

ing as the metal is discharged into the side of the mould cavity. This system usually offers the best compromise between moulding convenience and the ideal gating arrangement. Employing side-gating, uphill casting canditions can be achieved by progressive mould filling achieved by tilting the mould towards ingates.

The side gates may contain devices such as skim bobs or relief sprues to collect dross or slag or relieve pouring pressure. The various designs of side gate are illustrated in Fig. 7.10. The design shown in Fig. 7.10 (a) shows the use of skim bob in cope, whereas the design in Fig. 7.10 (b) shows the use of skim bob made in cope as well as in drag. Fig. 7.10 (c) represents the use of relief sprue. In design shown in Fig. 7.10 (d) the pouring basin serves the function of collecting the slag or dross. Design shown in Fig. 7.10

(e), represents the use of a shrink bob which serves the dual function, i.e. to act as a collector of slag or dross and to act as a reservoir to feed the casting as it shrinks. Fig. 7.10 (f) illustrates the use of strainer core in the gate, whereas Fig. 7.10 (g) illustrates the use of splash core in pouring basin. All these designs are intended to break-up the metal stream to avoid turbulence and erosion action. The side gate is used for large size and long castings, such, as tubes and hollow cylinders.

One variation of side gate is whirl gate which is shown in Fig. 7.10 (h). This also accompolishes somewhat the same purpose as the other side gates shown in Fig. 7.10.

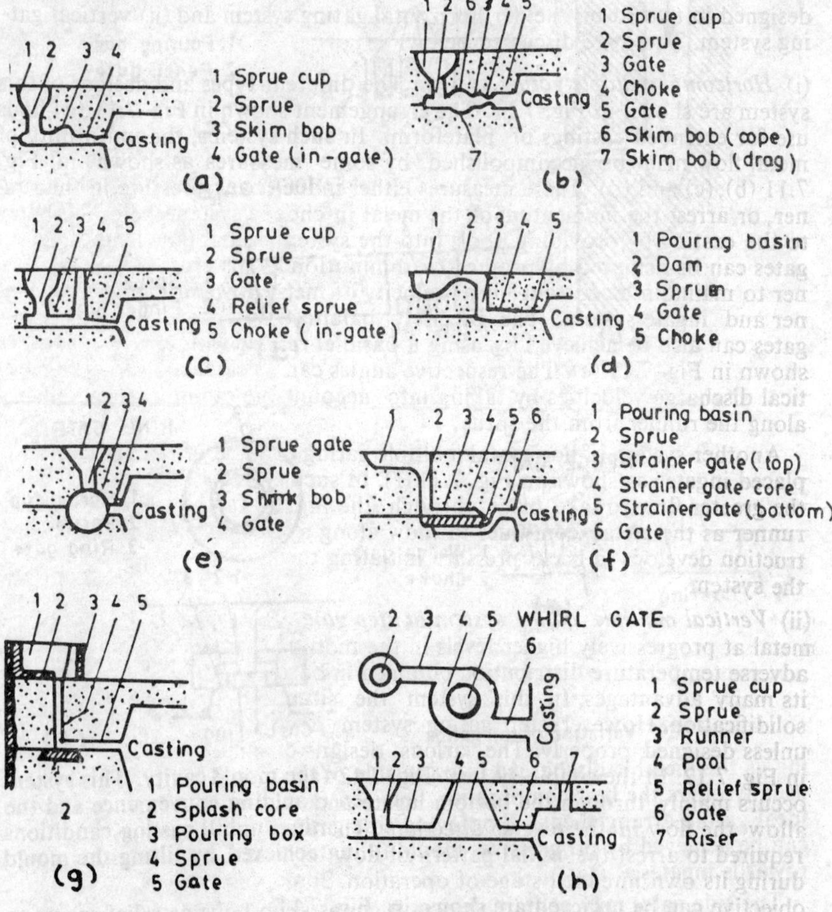

Fig. 7.10 Various designs of side gates

7.4.2 Multiple Gates

The small castings can be usually poured or gated through a single sprue and ingate, but this method becomes unsuitable for castings of larger dimensions due to the danger of overheating the mould adjacent to the metal entry point and low flow distances within the mould cavity. The alternative is the use of more complex gating systems in which the metal is directed through separate gating elements to different parts of the mould. Therefore, a multiple gating system can be devised to introduce metal either at widely separated points in the same horizontal plane or to obtain flow at progressively higher levels as the mould fills (step gating). The choice of the gating system, however depends on the shape and orientation of the casting. In multiple gating system it is essential to control the distribution of metal between the separate elements of the system. The multiple system may be designed in two forms, i.e. (i) horizontal gating system and (ii) vertical gating system. These are discussed below.

(i) *Horizontal multiple gating system* The different types and designs of this system are shown in Fig. 7.11. The arrangement shown in Fig. 7.11 (a) finds use for extended castings of plateform. In such systems, the uniformity of metal flow may be accompolished by some measures as shown in Fig. 7.11 (b), (c) and (d). These measures either induce back pressure in the runner, or arrest the momentum of the melal in choked systems before its entry to the casting by providing pools into the system. Equal flow from different gates can be achieved by progressive diminution in the cross section of runner to maintain a constant linear velocity of metal flow throughout the runner and ingate system. The precisely balanced flow rates from multiple gates can also be achieved by using a parallel runner with angled ingates as shown in Fig. 7.11 (e). The respective angles can be calculated to give identical discharge velocities by taking into account the diminution in velocity along the runner from the sprue.

Another system of horizontal multiple gating consists of circumferentially placed ingates as shown in Fig. 7.11 (f). In such systems the tendency is for the greater flow to take place through the ingates farthest from the down runner as the metal continues to flow along a straight path unless an obstruction develops a back pressure initiating the flow in other elements of the system.

(ii) *Vertical multiple gating system or step gate* This system introduces the metal at progressively higher levels in the mould. This system eliminates the adverse temperature distribution obtained in bottom gating whereas retains its many advantages. In this system the situation promotes directional solidification. However, step gating system does not function as desired unless designed properly. The various designs of step-gates are illustrated in Fig. 7.12. In the simple system (Fig. 7.12 (a)), filling of the mould cavity occurs mainly through the bottom ingate and the upper gates may even allow the flow in the reverse direction. Therefore, further measures are required to arrest the initial pattern of flow to permit each gate to function during its own intended stage of operation. Some methods by which this objective can be achieved are shown in Fig. 7.12, i.e. inclined ingates with or without common junction and reversed downrunner. The step gating is relatively more costly and cutting of the gates from casting is more difficult. This type of system is used for large cylindrical castings.

7.4.3 Special Gating System

Another group of gating methods in common use consists of a single long and continuous ingate of comparatively small breadth and connected to the

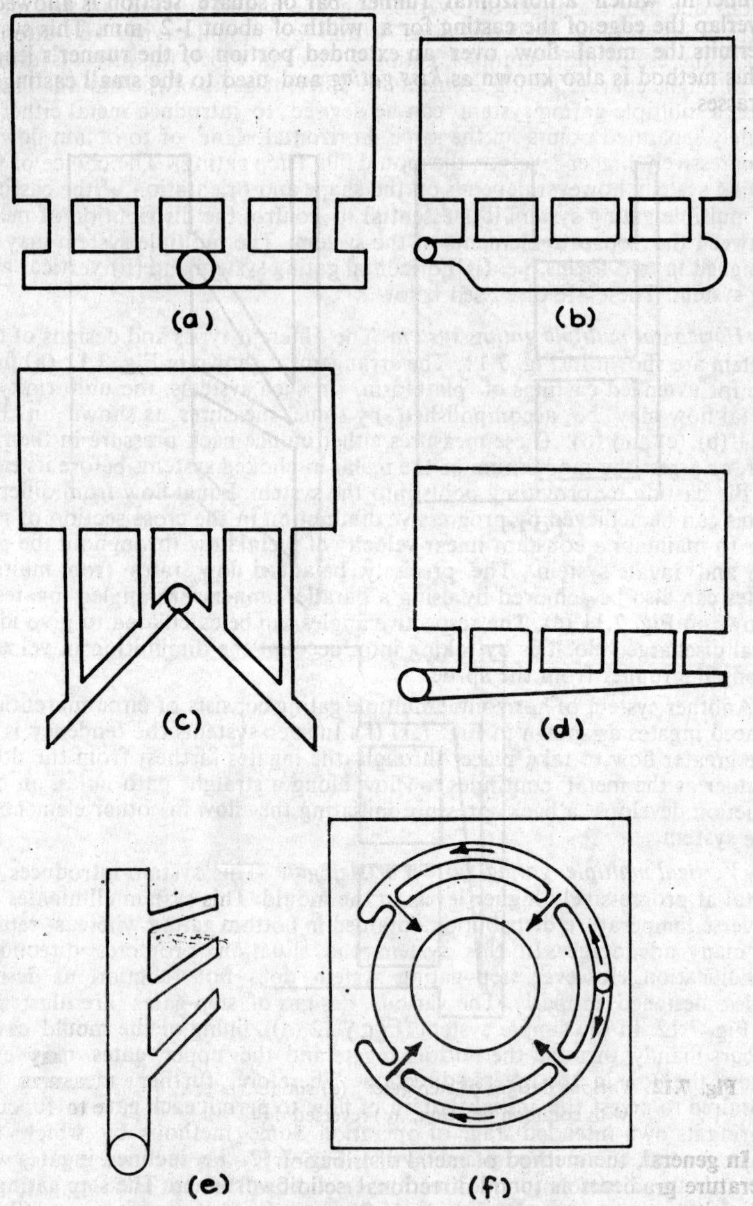

Fig. 7.11 Different types and designs of horizontal multiple gating system: (a) parallel fingure ingate, (b) pool-in system, (c) back swept runner (d) tapered runner, (e) parallel runner with angled ingates and (f) circumferentiallv faced ingates

runner in which a horizontal runner bar of square section is allowed to overlap the edge of the casting for a width of about 1-2 mm. This system permits the metal flow over an extended portion of the runner's length. This method is also known as *kiss gating* and used to the small castings of brasses.

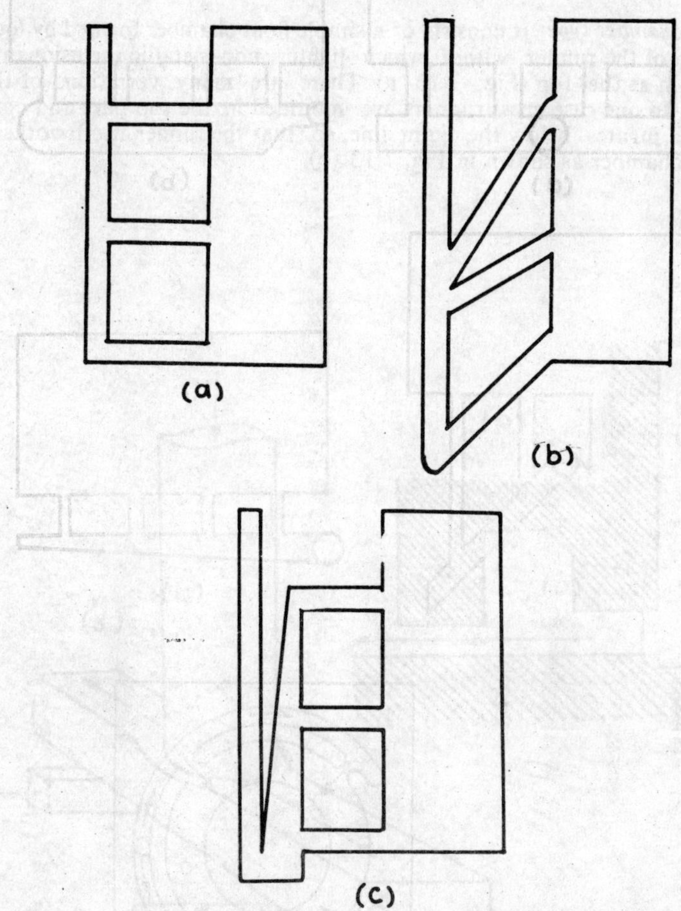

Fig. 7.12 Various designs of step gate: (a) simple system, (b) inclined steps with common junction, (c) reversed down runner

In general, the method of metal distribution is used so develop such temperature gradients as to aid directional solidification, but in some cases the opposite purpose may be required, i.e. minimum temperature gradients within the casting. This practice is particularly needed for pressure tightness in castings prone to leakage at hot spots. The usual method of gating employed is either multiple gating or gating into thin sections, which promote uniformity of temperature and thus equilising the freezing rates.

7.4.4 Traps or Gating Grids

Traps are the devices used for the separation of nonmetallic inclusions and form an important part of gating system. The various traps employed in practice can be grouped in three categories i.e. (i) float chamber trap (ii) baffle or skimmer core and (iii) filteration method.

(i) *Float chamber type* It consists of a simple float chamber formed by local expansion of the runner within which lighter non-metallic inclusions rise and remain at the top (Fig. 7.13 a). There are many variations of this principle. In one case cross runners are moulded in the top part and combined with ingates below the joint line, so that the runner itself works as the float chamber as shown in Fig. 7.13 (b).

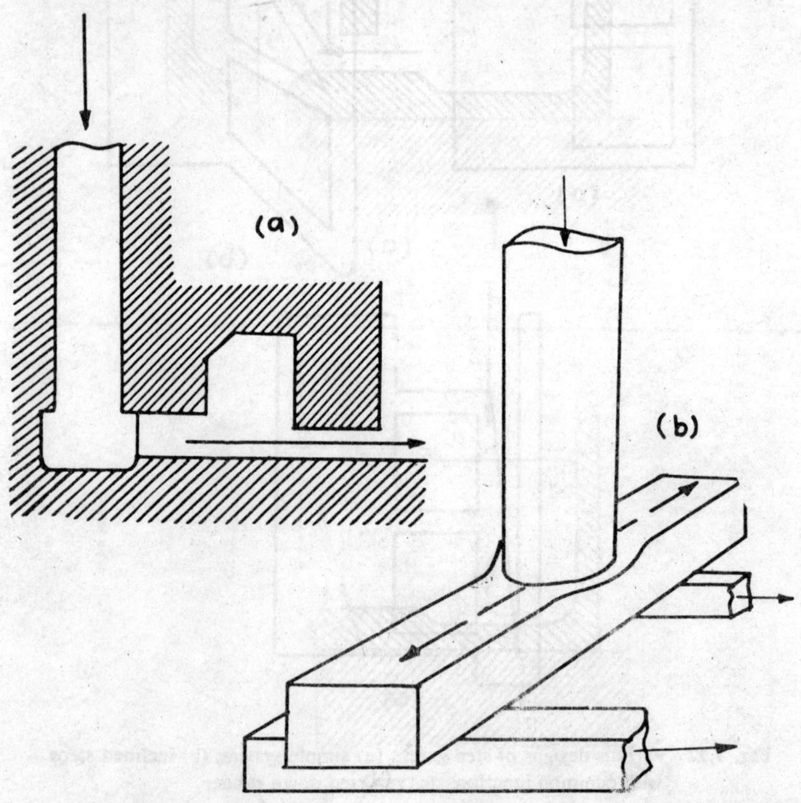

Fig. 7.13 Float chamber type traps: (a) trap in gate member, (b) underslung gates

(ii) *Baffle or skimmer core* In this case baffle or skimmer cores are placed across the runner basin as shown in Fig. 7.14 (a). In this case opportunity should be given to metal head to accumulate in an open basin in order to separate the inclusions by flotation. This purpose can also be achieved alongwith steady flow into the sprue by using a ball plug device as shown in Fig. 7.14 (b).

(iii) *Filtration method* In this system major inclusions are separated by filtration. This mothod is commonly used for light alloys where the density difference between alloy and inclusion is small. The filtration method is also employed for more fluid alloys of heavier metals. The most common device is the strainer core which is a perforated disc of core sand or pre-

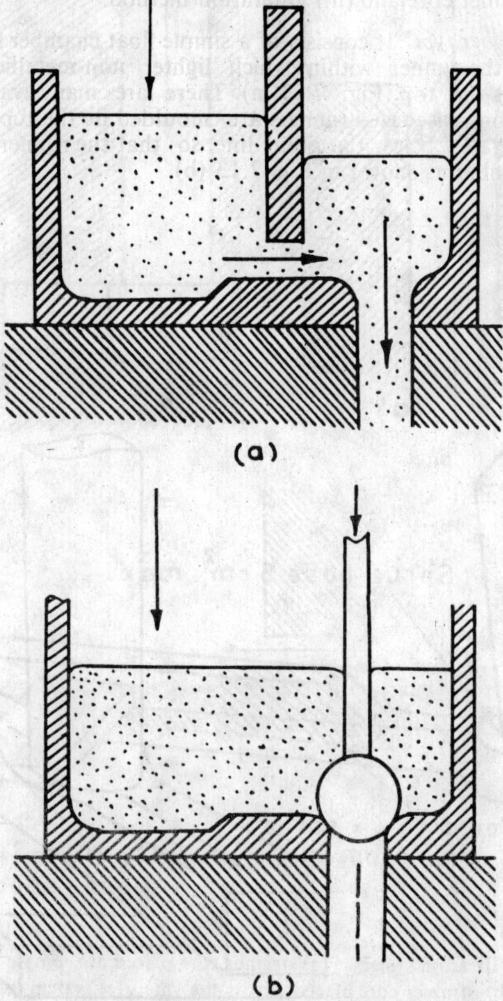

Fig. 7.14 Baffle or skimmer devices: (a) baffle core; (b) ball plug

fired refractory. The strainer core is placed across the system either at the base of the basin (Fig. 7.15 a) or across the gate (Fig. 7.15 b). The use of strainer core serves the dual purpose of arresting dross inclusions and reduction in velocity of metal and thereby reducing turbulance in the system.

In place of strainer cores, thin sheet metal screens can be used in the casting of light alloys. One such system has been developed by Flemmings and

Poirier for magnesium alloys (Fig. 7.15 c). In another such system described by Elliott and Mezoff, the metal is filtered through a column of steel wool retained in a cylindrical tinned steel screen. Use of screens and steel wool is highly effective in preventing the entry of dross into casting.

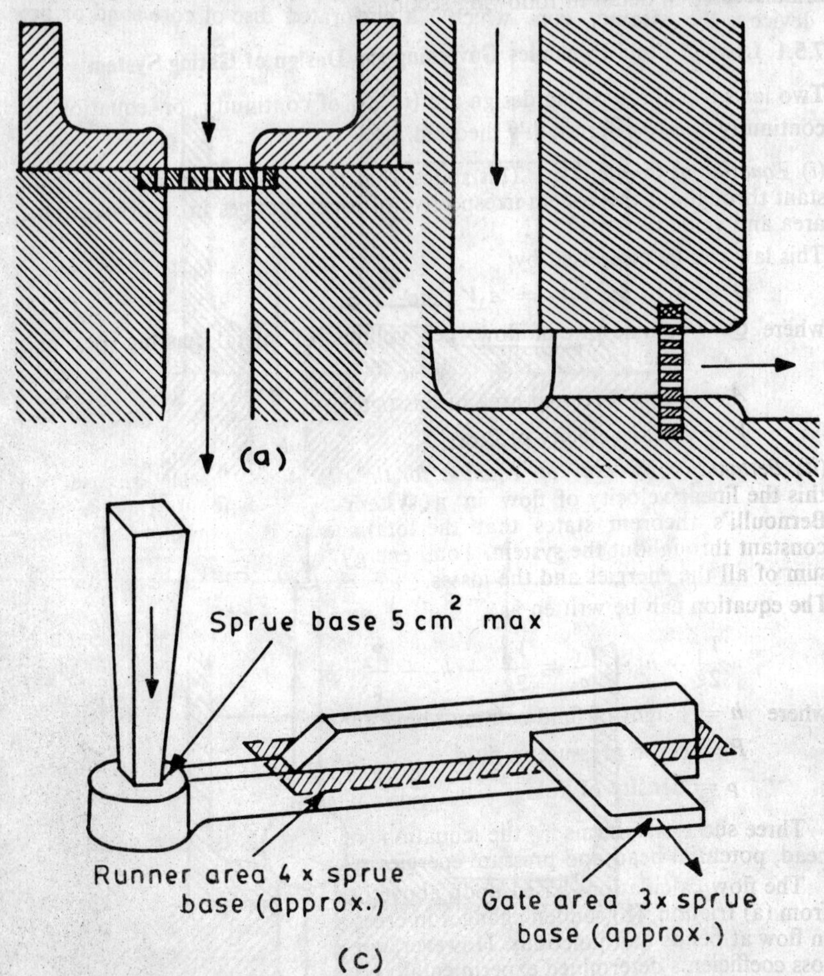

Fig. 7.15 Filtration systems: (a) strainer core placed at the base of the basin (b) strainer core placed across the gate, (c) system using metal screen

7.5 DESIGN OF GATING SYSTEM

The gating system principally consists of three elements, i.e. sprue, the runner, and the ingates. With the assumption that same quantity of metal is passing per unit time through each of the elements, the dimensions of the gating system can be found out by applying the two laws of fluid dynamics.

Once the dimensions are found out, suitable gating system can be designed by taking into account various considerations such as mould, material, metal to be cast, force acting on the mould, design of casting, etc. These are discussed in detail in following sections:

7.5.1 Laws of Fluid Dynamics Governing the Design of Gating System

Two laws useful for gating design are (*i*) law of continuity or equation of continuity and (*ii*) Bernoulli's theorem.

(*i*) *Equation of Continuity* This states, that the volume rate of flow is constant throughout the system irrespective of local changes in cross-sectional area and velocity.

This law can be expressed by:

$$Q = A_1V_1 = A_2V_2 = A_3V_3, \text{ etc.} \tag{7.14}$$

where Q = volume rate of flow, *i.e.* volume of metal passing per unit time.

A = Cross sectional area of passoge, and

V = linear velocity of flow.

(*ii*) *Bernoullils theorem or equaton for the energy of flowing stream* In this the linear velocity of flow in a system is related to other factors. Bernoulli's theorem states that the total energy of unit weight of fluid is constant throughout the system. Total energy of a flowing stream is the sum of all the energies and the losses.

The equation can be written as

$$\frac{V_1^2}{2g} + h_1 + \frac{P_1}{\rho} = \frac{V_2^2}{2g} + h_2 + \frac{P_2}{\rho} \text{ etc.} \tag{7.15}$$

where h = height of fluid column above the datum plane.

P = static pressure in fluid.

ρ = density of fluid.

Three successive terms in the equation represent the kinetic (velocity) head, potential head and pressure energies respectively.

The flow calculations based upon above laws are subject to errors arising from (a) friction, (b) sudden changes in cross-section, and (c) sharp changes in flow at bends and junctions. However, corrections can be applied using loss coefficients determined experimentally.

Taking into consideration the losses (other than heat losses) the equation can be written as following:

$$\frac{V_1^2}{2g} + h_1 + \frac{\rho_1}{\rho} = \frac{V_2^2}{2g} + h_2 + \frac{\rho_2}{\rho} + \frac{V_2^2}{2g}\left(K_B + \frac{f_l}{d}\right) \tag{7.16}$$

where $V_2^2/2g\,(K_B + (f_l/d))$ is frictional loss of head, in which K_B = loss coefficient for fllow velocity due to bends, constriction etc., f = friction coefficient, 1 = length of the channel, and d = diameter of the channel.

The velocities of flow in individual system ean be estimated by the use of above equations. For example, the velocity in down runner can be found from consideration of the linear velocity attained in a metal stream falling from an initial point of rest. In this case. under atmosphre, the potential energy at rest may be equated directly with kinetlc energy at any point in the moving stream. Thus using the point in the stream as datum, the equation (7.16) becomes

$$h = \frac{V^2}{2g}\left(1 + K_B + \frac{f_l}{d}\right) \tag{7.17}$$

or

$$V = (2gh)^{1/2}\left(1 + K_B + \frac{f_l}{d}\right) \tag{7.18}$$

The term $\left(1 + K_B + \frac{f_l}{d}\right)^{1/2}$ can be represented by μ representing coefficient of losses. Finally velocity of metal flow can be represented as

$$V = \mu\sqrt{2gh} \tag{7.19}$$

In practice the losses are not so high and in many cases negligible. Therefore, velocity is simply related to the distance of fall from the initial point of rest.

In free fall or in a parallel downrunner, the metal gains velocity during descent and thus metal should contract inwards or away from the walls (according to law of continuity .i.e eqn 7.14). This situation will cause aspiration of air in a long downrunner through the permeable walls of the mould and this forms the basis for adoption of downrunner progressively tapered. The diminishing cross-sectional area of downrunner accompanies the increasing linear velocity of the metal to provide a constant volume rate of flow. If the losses are neglected, the dimensions of the downrunner required to meet this requirement can be defined by the following equation derived from eqns 7.14 and 7.15.

$$\frac{A_2}{A_1} = \left(\frac{h_1}{h_2}\right)^{1/2} \tag{7.20}$$

Where A_1 and A_2 represent the cross-sectional areas and h_1 and h_2 are the depths of metal at the entry to the downrunner, and at any point within it, respectively.

If weight of casting is W in grams, then

$$W = \rho.A.t.\mu\sqrt{2g\,h_p} \tag{7.21}$$

or

$$A = \frac{W}{\rho.t.\mu\sqrt{2g\,h_p}} = \frac{W}{\rho t V} \tag{7.22}$$

where t = pouring time in seconds and

In equation (7.22), t and h_p or V are unknow-n since pressure head changes constantly due to rise of metal in the casting. The calculation of V or h_p is somewhat complex. However, the average value of h_p can be calculated from the formula.

$$h_p = h_{sp} - \frac{h^2}{2c} \tag{7.23}$$

where

h_{sp} = effective height of sprue,

h = height of casting above the point of metal entry into the mould in cms. and

c = height of casting in cms.

The discharge of metal from the ingates may occur in three positions (depending upon the gating system), i.e. (i) on the top of the mould, (ii) from the bottom of the mould, and (iii) at the parting line of the mould.

In the first case $h = 0$, and hence $h_p = h_{sp}$, in the second case $h = c$ and hence $h_p = h_{sp} - c/2$, and in the third case $h = c/2$ and hence $h_p = h_{sp} - (c/8)$.

7.5.2 Various Head Losses

The total head lost, $\sum \Delta h$ is compounded of a number of losses such as (i) entrance loss (Δh_0), (ii) discharge loss (Δh_d), (iii) head loss due to contraction (Δh_c), (iv) head loss due to enlargement (Δh_e), (v) head loss due to bends (Δh_b), and (vi) head loss due to wall friction (Δh_f).

$$\sum \Delta h = \Delta h_0 + \Delta h_d + \Delta h_c + \Delta h_e + \Delta h_b + \Delta h_f \quad (7.24)$$

Most of these losses are approximately proportional to the square of velocity, i.e.

$$h_0 = k_0 \frac{V^2}{2g},\ h_d = k_d \frac{V^2}{2g};\ h_c = k_c \frac{V^2}{2g} \text{ etc.} \quad (7.25)$$

where k_0, k_d, k_c, etc, are variable constants and can be determined experimentally.

The wall friction loss (Δh_f) is expressed in different way, i.e.

$$\Delta h_f = \frac{fl}{d} \frac{V^2}{2g} \quad (7.26)$$

where f = coefficient of friction, l = length of the channel, and d = diameter of the channel.

The constant f is a complicated function of many factors such as roughness of the channel walls, channel perimeter, and its length, velocity of metal flow, and viscosity, and hence the formula given in equation 7.26 is only approximate. For laminar flow

$$f = \frac{64}{R_e} \quad (7.27)$$

and the value of f calculated from this formula has been found to be in good agreement with the experimental values.

Some empirical formulae have been developed for the estimation of f. These are

$$1/\sqrt{f} = 2.0 \log_{10} (Re\sqrt{f}) - 0.8 \quad (7.28)$$

for $Re > 2000$

$$1/\sqrt{f} = 1.95 \log_{10} (Re\sqrt{f}) - 0.55 \quad (7.29)$$

for higher values of *Re*

$$1/\sqrt{f} = 2.0 \log_{10} (r/k) + 1.74 \quad (7.30)$$

for very rough channel, where r/k = roughness factor, r = channel radius, and k = mean height of roughness projections.

The velocity of metal flow constantly changes due to the loss of head at each section and thus the velocity of metal flow and loss of head should be estimated at different sections. The total head will be $h = \frac{V^2}{2g} + \sum \Delta h$

7.5.3 Calculation of pouring time

If the total loss coefficient is known, the pouring time can be calculated. Calculation of pouring time in bottom gating system shown in Fig. 7.16 can be made as following:

Let the metal height in the mould at any time t after beginning the pouring be H and after a short time dt, the height of metal becomes $H + dH$. If the cross sectional area of mould cavity in the horizontal plane is x, then the volume of metal poured in time dt will be equal to xdH.

If the cross section area of ingates is A, and the instantaneous velocity of metal is V' then metal delivered by the gate in time dt will be equal to $AV'dt$. The volume of metal poured in time dt will be equal to metal delivered by the gate to the casting in time dt, i.e.

$$xdh = AV'\,dt \quad (7.31)$$

If h is the total head, then the affective head at that instance will be $h - H$, then according to equation (7.19),

$$V' = \sqrt{2g\,\mu\,(h - H} \quad (7.32)$$

therefore, $xdH = AV'\,dt = A\sqrt{2g\mu\,(h - H)}\,dt$

or
$$\frac{dH}{\sqrt{2g\mu\,(h - H)}} = \frac{A}{x}\,dt \quad (7.33)$$

The equation (7.33) can be integrated between 0 and t_0 and 0 and H_0 where t_0 = pouring time, and H_0 = height of the mould cavity.

Thus
$$\int_0^{H_0} dH/\sqrt{2g\mu\,(h - H)} = \int_0^{t_0} A/x\,dt \quad (7.34)$$

or
$$t_0 = \frac{2x\,\mu}{A\sqrt{2g}}\,[\sqrt{h} - \sqrt{h - H_0}] \quad (7.35)$$

Hence pouring time varies linearly with the term $[\sqrt{h} - \sqrt{h - H_0}]$.

There is an optimum pouring time for each casting. If the pouring is slow, the metal may freeze too fast to fill out the mould and would result cold shuts in the castings. Whereas, rapid filling of the mould may cause many problems such as erosion of the mould wall, rough surface, excessive shrinkage, etc. The optimum rate is thus established for each casting by

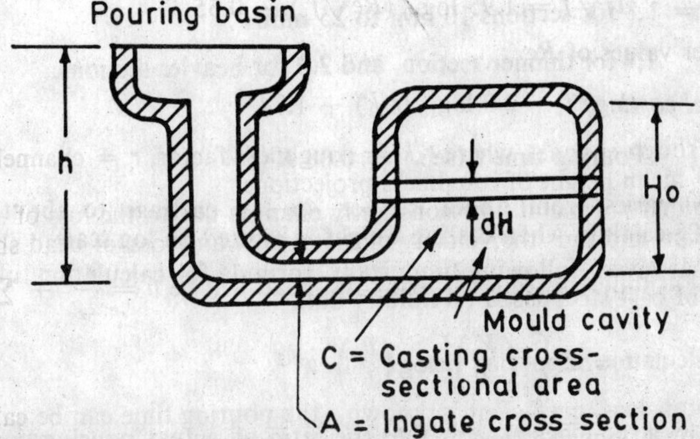

Fig. 7.16 Bottom gating system for calculation of pouring time

experience. In casting methods when metal is forced into a mould under pressure, the optimum pouring time depends on the fluidity of metal. In case of green sand moulding, pouring time also depends on wall thickness, shape and weight of the casting. Pouring times for grey cast iron and steel casting can be determined as following:

(i) *Grey Iron Castings*

Cast irons are less sensitive to pouring rate compared to other metals. However, it is advisable to determine the optimum pouring time which is a function of shape and size of the casting. For grey iron castings the pouring time t may be given as following:

$$t \text{ (in seconds)} = S\sqrt{W} \tag{7.36}$$

where S = co-efficient taking into account the thickness of the casting wall.

and W = weight of casting with gates and risers.

For gray cast iron casting weighing more than 450 Kg and green sand moulding, the pouring time can be given by

$$t = K\left(1.42 + \frac{T}{1.44}\right)\sqrt{W} \tag{7.37}$$

where K is a fluidity factor which is equal to fluidity value in cms. divided by 100, T = thickness of casting in cms and W = weight of casting in Kgs.

For grey iron castings weighing more than 450 Kgs, the pouring time in seconds (t) can be determined by

$$t = K\left(1.16 + \frac{T}{1.80}\right) 2\sqrt{W} \tag{7.38}$$

(ii) *Shell moulded ductile iron*

Pouring time t (Sec.) $= 0.68\ K\sqrt{W}$ (7.39)

where K = 1.8 for sections 10 mm to 25 mm.

1.4 for thinner section, and 2.0 for heavier sections.

(iii) *Steel castings*

$$\text{Pouring time } t \text{ (sec.)} = 0.08\, K\sqrt{W} \quad (7.40)$$

where K varies from about 1.2 for 45 Kg castings to about 0.4 for 45,000 Kg castings when casting weight is plotted on log scale.

Hess has given following important formula for calculation of pouring time (t) of cast steel based on fluidity data.

$$t = Z \frac{G}{G_z} \left(\frac{V_c}{A_c}\right)^2 \log\left(\frac{\theta_p}{b}\right)^a \lambda^{0.2} \quad (7.41)$$

where Z = a co-efficient equal to the ratio of actual pouring time to the theoretical pouring time, it can be given by

Z = 0.5 to 0.7 for castings with uniform wall thickness.

= 0.4 to 0.7 for castings with varying wall thickness

and = 0.6 to 0.8 for castings having projections.

G = Total weight of casting

G_z = portion of casting weight through which metal front should flow for the largest time without being topped with fresh hot metal.

a, and b are constants depending on carbon content of steel. The value of a being 1405 to 1434, and of b 7.9 to 12.5. This formula has been found to be valid for low and medium carbon steels upto 0.5 pct. C.

(iv) *Copper alloys*

The pouring time (t) in sec. for copper alloys can be given by

$$t = S\sqrt[3]{TG} \quad (7.42)$$

For bottom gating, the value of S is 1.3 to 1.9, lower value for brass and higher values for tin bronzes. For top gating the values of S are 1.8 to 2.8, lower values for brass and higher values for tin bronzes.

7.5.4 Pouring Basin Design

The design of pouring basin should be such that a proper flow system is established as rapidly as possible. This means that, when the metal enters the sprue, it should flow under the conditions equivalent to those present under full flow. These conditions can be achieved by:

(i) Streamlining the pouring basin and providing a dam which helps to establish constant conditions.

(ii) Using a strainer core in the pouring basin, and

(iii) Using a delay screen or sprue plug.

The design of pouring basin is governed mainly by the ease with which it can be filled and made. The diameters of the basin should be sufficiently large so that sprue remains full of metal without any splashing.

7.5.5 Sprue Design

When metal passes through the sprue, it gains velocity and loses its pressure energy or head. The loss of head in a sprue may cause a tendency to form a vortex on the metal in the sprue, or a negative pressuse effect in the metal column resulting into sucking of gases from the mould into the metal stream. Therefore, the sprue opening should be tapered which also helps in reduction of mould erosion and metal turbulences. The sprue size is usually selected as to control the pouring rate which has the advantage of early establishment of proper flow characteristics and reduction in flow rate of metal entering the mould cavity from the gates.

In case of light metal alloys prone to drossing, the use of restricted area to reduce the velocity of metal may result in drossing of the metal as it enters the runners. Therefore, many precautions are necessary for ideal gating and sprue design. Cast iron and steel are not prone to dross and thus no special precaution is necessary. Steel is poured at high temperature and solidifies very fast and thus the factor controlling the sprue design is not drossing and quiet entry of metal into the mould, rather it is necessary to get the metal into the mould rapid enough to avoid excessive chilling. Thus, the design and sprue-runner ratio may differ drastically from those needed for the light metal alloys.

7.5.6 Choke Area Calculation

Choke area is the smallest area in the feeding channels which controls the flow rate of metal into the mould cavity and in turn controls the pouring time. In general, the choke area occurs at the bottom of the sprue and it can be calculated by applying a formula based on the use of Bernoulli's theorem. The choke area A in sq. cm. can be given by

$$A = \frac{13.8\ W}{dt\ C\sqrt{2gH}} \tag{7.43}$$

where A = Choke area in cm^2.

W = Casting weight in Kgs.

d = specific gravity of metal.

H = effective height of metal head in cms.

C = efficiency factor or nozzle coefficient which is a function of gating sytem employed.

g = Acceleration due to gravity in cms/sec^2.

t = pouring time in seconds.

7.5.7 Importance and Determination of Dimensions of Passages (Gating Ratio)

In an individual gating system the shape and dimensions of the passages are the most important characteristics which determine the rate and type of flow and the position at which the metal enters the mould cavity. The dimensional characteristics of any gating system are expressed in terms of gating ratio, $a{:}b{:}c$ where a = cross sectional area of sprue or down runner, b = total cross sectional area of runners, and c = total cross sectirnal area of ingates. The gating ratio used in industries varies widely. It mainly emphasises the general feature of gating system, i.e. whether the system is pressurised in which cross section diminishes towards the casting providing a choke effect (e.g. gating ratio 1:2:4), or non pressurised in which the area increases giving a reverse choke effect (e.g. gating ratio 1:2:1).

In pressurised systems, the rate and distribution of metal flow are more predictable but the metal tends to enter the casting at high velocity giving a jet effect. On the other hand, unpressurised system cause irregular flow of metal and aspiration of air. These factors should be considered with respect to metal cast. However, in general, an ideal system for all purposes would be one in which pressure is sufficient to maintain all passages full just to avoid aspiration of air. The location and dimensions of the gating system are not only important to avoid turbulence, but mould filling is also an important objective, particularly for thin castings. In case of thin castings, the gating system should be so designed as to flush metal rapidiy but progressively through sections, otherwise laps and misruns may be caused. However, in case of thicker castings preventing of turbulence is more important to avoid mould erosion and to obtain proper temperature gradients. Gating system can be computed by the following methods:

(i) For light metals and alloys such as aluminium, magnesium, titanium, etc. and their alloys, the gating system can be designed on the basis of feed metal criterion K_m which is defined by

$$K_m = W/t \tag{7.44}$$

where W = weight of casting in Kg. with gates and risers

and t = time of pouring in seconds.

Experiments show that feeding is adequate for

(i) aluminium alloys when $K_m > 0.3\sqrt{W}$ Kg/sec and

(ii) magnisium alloys when $K_m > 0.2\sqrt{W}$ Kg/sec.

Based on feed metal criterion, the choke area

A_{choke} can be given by

$$A_{\text{choke}} = \frac{K_m \times 1000}{\mu\sqrt{2gh}} \text{ cm}^2 \tag{7.45}$$

If the rate of metal flow is less than 1·7 Kg/sec., one sprue is sufficient, whereas for large castings if the rate exceeds, more than one sprue should be used.

(i) Empirical methods based on stastical treatment of plant data can be used to compute the gating system. These methods establish the re lationship between the velocity of metal flow, pouring time, and choke area. One such system given by Dietert is as following:

Pouring time, $t = A\sqrt{W}$ sec. (7.46)

where A = Constant, and W = weight of casting and total ingate area

$$A_{ing} = \frac{B\sqrt{W}}{\sqrt{E}} \tag{7.47}$$

where B is another constant, E = effective sprue height.

However, these equations give much error under the present improved gating design.

(iii) Hydrodynamic methods can be used in design of gating system. In this case velocity of metal flow (V), time of pouring (t) and area of choke (A_{choke}) are evaluated with the help of fluid mechanics. This is already discussed in the previous sections.

8

Risering and Feeding and Castings

The main aim of designing the feeder head or riser is to produce sound casting by compensating for the shrinkage cavity. However, the metal consumption should be minimum to achieve high casting yield and minimum cost. Sometimes the riser is considered as a part of the gating system. The riser requirements are appreciably governed by the metal being cast, e.g., grey cast iron requires less of feeding than other alloys. During the final stages of solidification of grey cast iron graphitisation occurs causing an expansion which counter acts metal shrinkage.

Shrinkage cavity can be compensated by the progressive solidification of the casting towards the riser. The design of feeding system requires the establishment of shape, size and positions of the risers. The size and shape of each feeder head must satisfy the following two requirements.

(a) The feeder head should freeze sufficiently slowly to ensure the availability of liquid metal throughout the freezing of the casting. This condition will enable the directional solidification to continue from the casting into the feeder head. This may be referred as freezing time criteria.

(b) The feeder head should supply sufficient quantity of liquid metal to compensate for shrinkage. Any resulting shrinkage cavity or pipe should be clear of the casting. This may be referred as volume feed capacity criteria.

Table 8.1 indicates that the volume of metal required to compensate for shrinkage is relatively small (usually less than 7 pct. for most casting alloys). If it were possible to maintain completely liquid metal in the reservoir, the risering is fairly simple and only small heads are necessary to compensate for the shrinkage. This ideal condition is quite closely approached in either of the following situations and much higher yields are obtained.

(a) When solidification in the riser is retarded by exothermic compound or other means.

(b) When solidification of the casting is accelerated by heavy chilling or by an extremely low volume/area ratio (e.g. in thin plates).

Table 8.1 Approximate volume shrinkage on freezing of some foundry alloys

Metal/alloy	*Volumetric shrinkage*
Medium carbon steel	2.5 – 3.3
1 Pct. carbon steel	4.0
Pure aluminium	6.60
Pure copper	4.92
Gray cast iron	1.90 to negative depending on graphitisation and composition
White cast iron	4.0 – 5.50

However, in most cases solidification proceeds at similar rates in both the feeder head and the casting and thus for satisfactory feeding, solidification time for the head should be more than that for the casting by a safe margin. The indirect comparison of solidification times for appropriate geometric factors provides the quantitative basis for designing feeder heads to satisfy freezing time criterion. The correction factors such as local chilling or insulation may be incorporated to obtain change in freezing rate or increasing the margin of safety to ensure a sound casting.

In case of compact and a simple shaped casting, a single feeder head of appropriate dimensions (dimensions calculated on the basis of the above mentioned principle) may be provided, whereas complicated design may require additional feeder heads.

Other factors involved in the effective functioning of a riser are, (a) maintenance of pressure differential for feeding, and (b) location of the head in a manner to enable the flow of liquid metal throughout freezing. Another factor is freezing range which becomes a limiting factor in feeding metal from a single head to the casting having extensive walls of parallel section. For this, number of heads should be provided for the various effective freezing ranges. This system is referred as multiple head feeding. The converse case to multiple head feeding is a single head feeding of a number of separate castings.

In general, the shape and size of the feeder head are governed by three principal factors, i.e. freezing time, feed volume, and feeding range.

8.1 CONSIDERATIONS FOR FEEDER HEAD SHAPE

Solidification time is directly related to the ratio of volume to surface area. Table 8.2 illustrates the comparison of volume/surface area (V/A) ratios for a number of simple shapes of identical volume together with relative solidification times as calculated according to Chworinov relationship, (Solidification time (t_s) = $B\ (V/A^2)$. From Table 8.2, it may be observed that a spherical riser will have the largest V/A ratio and thus a spherical feeder head of given mass will remain liquid for the largest period. However, spherical heads have not been widely used due to two disadvantages, (a) inconvenience of moulding, and (b) solidification of head towards thermal centre unlike the thermal axis of other shapes. This centre is inaccessible to external feeding aids. In general. feeder heads are made with shapes having parallel walls e.g. squat cylindrical form. The use of squat cylindrical head facilitates the withdrawal of pattern piece either through the upper surface of the mould or towards the parting plane.

Table 8.2 Module and freezing times of constant volume bodies (1000 cm^3)

Shpere	*Ruling dimension* cm	*Modulus* $M = V/A$ cm	M^2 Cm^2	*Freezing time* $2.1\ XM^2$ min.	*Freezing time as % of the of sphere*
Sphere	$D = 12.41$	2.068	4.277	9.0	100
Cylinder $H = D$	$D = 10.84$	1.806	3.26	6.8	76
Cylinder $H = 10D$	$D = 5.003$	1.198	1.44	3.0	38
Cube	$T = 10$	1.667	2.78	5.8	65
Square bar					
$L = 10\ T$	$T = 4.64$	1.101	1.23	2.6	29
Plate or slab					
$L = 2T$	$T = 6.30$	1.575	2.48	5.2	58
$L = 5T$	$T = 3.42$	1.221	1.49	3.1	35
$L = 10T$	$T = 2.15$	0.898	0.81	1.7	19
$L = 20T$	$T = 1.35$	0.617	0.38	0.8	9

Narrow rectangular heads give the least satisfactory V/A ratio. This type of head is satisfactorily used in the form of ring head on large annular castings carrying a centre core. Though the efficiency of ring head is poor, it can be removed easily by machine parting.

However, the shape of feeder head is not solely dependent on solidification time. Other factors affecting the shape and size of feeder metal are (a) the shape of shrinkage cavity in the head, and (b) permissible area of casting and feed head junction which should be minimum possible for low cost of fettling.

8.2 CONSIDERATIONS FOR FEEDER HEAD VOLUME (SIZE OF RISER)

In order to ensure directional solidification and a sound casting, requirements for feezing time as well as feed volume should be met by the feeder head. One of these requirements may be the critical factor controlling the minimum size of riser. In most of cases where head size is governed by the freezing time criterion, a purely geometric comparison may be adopted to ensure that the head will solidify last and a separate estimation of freezing times of head and casting may be unnecessary. However, in case of differential cooling through the use of chills or exothermic materials, the freezing times of head and casting should be estimated separately.

8.3 CALCULATIONS OF RISER DIMENSIONS

The solidification time of riser (t_r) should be greater than that of the casting (t_c). In general, $t_r/t_c = 1.5$ gives satisfactory condition of feeding. According to Chworinov's rule $t = B\,(V/A)^2 = BR^2$ and therefore

$$\frac{t_r}{t_c} = \frac{R_r^2}{R_c^2} = 1.5 \tag{8.1}$$

or

$$\frac{R_r}{R_c} = 1.2 \quad \text{or} \quad R_r = 1.2\,R_c \tag{8.2}$$

The shape and dimensions of the feeder base are usually determined by the shape and dimensions of the casting section on which the riser is to be located. The simplest method of determining the riser dimensions is by the use of the relation established by practice.

As a general rule, the base of riser is 1.2-1.5 times the base of casting and the height 0.7 to 1.5 times the height of the casting section fed. Height of the riser may also be calculated on the basis of 1.0 to 1.5 times the base of the feeder (diameter).

In general, the calculations for riser dimensions can be made under three groups. i.e. (a) methods based on solidification time, (b) methods based on volume shrinkage and (c) combined methods. These are discussed below:

8.3.1 Methods Based on Solidification Time

(i) *Caine's Method.* This method is based on experimentally determined hyperbolic relationship between relative volumes (Y) and relative freezing rates (X) of riser and casting to produce a sound casting. Heat dissipation is taken as a function of the surface area of the casting, whereas heat content as a function of volume. The freezing ratio (X) is defined as following:

$$\text{Freezing ratio } (X) = \frac{\text{Cooling rate of the casting}}{\text{Cooling rate of the riser}} = \frac{A_c/V_c}{A_r/V_r} \tag{8.3}$$

The basic equation obtained for the hyperbolic relationship was of the form

$$X = \frac{a}{Y - b} + C \tag{8.4}$$

where X = freezing ratio, Y = relative volume (V_r/V_c), a = freezing characteristic constant, b = liquid – solid solidification contraction, and C = relative freezing rate of riser and casting.

For steel castings, the Eq. (8.4) is modified to

$$X = \frac{0.10}{Y - 0.03} + C \tag{8.5}$$

The value of C differs for different conditions of casting, e.g. $C = 1$ for ordinary sand castings, 0.05 for exothermic sleeves, and 0.8 for slightly exothermic antipiping addition to the top surface of the riser.

This relationship is illustrated in Fig. 8.1 for steel castings under different conditions (different values of C). It is pointed out that if the freezing rates of riser and casting were identical, the riser volume required would theoretically be infinite. When the casting freezes increasingly rapidly relative to the riser, the required riser volume decrease towards minimum approaching the shrinkage volume (condition approaching in case of extended and thin walled casting of very low V/A ratio).

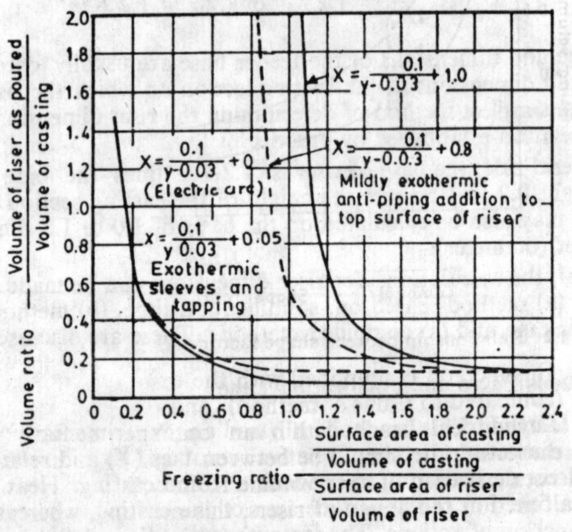

Fig. 8.1 Relationsoip between freezing ratio and relative volume for steel castings

Caine's relationship, together with additional information provided by the curves in Fig. 8.1 has contributed considerably to the basic calculation of risers. However trial and error calculation would be required to arrive at the suitable riser size.

(ii) *NRL Method* This is the simplified development of Caine's approach given by Bishop et. al. In this case surface area to volume ratio is replaced by the shape factor which is given as

$$S = \frac{L + W}{T} \tag{8.6}$$

where L = length of the section concerned

W = breadth of the section concerned

and T = thickness of the section concerned.

The relationship between shape factor (S) and riser volume requirements is illustrated in Fig. 8.2. The empirical graph is essentially similar to that of Caine. As the value of shape factor increases i.e. towards thinner and more extensive shapes of casting, the feeder head requirement diminishes towards the limiting level at which shape factor is no more controlling factor but the volume of feed metal required becomes the controlling factor.

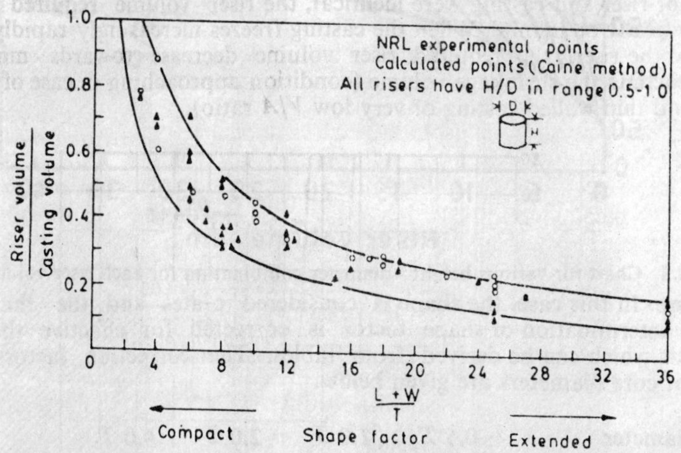

Fig. 8.2 Relationship between shape factor and riser volume requirements

Further, this method is simplified with the omission of riser geometry as a variable (compared to Caine's method) and the plotted relationship is restricted to cylindrical heads within an optimum range of V/A ratios (height to diameter ratio should be between 1 and 0.5). Thus this method permits direct derivation of riser volume from consideration of the casting geometry alone. For the design of risers, this method involves three following steps:

(a) Determination of shape factor ($L + W/T$) for the casting using the dimensions of the main section of the casting or of the section being considered

(b) Obtaining the $V_{riser}/V_{casting}$ ratio from the graph

(c) Calculation of V_{riser} from the volume of the casting or section.

A further chart (Fig. 8.3) is given for various alternative height diameter combinations for each riser volume.

Procedures have also been evolved to take care the effects of design features such as cores, ribs, thicker projection, etc. A thin rib or bracket freezes rapidly and may even accelerate the cooling of the casting by decreasing the V/A ratio and thus it may require little additional feed metal. However, thicker projections increase the freezing time and thus risers should be enlarged accordingly.

In case of hollow castings, the cooling rate in the neighbourhood of core is retarded to a value equivalent to that of thicker section containing no

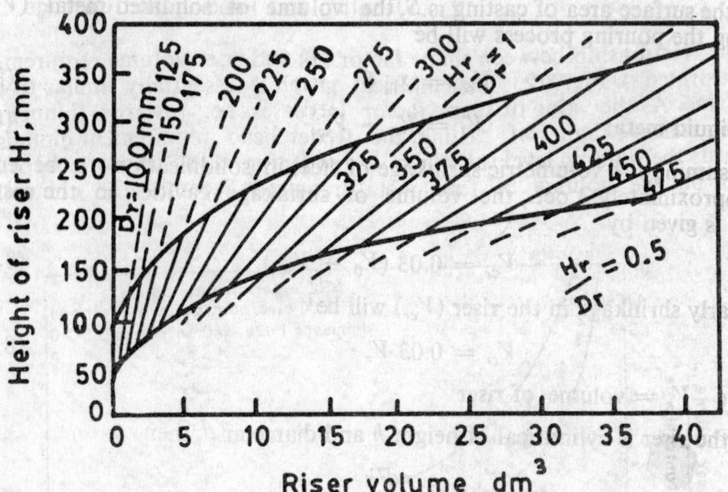

Fig. 8.3 Chart for various height—diameter combination for each riser volume

such core, In this case. the shape is considered plates and the thickness **used** in determination of shape factor is corrected for effective thermal **thickness** which can be derived from tables. The correction factors for **different** core diameters are given below:

Core diameter	0.5 T	1.0 T	2.0 T	4.0 T
Correction factor (K)	1.17	1.14	1.02	1.0

where T = true wall thickness,

Hence effective plate thickness

$$(T_e) = K.T.$$

The length of the hollow casting is considered to be equal to circumference at point of outside radius 0.5 thickness.

8.3.2 Methods Based on Shrinkage

(i) *Popov's Method*

Popov developed a method for calculating the riser size for steel castings. He based his calculations on the square root equation of Chworinov. Although there is a certain error in the calculation of cylinderical risers, it does not affect much the determination of their size. According to Chworinov, the mean thickness of the skin solidified (x_m) during the pouring time, t is

$$x_m = \frac{k\sqrt{t}}{2} \tag{8.7}$$

where k = constant.

If the surface area of casting is S, the volume of solidified metal (V_{sol}) during the pouring process will be

$$V_{sol} = x_m S \tag{8.8}$$

and liquid metal

$$V_l = V_0 - V_{sol} \tag{8.9}$$

Assuming the volumetric shrinkage of steel in solidification to be equal to approximately 3 pct., the volume of shrinkage cavities in the casting (V_{cc}) is given by

$$V_{cc} = 0.03\,(V_0 - V_{sol}) \tag{8.10}$$

Similarly shrinkage in the riser (V_{cr}) will be

$$V_{cr} = 0.03\, V_r \tag{8.11}$$

where V_r = volume of riser

If the riser is cylindrical of height h and diameter d, then

$$V_{cr} = 0.03\,\frac{\pi d^2 h}{4} \tag{8.12}$$

The total volume of the shrinkage cavity in the riser ($\sum V_{sc}$) will comprise the volumes of shrinkage cavities in the casting and the riser. Hence

$$\sum V_{sc} = V_{cc} + V_{cr} = 0.03\,(V_0 - V_{Sol}) + \frac{0.03\,\pi\, d^2 h}{4} \tag{8.13}$$

The shape of the shrinkage cavity in riser for steel castings obeys a parabolic law and is of conical shape. If n is the ratio between cone base and the riser diameter, then

$$\sum V_{Sc} = \frac{\pi\,(nd)^2\, h}{12} = 0.03\,(V_0 - V_{sol}) + 0.0075\,\pi d^2 h \tag{8.14}$$

with bottom gating cone base is 0.5 d and with top gating the cone base is 0.6 d.

Equations (8.7) and (8.8) can be used to solve equation (8.14) for h,

$$h = \frac{0.0575\,(2V_0 - k\,S\sqrt{t}}{(n^2 - 0.09)\,d^2} \tag{8.15}$$

When pouring time is quite short (for ball or cube shaped castings with high V/A ratio), the expression $kS\sqrt{t}$ tends to zero and can be neglected. Then the equation (8.15) gets simplified as

$$h = \frac{0.115\, V_0}{(n^2 - 0.09)\, d^2} \tag{8.16}$$

In case of large castings of small V/A ratio, the term $kS\sqrt{t}$ becomes nearer to $2V_0$ and the numerator of Eq. (8.15) becomes zero. This will also occur at slow rates of pouring and hence risers are no longer required.

The actual height of the riser for steel castings may be calculated with the help of Fig. 8.4 as following

$$H_r = h + h_{sc} + h_{br} + h_{con} \quad (8.17)$$

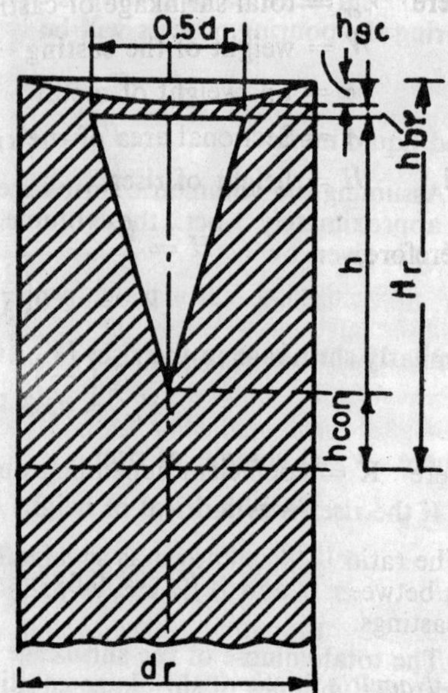

Fig. 8.4 Diagram illustrating the actual height of the riser for steel castings

when H_r = total height of the riser

h = height of the shrinkage cavity in the riser

h_{sc} = height of the spherical concavity

h_{br} = thickness of the bridging (solid metal) over the cavity

h_{con} = reserve height of the riser for loose metal and nonmetallic contamination under the tail of the shrinkage cavity

As a rule $h_{sc} + h_{br} + h_{con} = 4$ cm.

Hence $$H_r = \frac{0.0575\,(2V_0 - kS\sqrt{t})}{(n^2 - 0.09)\,d^2} + 4 \text{ cm} \quad (8.18)$$

This equation should be modified for different gating systems, e.g.

$$H_r = \frac{0.7\,V_0}{d^2} + 4 \text{ cm (heavy castings bottom gated)} \quad (8.19)$$

$$H_r = \frac{0.15\,(3V_0 - S\sqrt{t})}{d^2} + 4 \text{ cm (Top gated sand moulds)} \quad (8.20)$$

$$H_r = \frac{0.45\,V_0}{d^2} + 4 \text{ cm (Heavy castings top gated into sand moulds)} \quad (8.21)$$

(ii) *Ryzhikov's Method*

A.A. Ryzhikov developed the following equation for cylindrical risers

$$\frac{W}{\sigma}\,V_{cr} + AH\cdot V_{cr} = \frac{1}{3}\,A.H \quad (8.22)$$

where V_{cr} = total shrinkage of casting and riser
W = weight of the casting
σ = Sp. weight of metal
A = Sectional area of the riser base
and H = height of riser.

Therefore,
$$H = \frac{W}{A} \cdot \frac{V_{cr}}{\sigma\left(\frac{1}{3} - V_{cr}\right)}$$
$$= K\left(\frac{W}{A}\right) \qquad (8.23)$$

Where K = coefficient depending on σ, V_{cr} and the configuration of a casting.

The ratio W/A is known as specific coefficient of casting feed. The relation between H and W/A can be found experimentally for various groups of castings.

(iii) *Namur's Method*

R. Namur developed a precise but cumbersome method for the calculation of risers. The riser volume (V_r) can be determined from the equation
$$V_r = AR_c^3 + 3\alpha V_c \qquad (8.24)$$

where $A = \frac{V_c}{R_r^3}$ is the coefficient depending on the shape of riser. (e.g. A = 280 for semispherical riser)
V_c = volume of the casting fed by riser
R_r = referred thickness of the riser
R_c = referred thickness of the casting or its heavy section connected with the riser
α = coefficient of volumetric shrinkage of steel (α = 0.06 for low carbon steel)

This method indicates that the riser supply should be atleast three times of the feed metal required and not equal to Caine's theory.

(iv) *Pribyl's Method*

Pribyl proposed the following relation to calculate the riser volume (V_r)
$$\beta V_r = a_{cr}(V_c + V_r)$$
or
$$V_r = \frac{a_{cr}}{\beta - a_{cr}} V_c \qquad (8.25)$$

where β = ratio of the cavity volume in the riser to the riser volume (usually 0.1 to 0.13)

V_c = volume of casting

a_{cr} = combined shrinkage of casting and riser

8.3.3 Combined Methods for Calculation of Riser Size

(i) *Merchant's Method*

H.D. Merchant proposed an improvement on the Caine's or N.R.L. methods by introducing solidification shrinkage and shape factors for the risers. He obtained the general relation as follows:

$$\frac{25\,(4P+1)}{A_c} = \frac{PD}{\alpha V_c} - \frac{1.275}{nD^2} \tag{8.26}$$

where P = diameter to height ratio of riser (H/D)

n = number of risers needed

and α = solidification shrinkage in pct.

(ii) *Heine's Method,*

R.W. Heine developed a suitable method of riser calculation for malleable iron castings. The method is based on the actual measurement of the pipe in a casting and riser system which produces a sound casting. The actual measurement of the pipe can be carried out by (a) filling the pipe with water and measuring the volume, (b) comparing the weight of the riser with the expected weight calculated from the external diamensions, or (c) based on prior knowledge of the casting characteristics of the alloy. The procedure of calculating the riser size is outlined as below taking reference of Fig. 8.5.

(a) Find the weight of the casting.

(b) Find the volume of the feed metal required by the castings which is given by

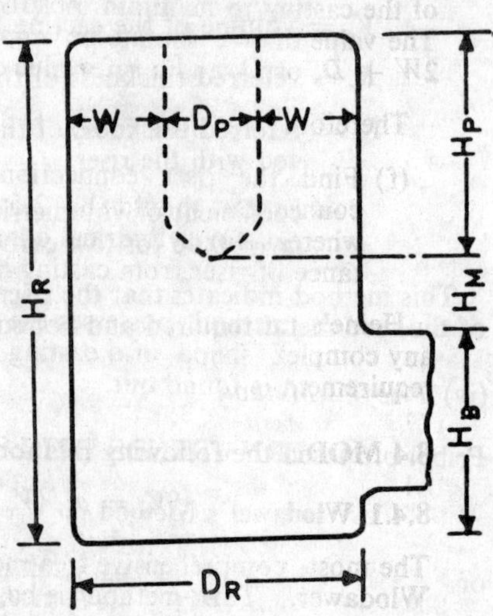

Fig. 8.5 Illustration of the calculation of riser size for maleable iron castings

$$V_f = \frac{\text{casting weight}}{\text{density of metal}} \times \alpha \tag{8.27}$$

where α = solidification shrinkage in pct.

(c) Find the values of pipe diameter (D_p) and height of pipe (H_p).

$$V_f = \frac{\pi D_p^2}{4} H_p \tag{8.28}$$

The required feed metal volume may be supplied by a number of combinations of D_p and H_p. However, in practice H_p is usually 1.5 to 3 times of D_c and most common ratio of H_p/D_p is 2.5. The equation (8.27) can then be solved by substituting the values of D_p and H_p.

(d) Find the diameter of the riser (D). From Fig. 8.5, D may be given as

$$D = 2W + D_p \tag{8.29}$$

where W = representative section thickness of the casting (W is the effective plate thickness, for bars W = 0.5 diameter of bar and for cubes $W = 0.35 \times$ edge length of cube).

(e) Find the height of riser (H_r). The total height of the riser should be equal to

$$H_r = H_p + H_m + H_b \tag{8.30}$$

where H_m = pressure section of the riser required to keep pressure on liquid metal during the final stage of feeding.

and H_b = the bottom section height of the riser through which neck down connection is made to the casting.

H_m depends on the location of the height point to be fed in the casting. For white cast irons, H_m should be usually 2.5 cm above the highest point of the casting to maintain positive pressure in the final stages of feeding. The value of H_b should be minimum of $2W$ and need not be more than $2W + D_p$ or D_r and a safe value of H_b is $3W$.

Therefore $$H_r = H_p + 3W + 2.5 \text{ cm} \tag{8.31}$$

(f) Find the riser connection. For rectangular connections the riser connection should be 2.5 W to 0.8 D_r in width and W in height whereas for cylindrical connections it should be $2W \times 2W$. The distance of riser from casting is 0.5 to 1.5 W.

Heine's method possesses the advantage that risering can be done for any complex shape of a casting in a simple manner once the feed metal requirement is found out.

8.4 MODERN TRENDS IN DESIGN OF RISERS

8.4.1 Wlodawer's Method

The most comprehensive treatment of riser dimensioning has been that of Wlodawer. This method is based on the original approach of Chworinov (Chapter 6). This method is also extended to include systematic considerations of padding, chills, exothermic materials, and other aids to directional solidification.

Although Wlodawer's methods have much in common with those of Caine and Bishop et. al, the later was criticised for not offering any real advantage of shape factor over Chworinov ratio of volume to surface area or cooling modulus. In the Wlodawer's method, the riser requirement is reduced from the cooling modulus for the casting. A complicated casting may be subdivided into smaller basic shapes for each of which the modulus (V/A ratio) can be easily calculated. In this system, only the cooling surfaces are considered in the calculation and the imaginary contact surfaces are omitted. The riser is then selected on the basis that it should have a modulus value of 1.2 times that for the casting. Since riser requirement for a slender and extensive cast shape also depends on the volume of feed metal

Shape	Volume V	Area A	Modulus V/A
Sphere	πt^2	$\frac{\pi t^3}{6}$	$\frac{t}{6}$
Cylinder h = t	$\frac{\pi t^3}{4}$	$\frac{3\pi t^2}{2}$	$\frac{t}{6}$
Cube	t^3	$6t^2$	$\frac{t}{6}$
Bar (square semi infinite)	$t^2 l$	$4tl$	$\frac{t}{4}$
Bar (cylindrical semi infinite)	$\frac{\pi t^2 l}{4}$	$\pi t l$	$\frac{t}{4}$
Plate (semi infinite)	At	$2A$	$\frac{t}{2}$

Fig. 8.6 Moduli of the simple shapes

required, a further check is necessary to verify that the feed volume from the proposed riser will be sufficient under the given conditions.

The modulus calculations for intricate castings are simplified by employing the principle of substitute bodies. The plain shapes are substituted for more elaborate shapes of equivalent modulus on the basis that two bodies of equal modulus will solidify in the same time. Wlodawer has shown that angular and curved bodies of the same ruling dimension have the same modulus as shown by sphere, cylinder, and cube (Fig. 8.6). This figure shows that calculation can be simplified by replacing an irregularly shaped body by one having approximately equivalent mass. Similarly, rings and hallow cylinders can be treated equivalent to bars or plates which they would form on opening out. A number of generalised formulae tables and charts can be used to determine the moduli of the simpler shapes resulting from the break down of a complicated shape.

As shown in Fig. 8.6, the modulus of the most compact bodies (sphere, cube and equiaxed cylinder) is same, i.e. $d/6$ for a ruling dimension d. In case of bars and plates having the same ruling dimension, d, the ratio of modulus to d increases from the minimum represented by most compact bodies towards a limiting value at which end effects become negligible. The progressive change in the modulus with decreasing thickness is represented in Fig. 8.7.

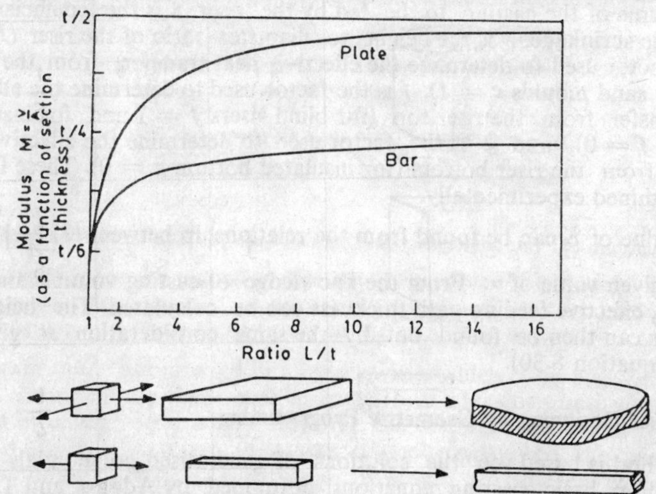

Fig. 8.7 Progressive change in the monulus with deceasing thickness

If end effects are absent, the modulus of a bar of given section can be calculated from the general formula

$$M = \frac{\text{cross sectional area}}{\text{perimeter}} \tag{8.32}$$

or

$$M = \frac{a\,b}{2\,(a + b)} \tag{8.33}$$

where a and b are the cross-sectional dimensions.

The values of M for bars of more complex cross section can be found directly from the general formula given in relation (8.32). For rectangular bars of all dimensions, the values of M can be read directly from the chart prepared from the bar formula.

8.4.2 Riser Design Based on Heat Transfer

Adams and Teylor, were the first to work on the design of risers based on heat transfer. The heat lost from a riser is caused due to convection, radiation, and conduction. Conduction and radiation apply to the heat lost from the tops of open risers, whereas condition applies to the surfaces in contact with moulding materials Adams and Teylor developed an expression for total heat lost by the riser and recommended the use of correction factor for the suitable designs of riser. Based on the data of Adam and Teylor, researchers in Wisconsin developed the following equation for risering:

$$\frac{4e\,x + f + g}{x\,(1 - \alpha)}\left[K\left(\frac{V}{A}\right)_F\right] = D - \frac{4\,\alpha}{\pi\,x\,(1 - \alpha)}\frac{V_0}{D^2} \qquad (8.34)$$

where K is shape factor based on heat transfer calculations, $(V/A)_F$ is the effective ratio of volume to surface area of thc feeding path of the casting to which the feeder head is connected, D is diameter of the riser, V_0 is the total volume of the casting to be fed by the riser, α is the coefficient of volumetric shrinkage, x is the height to diameter ratio of the riser (H/D), e is the factor used to determine the effective heat tranefer, from the riser sides (for sand moulds $e = 1$), f is the factor used to determine the effective heat transfer from the riser top (for blind risers $f = 1$ and for insulated top riser $f = 0$), and g is the factor used to determine the effective heat transfer from the riser bottom (for insulated bottom $g = 0$). These factors are determined experimentally.

The value of K can be found from the relationship between $K\left(\frac{V}{A}\right)^F$ and D for a given value of α. From the knowledge of casting volume and riser diameter, effective feeding path thickness can be calculated. The height of the risers can then be found out by the same consideration as given by Heine (equation 8.30).

8.4.3 Riser Designing by Geometric Programming

This method is based on the solutions of generalised polynomials which are based on basic risering equations developed by Adams and Taylors. Based on the problem of minimising the riser height and subject to the constraint $t_r > t_c$ (t_r and t_c are the solidification times for riser and casting respectively), computer solutions are obtaincd.

8.5 FEEDING CAPACITY OF THE RISERS

The use of the factor 1.2 in deriving the modulus of the riser is intended to take care of the unfavourable change in the riser modulus due to shrinkage during solidification. The shape of the shrinkage cavity also decides the actual proportion of the initial riser volume available for feeding.

Fig. 8.8 illustrates the metal utilization capacity of riser as a proportion of initial volume. In cylindrical riser under normal cooling conditions, the shrinkage cavity takes up a general form of Fig. 8.8(a) Although the shape of cavity will depend on the time required for feed metal, it has been shown that the increased surface area due to cavity formation reduces the final modulus to a value approximately 75% of that before the occurence of shrinkage. Thus the factor 1.2 is employed to offsst the consequent reduction in solidification time.

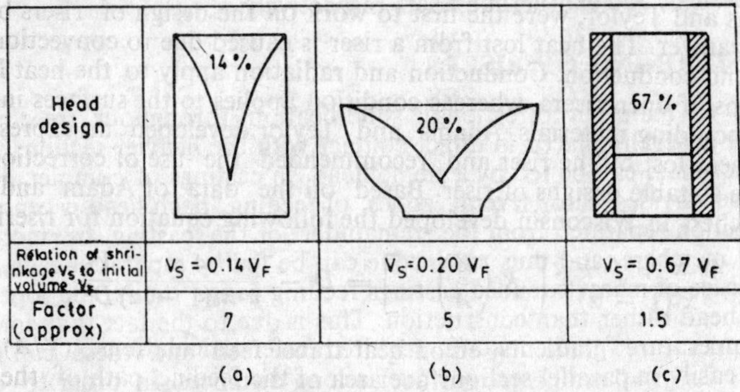

Fig. 8.8 Metal utilization capacity of riser as a proportion of its initial volume: (a) cylinderical, (b) hemispherical (c) with exothermic sleeve

The volume of the shrinkage cavity represents the maximum volume of metal available for feeding. For cylinderical riser under normal cooling, this comprises roughly 14% of original volume of riser. Hence, the maximum volume of casting (V_{max}) which can be fed by a cylindrical head under normal cooling (without exothermic feeding) may be given by the following relation

$$\underset{\text{(casting)}}{V_{max}} = \underset{\text{(riser)}}{V} \times \frac{14 - \alpha}{\alpha} \qquad (8.35)$$

where α = coefficient of shrinkage (pct.)

In hemispherical riser (Fig. 8.8(b)) metal utilization is more efficient as the normal shrinkage cavity represents 20 pct. of the initial volume. In this case the relation of V max (Casting) and $V_{(riser)}$ can be given by

$$\underset{\text{(Casting)}}{V_{max}} = \underset{\text{(Riser)}}{V} \times \frac{20 - \alpha}{\alpha} \qquad (8.36)$$

Therefore, the hemispherical shape of riser is recommended for those cases where volume feed capacity is the governing factor for the riser size.

However, the most efficient utilization of riser metal is achieved with the use of exothermic sleeves (Fig. 8.8 c). In this case, the metal surface falls uniformly and pipe is practically absent and a utilization of 67 pct, is feasible, hence

$$\underset{\text{(Casting)}}{V_{max}} = \underset{\text{(Head)}}{V} \times \frac{67 - \alpha}{\alpha} \qquad (8.37)$$

The modulus value for an exothermic head is an effective value which can be derived from geometric modulus and using a conversion factor determined experimentally. The conversion factor will depend on the nature and thickness of the exothermic material. The maximum casting yield attainable depends on the feeding capacity of riser. Based on the above values (equation 8.35 to 8.37), an alloy with 6 pct. shrinkage would give maximum casting yields of 57, 70, and 71 pct., with plain cylindrical, hemispherical, and exothermic risers respectively.

8.6 POSITIONING OF RISERS

In order to maintain directional solidification throughout freezing, the risers are generally placed in direct contact with the heavier sections. Since a compact mass can be fed by a single head, in castings of complex designs, the shape divides into several zones of feeding. Each head is centred on heavy section separated from the remainder of the casting by more constricted members and thus each zone can be fed by separate heads. However, in case of many extended castings feeding range limits the operation of each head rather than consrtiction. This is due to the fact that considerable temperature gradients are required to feed and these can not be induced easily in parallel sections over a long distances. In order to obtain castings free from dispersed porosity, gradients of 0.2 to 12 °C/ cm may be required depending on the shape and alloy of casting. For example, gradients of 1.35 – 2.7 °C/cm for bars and 0.2 – 0.4 °C/cm for plates are necessary for steel. The gradients of 5.5 — 13°C/cm are required for some non-ferrous alloys having long freezing range.

8.6.1 Feeding Distances for Simple Shapes

In general 9.5 $\sqrt{T}$ cm for bars, 4.5 T cm for plates can be fed on either side of the riser with a proper thermal gradient (as given above). When the plates are joined at their edges to form stepped members, conjunction of two plates differing in thickness by a factor exceeding 1.4 increases the feeding distance in the thinner plate as compared to its casting along. The empirical formula for stepped casting developed by Bishop et al, is represented in Fig. 8.9. When a very thin section (about 3 mm plate) is attached to a heavy section, the feeding distance in heavy section is not reduced as the thin section slows down the cooling of end plate.

8.6.2 Feeding Distances for Complex Designs

Any change in a section leads to a change in cooling rate. The changes in section are usually of two types, i.e. (i) changes in direction and (ii) presence of intersections. The areas corresponding to change in direction and common in intersections are the areas of hot spots leading to hot tearing and residual stress. These problems cnn be overcome in following three ways:

(i) Employing chlll (either internal or external).

(ii) Providing the suitable changes in design by making the casting lighter.

(iii) Attaching thin fin across the intersectional valley.

All these methods result in increase of cooling rate.

8.6.3 Effect of Chills on Feeding Distances

The chills increase the feeding distance and promote the directional solidification. The effect of chills on feeding distances is illustrated in Fig. 8.10. From Fig. 8.10(a) it may be seen that in case of bars the maximum feeding distance is $9.5\sqrt{T}$ out of which $0.5\,T$ to $2\,T$ is the riser contribution and $1.5\,T$ to $2\,T$ is the edge contribution. The edge contribution can be further increased by T by the use of chills (Fig. 8.10(b) i.e. total feeding possible would be equal to $6\sqrt{T} + T$. Whereas in case of plate, fceding distance is $5.4\,T$ out of which $2\,T$ is the riser contribution and $2.5\,T$ is the edge contribution. The use of chill will contribute further 5 cm and thus making the total feeding distance equal to $4.5\,T + 5$ cms. If a chill is placed in the middle of the casting and two feeders are provided, a distance of $9T + 10$ cm can be fed.

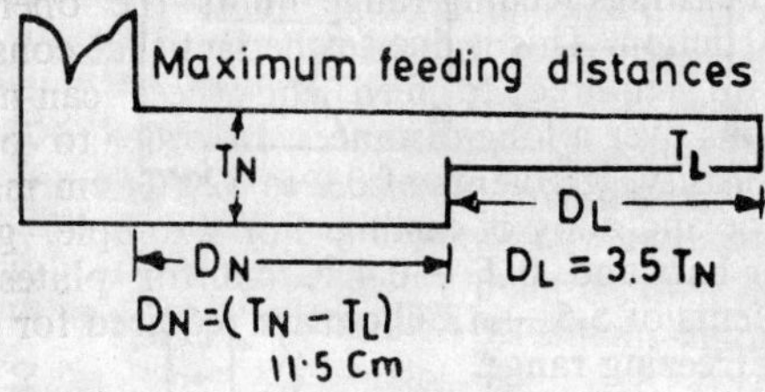

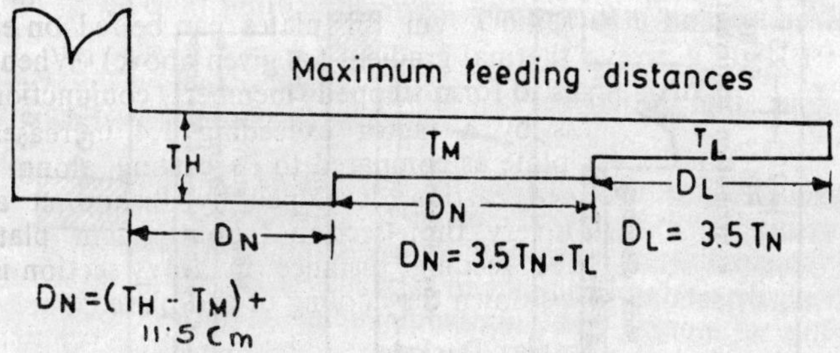

Fig. 8.9 Feeding distances for simple stepped castings

8.6.4 Effect of Freezing Range of Alloys on Feeding Distance

Most of the data on feeding distance has been designed from steel containing 0.25 – 0.3 pct. carbon and having moderate freezing range. The investigation on other alloys confirned that feeding distance diminishes with increasing freezing range and thus steep temperature gradients are required for feeding of alloys freezing over a long range, i.e. in pasty manner. In the extreme case of gun metal, satisfactory soundness of the casting is attainable only by the use of chills. In this, large feeder heads do not serve any beneficial purpose as they delay freezing resulting in lowering the lateral temperature gradients and these conditions provide a relatively sound

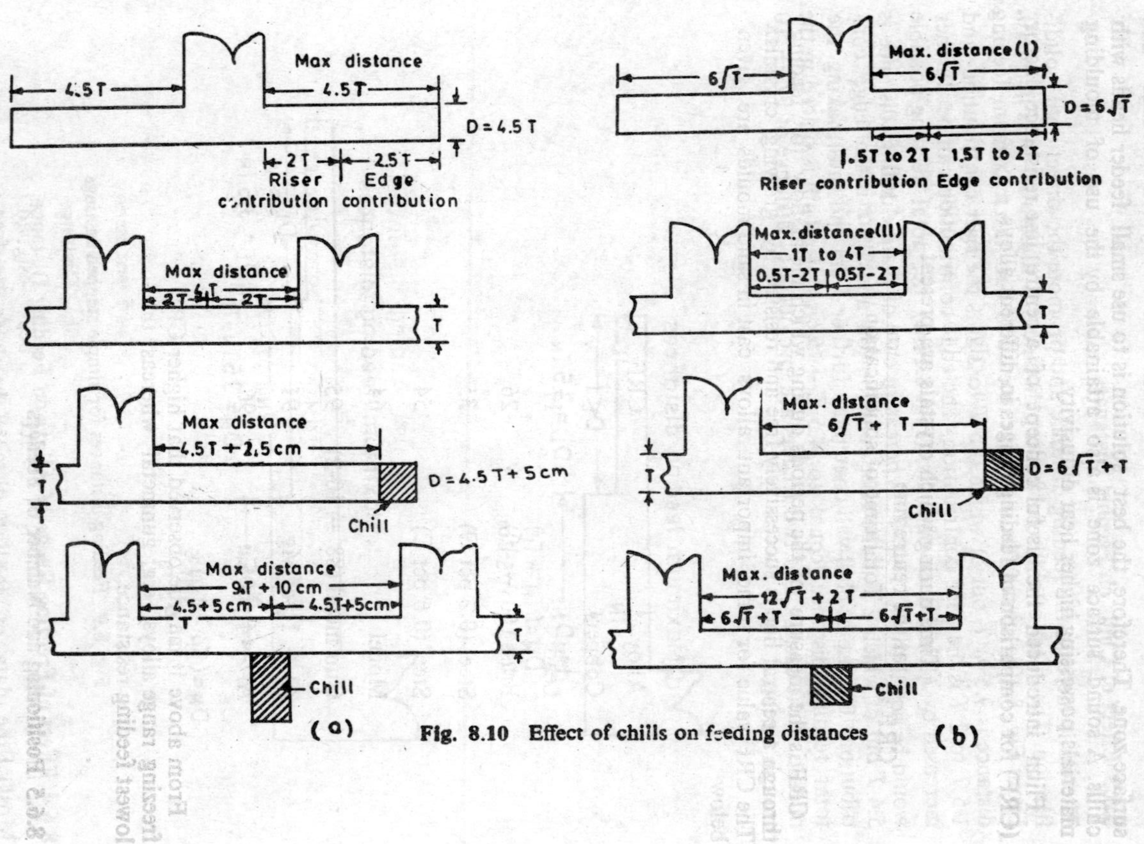

Fig. 8.10 Effect of chills on feeding distances

surface zone. Therefore, the best solution is to use small feeder heads with chills. A sound surface zone is also attainable by the use of moulding materials possessing higher heat diffusivity.

Flinn introduced the useful concept of a centre line resistance factor (CRF) for comparison of feeding ranges in different alloys.

$$\text{CRF} = \frac{\text{Time during which crystals are present at the centre line}}{\text{Total time of solidification}} \times 100$$

CRF is the measure of the periods during which metal flow for feeding through a central line is successively free and resisted by growing crystals. The CRF value for some important alloys cast in sand moulds are given below

Alloy	CRF
Copper	<1
Lead	17
Brass (60 – 40)	26
Steel (0.2 pct C)	35
Steel (0.6 pct C)	54
Monel	64
Gun metal (88 – 10-2)	95
Al – 1 pct Mg	91
Al – 4 pct Cu	96

From above it may be observed that higher CRF values are for the long freezing range alloys, e.g, gunmetal whereas the pure metals show the lowest feeding resistance.

8.6.5 Positioning and Number of Feeders

The main importance of feeding distance data lies in the ability to restrict the number of risers required. From the point of view of cooling moduls or freezing time criterion, few large risers should be provided rather than a large number of small as the V/A ratio is maximum for large risers.

For positioning of risers, a large number of factors should also be considered in addition to the freezing range (as discussed above). These are discussed below:

(i) *Generation of the pressure required for feeding.* In order to obtain maximum metallostatic pressure, the risers should be placed as high as possible with respect to casting. This consideration becomes more prominent in case of long freezing range alloys for which much taller heads are recommended as compared for steels.

(ii) *Moulding convenience.* For moulding convenience the risers should be located with existing parting surfaces as to avoid the need for additional cases and joints. The moulding difficulties can be overcome by the full mould casting. In this technique, the feeder head is shaped in foam plastic and rammed into the mould with the pattern. When molten metal is fed, the plastic is displaced and volatilised.

(iii) *Problems of feeder head removal and dressing.* The risers should be accessible for removal and as far as possible should be placed upon flat surfaces rather than curved surfaces. This will reduce the work in the fettling shop.

(iv) *Shared feeder heads.* A single head can be used to feed more than one casting. This practice is useful for intensive utilization of moulding boxes. However, it may not always provide a potential means for the improvement of casting yield. By the use of such feeder heads, the total volume of metal required becomes almost equal to minimum required to satisfy the volume demand criterion. However, the advantages of shared feeder heads are confined to those cases in which surplus feeding capacity is provided based on the freezing time criterion. These conditions are mainly associated with compact castings having low specific shrinkage.

8.7 SUMMARY OF RISER DESIGN PRINCIPLES

On the basis of the above discussion, the designing of feeder head systems can be summarised as following :

(i) The Casting is subdivided into zones for feeding by separate heads. The number of risers should be minimum for maximum yields. Advantage of sharing of risers between casting should be derived.

(ii) Risers should be placed adjacent to thick sections and advantage should be taken of favourable factors such as temperature gradients, metallostatic pressure and atmospheric pressure.

(iii) All possible means (gating technique, chills etc.) should be employed to induce favourable temperature gradients within the casting.

(iv) Using the various methods of calculation, the freezing time of each riser should be designed to exceed that of casting or feeding zone by a safe margin. This objective is achieved with minimum metal consumption by the risers having compact form.

(v) The volume feed capacity of each riser is determined on the basis of the behaviour of riser during solidification. The feeder head volume is fixed by the higher of the two estimates based on feed demand and freezing time.

8.8 TYPES OF RISERS

The risers can be designed in different shapes depending upon the functions to be performed. The various designs of risers are discussed below:

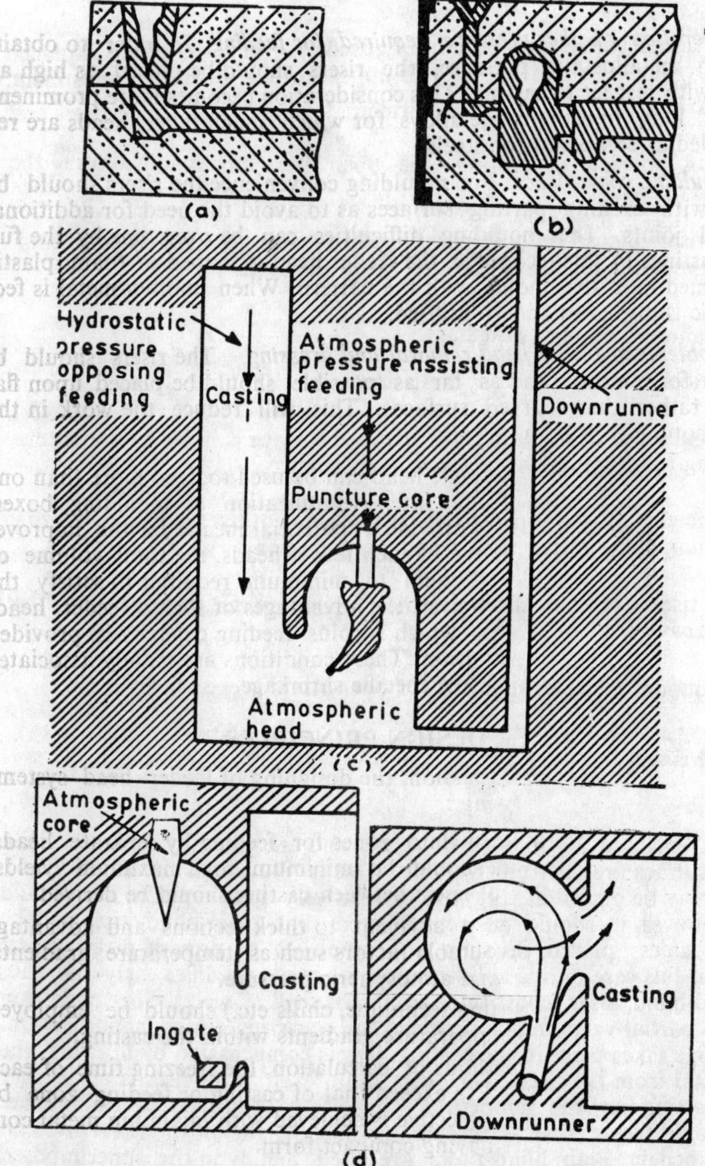

Fig. 8.11 Types of riser: (a) open riser (b) blind riser, (c) atmospheric or pressure riser. (d) whirl gate riser and (e) open top internal riser

8.8.1 Open Risers

This is the simplest and most commonly used type of risers which is placed at the top or side of a casting with its upper surface open to atmosphere. This is illustrated in Fig. 8.11(a). The Principal advantages of open risers are:

(a) During pouring of casting, the open risers function in venting the mould cavity.

(b) Open risers placed on heavy sections of the casting serve as collectors of non metallic inclusions floating up to the surface.

The disadvantages of open risers are:

(a) Heat loss to atmosphere by radiation.

(b) In the absence of insulation the exposed surface is liable to premature freezing resulting in loss of feeding efficiency.

(c) The height of the open riser is determined by the height of the moulding boxes rather than feeding requirements.

(d) Open risers are holes through which foreign matter may fall into the mould cavity.

(e) The amount of metal required for open risers is excessively high.

8.8.2 Blind Risers

These are enclosed by the sand mould and are generally designed for a minimum surfacs area per unit volume. This is illustrated in Fig. 8.11(b). The riser may be placed at the top or relatively low down in the mould cavity. However, it should be remembered that plain blind riser will be ineffective unless placed in such a position as to provide metallostatic pressure. In this case during solidification the sprue solidifies early and thus casting and blind riser constitute a closed shell of metal. This condition develops a partial vacuum due to shrinkage occurring during solidification. As shrinkage takes place in the casting, it gets compensated by drawing the molten metal from the riser. This can occur even if the riser is not higher than the casting but the temperature gradients should be favaurable, i.e. casting should freeze first. However, greater height of the riser gives an additional benefit. Plain blind risers are used mainly in the superimposed position to save metal which might otherwise be wasted in filling the mould to its upper surface.

The blind risers are usually made rounded at the top to save metal. The various advantages of blind riser are:

(a) The riser contains hottest metal whereas the casting contains coldest and hence direction solidification is achieved.

(b) Considerable flexibility in position of riser is possible.

(c) Blind risers are more easily removable from a casting.

(d) A blind riser is much smaller than the open riser.

The main disadvantage of blind riser is in feeding a section of the casting at a greater height when the skin of the casting breaks. This situation tends to equalise the pressure and thus reduce the risesr's effectiveness.

8.8.3 Atmospheric or Pressure Riser

This type of riser has been developed to overcome thc limitations of normal blind riser. When the skin of the casting is strong enough and the skin of the blind riser is pierced by a permeable core, the atmospheric pressrue will exert an influence on the liquid metal in the riser and the arrangement is called atmospheric pressure riser Fig. 8.11 (c). The atmospheric pressure forces the liquid metal into the shrinkage area developed due to solidification. Atmospheric pressure influencing in this system is equivalent to 1.30 m of steel, 1.15 m of copper and 4.34 m of aluminium and hence casting of greater height than itself.

The atmospheric risers are suitable for those alloys which form the required sealed layer of solid at the surface. In case of alloys associated with long freezing range, liquid channels tend to be retained in the surface zone and thus the residual liquid in the casting remains under direct influence of atmospheric pressure until a late stage of freezing.

Atmospheric risers arc mainly used to feed heavy sections of casting situated low in the mould cavity. In such cases these are used in combination with open risers to feed the upper most sections The atmospheric risers facilitates the feeding of skin forming alloys at inaccessible points or which are uneconomic for the use of open risers, or when simple blind risers are not operative.

8.8.4 Whirl Gate Riser

The whirl gate principle can be employed in conjunction with open or blind risers for lateral feeding. This is illustrated in Fig. 8.11 (d). In this system, the metal enters a sump at the base of a riser through a tangentially placed ingate which produces a swirling effect throughout pouring. The circulation of hot metal effectively heats the mould surrounding the riser and the neck between casting and riser and thus temperature differential between riser and casting becomes favourable for directional solidification. The main advantage of whirl gate riser is that it acts as trap for slag and inclusions which are forced to the centre of the risers and thus preventing from entering the casting.

8.8.5 Open Top Internal Riser

Risers surrounded in whole or part by the casting are known as ernal risers. These can be used inside circular or cylindrically shaped castings. Since the risers are surrounded by casting, their cooling rate is lower than that of risers located above or to one side of a casting. This condition favours the use of smaller risers than in the conventional case and thus increase the casting yield.

8.9 VARIOUS AIDS FOR RISER EFFICIENCY

The efficiency of a riser may be considered as the amount of feed metal supplied to the casting in relation to the total weight of metal initially present in the riser i.e.

Efficiency $E = [(V_i - V_f)/V_i] \times 100$ pct.

where V_i = initial volume of metal in the riser.

V_f = final volume of metal in the riser.

The efficiency of a plain riser is relatively low since the solidification in the riser as well as casting proceeds almost at the same time. The solidification rate of casting in relation to riser can be increased by number of measures which delay the solidification of metal or assists its feeding into the casting. Various measures to improve the riser efficiency are discussed in following sections:

8.9.1 Topping up and Rod Feeding

Topping up is the most direct method for extending the feeding period. It is referred to further addition of superheated molten metal after an interval in which some feeding has occurred. In this way the temperature differential and metallostatic head are restored. This method is particularly suitable to very heavy castings with long solidification times. In this case, the initial size of the riser can be reduced to increase the riser efficiency.

The influence of atmospheric pressure is maintained by rod feeding which mechanically disturbs the partially crystallised metal in the upper region of the riser.

8.9.2 Heating of Risers by Electric Arc or Current

The solidification of open riser can be delayed by heating of riser with electric arc or current. In case of electric arc heating a steel electrode is incorporated in the base of the riser and the arc is struck on to the liquid metal surface from an external graphite electrode. This technique is particularly suitable for risers used on very heavy steel castings. A modification of this technique uses consumable electrode (i.e. the principle of electroslagmelting). Similarly, induction heating or resistance heating techniques can be employed. However, the complications and costs of the procedures make their commercial applications unlikely when other simpler techniques are available.

8.9.3 Use of Iusulators

The solidification in the risers can be delayed by simple insulation. The insulating materials may be employed either as coverings to the surfaces of open risers or as linings for riser walls. The use of insulating materials to reduce radiation losses from the exposed metal surface is a long used practice. Efficient insulation can be achieved by lining the risers with moulded sleeves to reduce conductive heat losses through the mould. Adams and Taylor computed the freezing times of standard cylindrical heads with and without top and side wall insulation. The results for 10 ×

10 cm cylindrical heads may be summarised below:

Condition	Freezing time in minutes		
	Steel	Copper	Aluminium
1. No treatment	5.0	8.2	12.3
2. Top insulation only	13.4	14.0	14.3
3. Wall insulation only	7.5	15.1	31.1
4. Top and wall insulation	43.0	45.0	45.6

The above results indicate that the best conditions for prolonged feeding can be obtained by the combination of both types of insulation.

In case of sleels, during the initial short period of freezing, the cold walls of the riser extract the heat rapidly from the metal and then radiation losses predominate owing to the high casting temperatures and thus top insulation is most important in steel castings. On the other hand, for aluminium, use of insulating sleeves is most useful, since the conduction loss to the walls is proportionately greater at lower temperatures.

The insulating materials possess low heat diffusivity owing to their granular or porous structures. The materials used as top coverings may range from dry sand and powdered slag to organic substances (coal dust. chopped straw, etc.) which leave bulky ash on burning. Some antipiping compounds containing exothermic materials can also be used in place of insulating materials. In case of steel castings the carbonauous materials introduce carbon in the metal and thus locally freezing temperature gets depressed which prolongs the feeding without directly affecting the temperature of the riser's metal.

For the lining of riser walls, insulating materials may be used in the form of preformed sleeves. The use of plaster sleeves (made from feamed gypsum plaster) for non-ferrous castings has been reported quite frequently. Other materials for sleeve production are diatomaceous silica and vermiculite. The diatomaceous silica can be used for steel castings as it is thermally stable to a high temperature. Paper has also been reported effective as a riser lining material.

8.9.4 Use of Exothermic Compounds

These compounds utilize the thermic reaction which produce both metal and heat. The reaction takes place between oxides (oxides, of Fe, Mn, Cu, etc.) and reducing metal (Al, Ca, Mg or Si). In general, the reaction between iron oxides and aluminium forms the basis of exothermic feeding for steel castings. This reaction can be represented as following:

$$Fe_2O_3 + 2Al \longrightarrow Al_2O_3 + 2Fe,\ H_{298} = -\ 203.7\ \text{K cal}\ (-852.9\ \text{k J}) \quad (8.38)$$

$$3Fe_3O_4 + 8Al \longrightarrow 4Al_2O_3 = 9Fe,\ H_{298} = -\ 799.3\ \text{K cal}\ (-\ 3347\ \text{kJ}) \quad (8.39)$$

The above reactions indicate that the heat generated is more than required and infact some diluants such as bouxite, ash, fire clay, sand, etc. is used with thermit mixture. Exothermic feeding mixtures can be employed in following two ways.

(i) The exothermic mixture may be added directly to the riser in relatively large amounts. The control of feeding action is achieved by the amount and timing of the addition depending upon the dimensions of riser and casting. The superheated molten metal resulting from the exothermic reaction must be compatible in composition with that of casting. In some cases alloying elements may be added to produce the products approximate to the normal composition of the cast alloy[24].

(ii) The most significant advance in the use of exothermic mixtures is the introduction of mouldable materials which can be preformed into sleeves. This was first introduced by Finch and then it has been the subject of important investigations in recent years.. The exothermic reactants are blended with bonding materials (such as portland cement) and water to provide green and dry strength.

Permeability can be obtained by mixing relatively coarse materials. The specific substances are also added to delay the reaction. In this case, the main aim is not to produce molten metal but rather to retain the moulded shape of exothermic mixture throughout the solidification (this condition prevents the contamination of the casting). The sleeves are moulded and vented in the conventional manner. The properties of a typical compound are as follows.

Permeability	150
Dry compression strength	30 kg/cm^2 (2.75 MN/m^2)
Green compression strength	0.3 kg/cm^2 (27.5 KN/m^2)
Density	1340 kg/m^3
Ignition temperature	300 °C
Effectice calorific value	1650 Kcal/kg of metal (6.9 M J/kg)

The most important factor for an optimum feeding is the period during which the heat is generated. If the period is too large (i.e. compound is insufficiently sensitive), the reaction may exert its effect after solidification, particularly in case of small risers. On the other hand instantaneous reaction will liberate heat soon to achieve maximum feeding efficiency. Therefore, it is essential to control ignition and reaction rate. A sensitive exothermic compound is employed for low melting point alloys, whereas low sensitivity is preferred for high melting point alloys.

As already discussed in section 8.5, the powerful local heating effect of exothermic compounds produces a fundamental change in the character of riser shrinkage (Fig. 8.8). The metal level in the riser sinks almost uniformly across the section resulting into drastic shortening of the pipe.

Wlodawer has described the systematic scheme for the design of exothermic sleeves for steel castings. Riser dimensions can be selected with the help of charts given in Fig. 8.12. The suitable riser size is derived from the casting modulus and demand for feed metal.

The major consideration in using the exothermic compounds is the economic one. The material and labour costs should balance the metal savings. The greatest potential savings can be made in the case of expensive alloys involving high cost of melting and heavy melting losses. In the case of cheaper alloys usually top insulation is employed as the massive use of exothermic compounds to increase yield may be uneconomic in many instances. However, exothermic compounds provide specific technical advantages by reducing these defects which do not respond to other methods.

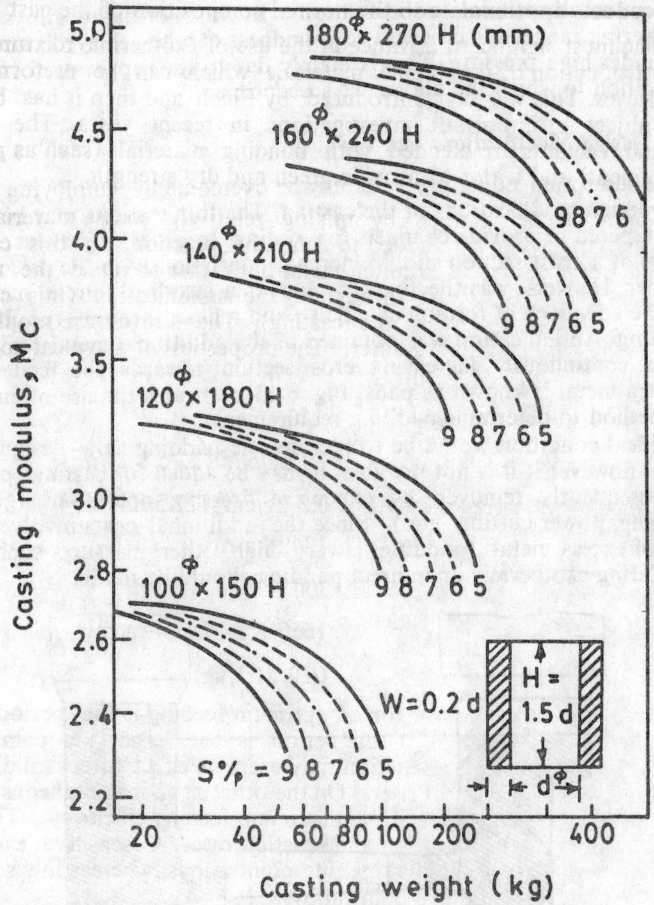

Fig. 8.12 Charts for the selection of riser dimensions

8.9.5 Formation of Pressure in Blind Risers (Pressure Feeding)

Effective feeding requires pressure differential which is generally provided either by gravity or by the atmospheric effect. The idea of using direct application of feeding pressure has received attention of many workers. In general, two methods can be used to increase the pressure in a blind riser, i.e. (i) use of pellets of gas forming compounds (lime stone + carbon powder) and (ii) direct application of compressed air. In the former method lime stone reacts with carbon at the temperature of molten metal and the gas evolved (CO and/or CO_2) exerts the pressure on the riser metal. It may be noted that evolution of gases should occur only after formation of solid skin of the mould of sufficient strength on the surface of the riser.

However, neither technique has been commonly used mainly due to difficulties in control and the associated dangers. It has been found that there is little advantage in case of skin forming alloys having strong tendency to directional solidification. In case of light alloys having long freezing range, some benefit to soundness can be achieved when solidified under high pressure. This is mainly due to the suppression of the gas precipitation to which the alloys are susceptible.

8.9.6 Design Modifications and Padding

In some cases the feeding problem may be overcome by employing fundamental design modifications in the casting. The thin sections may be thickened or tapered to provide channels for feeding, In order to prevent freezing, modulus of a cross-section shou!! increase continuously in the direction of the feeder. Heuvers was the fir.' to use the inscribed circle method to verify the adequacy of feeding channels from a head through the body of the casting. An indication was obtained of the additional metal required to ensure a continuously increasing cross-section towards the feeder head. This extra metal is known as pads. Fig. 8.13 illustrates the use of inscribed circle method to determine padding requirements.

The ideal condition would be to incroporate padding as a design modification, however if it is not possible, it may be added for casting purposes and subsequently removed by expensive dressing operations (grinding, machining, power cutting etc.). Since the additional costs involved in removal of excess metal (padding) is very high, other measures such as use of insulating exothermic or indirect padding should be made.

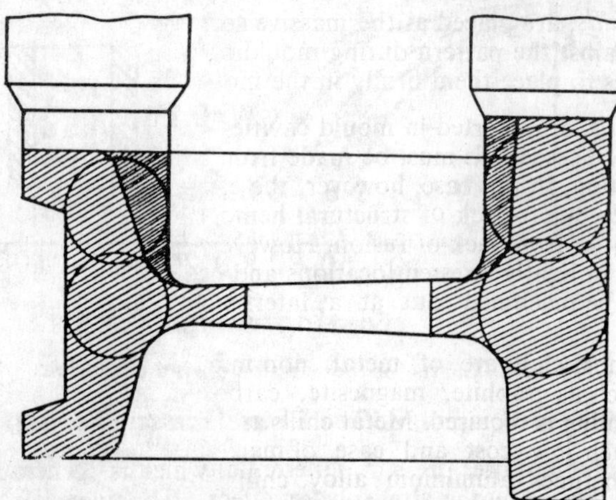

Fig. 8.13 Use of inscribed circle method for padding requirements

8.9.7 Use of Chills

The chills are massive metal inserts of increased heat capacity and thermal conductivity. The chills provide a highly effective means of promoting directional solidification. The surface chilling steepens the temperature gradient within the casting and as a result solidification proceeds by skin formation rather than in the pasty manner. The function of chills is of particular importance in improving the soundness of the long freezing range alloys, such as bronzes and gun metals. The close spacing of chills and risers is also used to produce high quality castings with high standard of soundness, fine microstructures and high level of mechanical properties.

However, in general, the chills are used to promote selective freezing of thicker sections of a casting by acceleration of heat extraction. Due to increased rate of cooling the thicker section can draw feed metal through the thin sections which remains liquid. Therefore, it results in (i) equalisation of freezing raies of heavy and thin sections and (ii) decreased volume of shrinkage cavity or its complete elimination. The effect of cooling equalisation is also an important factor in the prevention of hot tears. When chills and risers are placed on massive sections of the casting, the joint effect enables the production of sound castings completely free from shrinkage cavities. Metal chills may be employed either externally (external chills) or internally (internal chills) to the casting as shown in Fig. 8.14. External chilis are more commonly employed than internal chills.

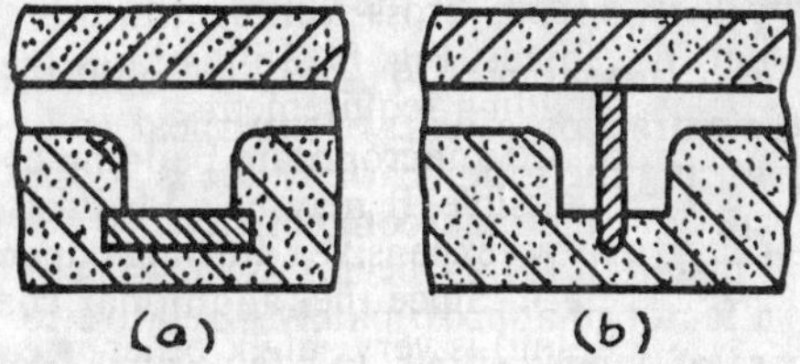

Fig. 8.14 Use of chills to the casting: (a) External chill and (b) internal chill

External Chills, are placed at the massive sections of a casting. These are positioned against the pattern during moulding and may be provided with hooks or wires to place them firmly in the moulding material.

Internal chills, are inserted in mould cavities which form massive portions of the casting. These chills must be made from alloys of same composition as that of casting. In this case, however, the castings loose credit for metallurgical quality due to lack of structural hemogeneity and the possibilities of discontinuities from lack of fusion. However, the internal ch;lls are less objectionable in lightly stressed locations and prcferable in situations where the chills can be machined out at a later stage, e.g. bosses intended for drilling.

Although most chills are of metal, non metallic substances of high heat diffusivity (such as graphite, magnesite, carborundum, etc.) may be used where less cooling is required. Metal chills are usually made of steel or cast iron due to their low cost and ease of machinability from bar or plate stock, but sometimes aluminium alloy chills are used for light alloy castings. In general, flat chills are preferred but shaped chills may be moulded and cast to fit irregular casting contours. Other factors affecting the function of chills are discussed below:

(i) A chill must be of sufficient mass in relation to the casting section to which it is to be applied. If the thermal capacity of chill is insufficient, its temperature will rise rapidly to a level at which the rate of heat extraction is similar to that of the remainder mould.

(ii) Chills must be placed in the sections from where heat can be extracted to useful cffect. If a chill is placed in the ingate region, it may get excessiveiy preheated due to passage of liquid metal across the chill surface during filling of mould. This situation may nullify or even reverse the influence of chill by the time the critical stage of feeding is attained.

(iii) Chills should be dry and free from oxide scale or grease, otherwise the presence of these contaminations should cause blowing defects.

(iv) The chills should be free from porosity and cracks as these may be the potential sources of gas defects. Therefore, chiles should be discarded when they become cracked or heavily oxidised.

(v) Differential contraction in the casting skin resulting by the use of single large chill may cause surface tearing of the casting near the edges of the chill. This danger can be reduced by the use of a number of small chills instead of single large chill. This reduces the strain concentration and permits the absorption of contraction stresses at large number of sites between the chills.

Calculation of Chill Dimensions. The use of chills reduces the time of solidification by the extraction of part of superheat and heat of fusion. The effect of chill in reducing the solidification time is a function of its diffusivity $(kc\rho)^{1/2}$, where k = thermal conductivity, c = specific heat and ρ density of the chill material. It has been shown that chills increase the solidification rate by a factor of approximately 4 relative to that obtained in sand moulds. In case of copper alloys, this factor has been indicated of the order of 14.

The solidification time of a chilled casting may be considered same as would be for a casting of smaller volume in the absence of chilling. According to Chowrinov, $t = k\left(\frac{V}{A}\right)^2$ where $V/A = M$ known as modulus. From this it can be seen that a factor of 1/4 applicable to freezing time is equivalent to a factor of 1/2 whcn applied to the modulus. This effect of the chill in reducing the modulus may also be expressed as an apparent increase in the total surface area of the casting. In conjunction with the actual volume of the casting, an effective modulus value can be determined for purposes of feeding calculations.

$$M_{\text{effective}} = V/A_{\text{apparent}} \tag{8.40}$$

for fully chilled castings

$$\frac{t_{\text{sand}}}{t_{\text{chill}}} = \frac{(M_{\text{sand}})^2}{(M_{\text{chill}})^2} \tag{8.41}$$

or

$$\frac{t_{\text{sand}}}{t_{\text{chill}}} = \frac{V/A)^2}{(V/A_{\text{apparent}})^2}$$

or $$A_{apparent} = A\sqrt{\frac{t_{sand}}{t_{chill}}} = A\sqrt{4} = 2A \quad (8.42)$$

For partially chilled casting the value of $A_{apparent}$ may be derived by the addition of area of sand surface to the area of chill multiplied by the suitable factor obtained from experiments on solidification.

$$A_{apparent} = A_{sand} + A_{chill}\,(t_{sand}/t_{chill})^{1/2}$$
$$= A + A_{chill}\,[(t_{sand}/t_{chill})^{1/2}]$$
$$(\text{Since } A_{sand} = A - A_{chill}) \quad (8.43)$$

Therefore, in the case of steel castings for which $t_{sand}/t_{chill} = 4$, it will require only the chill area to be added to the total area when calculating the modulus.

However, the factor of 2 is found to be applicable only in the presenee of air gap generally formed between a casting and a metal mould, whereas true metallic mould gives a factor of 3 for steel chill, In this case

$$A_{apparent} = A_{sand} + 3A_{chill}$$
$$= A + 2A_{chill} \quad (8.44)$$

Therefore $$A_{chill} = \frac{A_{apparent} - A}{2} \quad (8.45)$$

The effect of using chills together with design modifications of intersections for various designs is illustrated in Fig. 8.15. The differential cooling effect rendered by chiils is particularly suitable in the case of intersections extending for long distances, e.g. plate section reinforced with stiffening ribs. Small face chills are also invaluable for local applications to projections and isolated bosses where use of separate heads would be uneconomical.

Another method of using chills is to place the chill at an angle to the casting surface. In this case, the chill exerts its influence through a tapering layer of sand and thus the degree of cooling varies with the distance from the casting surface. Indirect chill can be used to promote directional solidification in a casting member of uniform section.

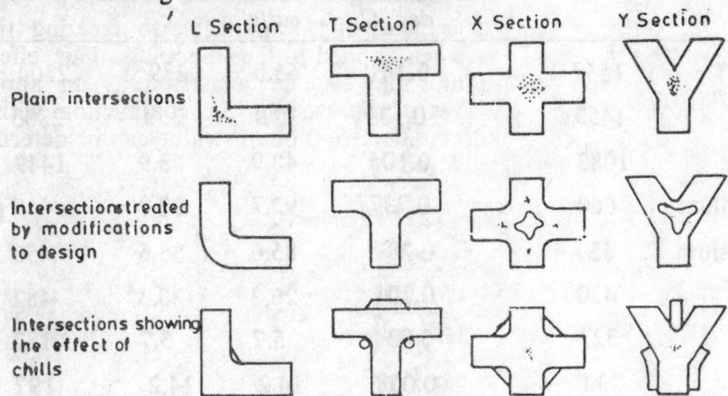

Fig. 8.15 Use of chills together with design modifications of intersection

9

Melting Furnaces

9.1 INTRODUCTION

In foundry practice, metal melting operation is the second in importance only to mould making. Metal melting process may be defined as tch process of producing liquid metal of the required composition at the desired rate and with sufficient amount of super heat in the most economical way. The basic requirements of metal melting depend upon their thermal properties which are given in Table 9 1 for the metals used in commercial casting alloys. In addition to the thermal properties of metals, selection of melting equipment depends upou the following factors:

Table 9.1 Properties and Heat Requirements for Metal Melting Total heat requirements for melting $l+s\,(t-20)$

Metal	*Melting point* t_m °C	*Mean sp. heat 20—*t_m C cal/g *deg. C*	*Latent heat of fusion cal/g*	*For 1 Kg.* Kcal	*For 1 litre* Kcals
Iron	1537	0.141	65.0	279	2204
Nickel	1453	0.133	72.1	259	2305
Copper	1083	0.105	48.9	48.9	1449
Aluminium	660	0.237	92.7	92.7	659
Magnesium	650	0.289	85.6	85.6	456
Zinc	420	0.101	26.3	26.3	482
Lead	327	0.034	5.7	5.7	187
Tin	232	0.058	14.2	14.2	197

(a) The compositional range of material to be melted and superheat required.

(b) Form and size of charge material.

(c) The fuel or energy used to melt the charge.

(d) The degree of refining and control over the process.

(e) The type and size of melting unit.

(f) Economic factors.

All these factors are inter connected and the melting operation should be viewed as a total system rather than each point in isolation. A melting system should be choosen that allows adequate flexibility for producing the range of composition and quantities required.

The form and size of the charge material have an important influence on the melting operation. This can be understood by the relationship between the surface area and the density of metal charged. Dense materials can be charged in the furnace at a fast rate because it has small surface area which gives low oxidation losses but low melting rate. Therefore, a compromise is preferred between these points (heavy and light scraps) but in general a charge o low surface area and relatively high density is preferred unless the charge is immerdiately immesed under liquid slag or flux. Light scrap (off cuts, sheets, etc.) can be used in compressed form whereas virgin metal may be supplied as small ingots or alloying elements as briquetts.

For melting of metals, electricity and solid, liquid and gaseous fuels are the main heat supply source although with improved technology nuclear energy will be used in future. Electrical energy can be used in following three ways:

(a) *Electrical resistance heating*. In this case melting unit incorporates the resistance wire around the crucible.

(b) *Induction heating*. In this case an electric current is induced in the metal to be melted by passing a current through a conducting wire surrounding the metal containing vessel. This produces a magnetic field in the metal charge in the vessel which in turn induces an electric current and thereby produces heating of the metal charge. This system is called coreless induction furnace. In another form a current is passed through a wire wrapped round on iron core which induces an electric current in liquid metal passing through a channel in the core. This acts as a transformer in which the liquid metal passing through the channel acts as a secondary and is therefore heated. This is called a channel induction furnace,

(c) *Electric Arc heating*. In this system the heat is produced in two ways, i.e., (i) by producing a high energy arc between inert electrodes (graphite) above the charge to be melted and the heat generated melts the charge. This system is known as indirect arc process and (ii) by striking an arc between the graphite electrode and the charge, this is called the direct arc process. The latter process is preferred since it gives better heating efficiency (more than 90%) due to direct put of electric currect into the charge.

Solid fuel usually takes the form of coke which is used in shaft type furnaces, e.g. cupola and blastfurnace or a pulverised coal form when burnt in air in hearth, crucible, or rotary type furnace.

Liquid and gaseous fuels can be conveniently used in the furnaces such as rotary furnace, reverberatory furnace, crucible furnace or any other furnace of similar designs. In case of liquid and gaseous fuels, desired atmosphere (oxidising, reducing or neutral) can be maintained by selection of proper burner and controlling the flow of fuel and oxidant (air).

In general, the metals obtained from smelting furnaces (blast furnace, steel making furnace, and other non ferrous smelting furnaces) are not directly cast into the desired shapes mainly because of two reasons, i.e., (i) the metals obtained from smelting furnaces are not usually in a suitably refined state and (ii) difficulty in pouring a huge quantity of molten metal in a large number of moulds. Therefore, most of the foundry melting is carried out on a batch basis from cold charges, the notable exceptions being cast iron production from continuously melting cupola and pig iron from blast furnace (pig iron is directly used for casting of ladles). A semi-continuous practice is also used in which of heel or aportion of liquid metal is retained in the furnace to provide a bath for the absorption of the next solid charge. In certain cases duplex or triplex practice is used involving separate furnaces for successive stages. Sometimes a holding furnace is used to provide a reservior of molten metal at the pouring temperature.

The principal foundary furnaces are shown in Table 9.2. The furnaces have been classified in to two main groups, based on fuel combustion and electrical heating. The details of some important foundary furnaces are discussed in this chapter.

9.2 CUPOLA

The Cupola is a shaft furnace (Fig. 9.1) employed for melting and refining pig iron along with scrap in the production of cast iron castings. The height of the shaft (4-6 m for cold blast cupola) provides a important function in absorbing the heat by the charge from the ascending gases. The charge materials i.e., metal, flux (lime stone) and coke are charged through a door at the top of the shaft. The cupola operates on a counter current principle, i.e. ascending hot gases heat the discending metal charge.

9.2.1 Construction of Cupola

Furnace is made up of a vertical steel shell (6-12 mm thick) lined with either fire brick or a rammed monolithic lining thickness of 150-225 mm for short runs and upto 300 mm for runs greater than one shift. The thickness of lining decreases to about 100 mm above the charging door. In case of short cupolas, the lining may be flared out to give increased area above tuyers to improve capacity and efficiency of operations. In some instances no lining is used in the melting zone and the shell is water cooled. This reduces the patching or relining work, but greatly increase the heat losses.

The air required for burning of coke is obtained from a blower and passed through a pipe called wind pipe. From the windpipe, the air passes to a circular jacket around the shell and then air is carried into the cupola through a number of tuyeres (openings) which are equally spaced around the circumference of the cupola to blow the air uniformly. The number of tuyeres required increases with increase in internal diameter of the

TABLE 9.2 Classification of Foundary Melting Furnaces

Melting Unit	*Fuel employed*	*Batch/ coutinuous melting rate in tonnes/hr*	*Temperature range °C*	*Melting efficiency*	*Heat transfer mechanism*	*Range of metals melted*	*Remarks*
1	*2*	*3*	*4*	*5*	*6*	*7*	*8*
Cupola	Coke	Continuous 2-20	1300-1500	High 50%	Conduction and convention of ascending gases	Cast iren	Main melting unit for cast iron, Pollution from exhaust gases, Some refining with slags, Low thermal efficiency for holding or superheating
Revorbaratory	Gas, oil- or pulverised coal	Batch 2-20	600-1650	Moderate	Radiation and Convection from combustion of fuel	Al, Cu, Ni, alloys, steel malleable C.I.	Considerable metal refining
Open hearth	Gas, oil	Batch 5-100	100-2000	High	—do—	Steel (Heavy)	Good refining through slags Production and refining rates can be increased by use of O_2
Rotary	Oil, gas or pulverised coal	Batch 0.1-0.5	800-1500	Moderate to high	—do—	Gunmetal brasses, bronze, cast iron and steel	Ideal for melting swarf, borings fines added under slag. Good mixing action in due to rotation
Crucible	Oil, gas coke	Batch 0.2-1.0	200-1400	Low 3.7% (Pit) 8-18% (Tilting)	Conduction through crucible	Al, Zn, Brasses cast iron	Tilting or pit type, Modern development of immerged crucible firing improves efficiency

Contd.

1	2	3	4	5	6	7	8
Converter	Oil, ggs	Batch 10-100	About 1650	High	Convection and conduction due to exothermic reactions in the melt	Mainly steels some Cu and cast iron	Bottom and top blown for steels, side blown for Cu, requtres molton charge, refining possible
Direct Arc	Electricity 3 Phase AC or DC	Batch 2-40	1250-1750	High (80%)	By arcing to charge	Steels, cast iron, Cu, Ni alloys	Refining is possible with basic or acid, oxidizing or reducing slags. O_2 lancing increases temrature and aids alloy steel production
Indirect Arc	Electricity single phase AC or DC	Batch upto 0.0-1.0	1000-1750	Moderate to high	By radiation and convection from arc and refractries	Special cast irons, bornzes, gun metals, Ni and Cu alloys	Rocking improves mixing and thermal efficiency. Metal refining is possible by using acid or basic slags
Coreless Induction	Electricity AC	Batch 0.5-3.0	600-1750	High (60%)	By induced current in melt and convective stirring action	Steels. alloy steels, high melting point alloys	High control of melting conditions, low metal loss and pollution. Ideal for compositional adjustments to melt
Channel Induction	Electricity AC only	Batch 0.5-2	600-1500	High (60%)	—do—	Al and low melting point alloys, some cast iron	Normally operates at or below 500 cycles/sec. No refining using slags. Cood mixing due to induced currents in melt
Resistor	Electricity AC or DC	Batch upto 1	1000-1750	Moderate to high	By radiation and convection from resistance rod or material and refactories	Al, Cu, Ni alloy, cast iron bronze and low melting point alloys	No refining, rod and crucible resistor furnaces, rocking in rod resistor furnace improves efficiency

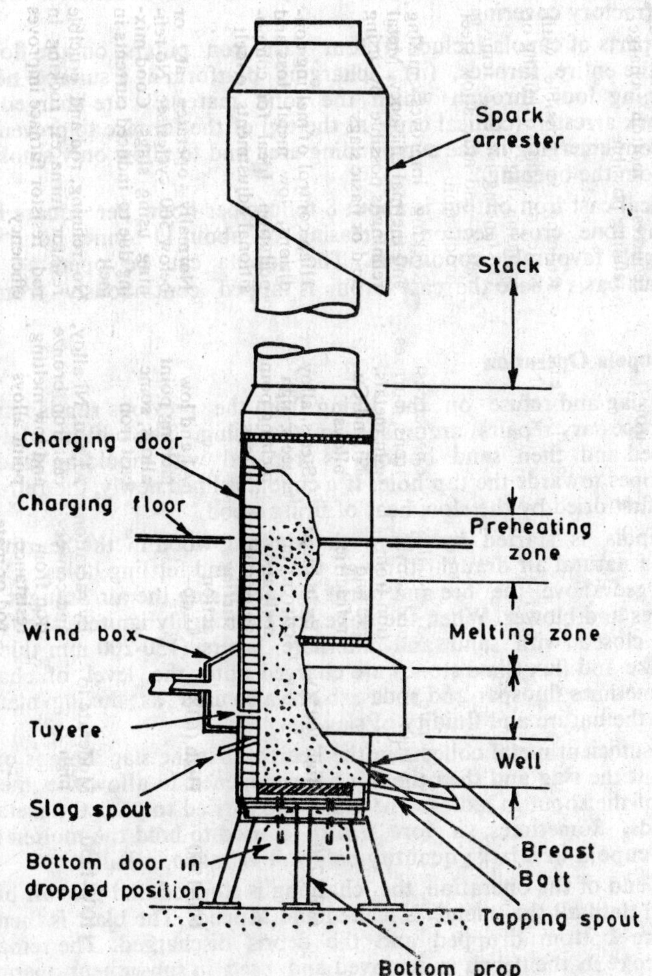

Fig. 9.1 The cupola furnace

cupola, e.g., 4 tuyeres for 0·5 m diameter to 10 tuyeres for an internal diameter of 2.0 m. The total area of air inlets (tuyeres) should be about 25 pct of the cupola plan area. The tuyeres are located at the height of about 500 mm from the bed of the cupola.

A small window with mica cover is provided opposite to each tuyere to inspect the combustion conditions. A tapping hole (spout) for molten metal is provided at the botton of the bed. Opppsite to the tap hole and some what above it, a slag hole is provided.

Most modern cupolas are of the drop bottom type in which plates (doors) are hinged under the hearth by a vertical rod known as *prop*. This system permits the botton to be lowered at the end of melting to facilitate cleaning operation. The bottom of the furnace is made up of thoroughly rammed moulding sand. Older cupolas may have a solid concrete bottom with a refractory covering.

Other parts of cupola include (i) four cast iron pillars on the floor to support the entire furnace, (ii) a charging platform at a suitable height, (iii) charging door through which the solid materials are charged, and (iv) a spark arrester (conical cap) at the top of the furnace to prevent hot sparks from emerging in the surrounding area and to allow only smoke to escape from the opening.

A typical cast iron output is about 8 tonnes per hour per square metre of melting zone cross section increasing to about 11 tonnes $hour^{-1}$ m^{-2} under highly favourable conditions. The cupola can be operated on a continuous basis where the cast iron is tapped continuously from the hearth.

9.2.2 Cupola Operation

First the slag and refuse on the lining from the previous run is removed and the necessary repairs are made in the lining. The bottom plates are then closed and then sand bottom is prepared with moulding sand such that it slopes towards the tap hole. If a cupola is lined newly, the refractory lining is first dried by the slow heat of firing wood.

The cupola is started by firing the kindling wood in the hearth while allowing a natural air draught through the slag and fettling holes. Coke is then charged above the fire and burnt by increasing the air draught using the tuyeres and blower. When the coke has thoroughly ignited, the fettling holes are closed with sand and alternate layers (150-200 mm thick) of metal, coke and flux (lime stone) are charged upto the level of charging door. Sometimes fluospar and soda ash are also used as fluxing materials to adjust the nature and fluidity of slag.

When sufficient metal collects in the hearth, first the slag hole is opened to flow out the slag and then the tap hole is opened to allow the metal to flowout of the shout in ladles. The ladles are carried to pour the metal into the moulds. Sometimes, a fore health is used to hold the molten metal from the cupola in a large quantity before it is taken in ladles.

At the end of the operation, the charging is stopped and the air blow is continued until all the metal is melted and tapped. The blast is then shut off and the bottom dropped and the debris discharged. The remaining unburnt coke in the debris is removed and used in subsequent operations. In the cupola having solid bottom, the debris is removed via the fettling hole. The cupola lining is repaired between the melts. Two or three cupolas may be employed to obtain continuous operation (one being repaird while the other maintains production).

9.2.3 Developments in Cupola

(i) *Use of Hot Blast*. The use of air blast heated between 350-500 °C results in following advantages.

(a) Higher temperature of molten metal can be obtained and thus it becomes possible to use steel scrap partly which enables the production of high grade cast iron.

(b) Operation efficiency is improved with reduced coke consumption.

(c) **Higher recovery of iron and alloying elements (silicon and manganese) is obtained with improved reduction of sulphur.**

(d) **It is possible to use inferior quality coke to have the same temperature and melting conditions.**

However, use of hot blast is economical only for long runs of large capacity (10 tonnes per hour and above) cupolas due to shortcomings such as higher cost of installation, maintenance and operation, higher rate of lining erosion, and greater control of equipment.

Preheating of blast can be carried out by two systems, i.e., (a) recuperator system and (b) external combustation system. In the former system, the sensible heat of exit gases is utilised to heat the blast, whereas in the latter case, cold blast is heated in a heat exchanger by hot gases produced by the combustion of oil or gas in a separate unit. The recuperator system lacks flexibility, since the temperature and quantity of hot blast will depend on the characteristics of exit gases (i.e. temperature and calorafic value), but it is most economical for continuously running large sized cupola. The external combustion system requiring less capital cost is more economical for smaller cupolas.

(ii) *Water cooling*. Large sized cupolas can be water cooled above the tuyere level upto the extent of melting and superheating zone, (i.e. about twice the internal diameter). Water cooling gives many benefits such as lower consumption of refractory lining, long melting periods, less time, and money required in repairing and larger melting rate obtained. In case of using water cooling, basic lining is provided only in the water cooled area.

Water cooling can be effected in the following two ways.

(a) By providing a jacket around the shell and circulating cold water through jacket. This is known as internal cooling which provides superior heat transfer and more efficient temperature control of the cupola melting operation.

(b) By the use of showers which drop water continuously on the outside of the shell and the water collects in a trough around the cupola. This is known as external cooling and it is simpler to install.

In both the cases, the water is recirculated after cooling. Water cooling is usually adopted for hot blast cupolas where a high temperature of metal can be maintained with uniform conditions of combustion. However, water cooling is economical for large sized cupolas running for larger durations.

(iii) *Basic Lined Cupolas*. In acid lined cupola, the slag produced is to be maintained of acidic nature which can not reduce sulphur during melting, rather than metal absorbs a certain amount of sulphur from the fuel. Whereas cupola using a basic slag and a basic lining provides improved melting control due to increased carbon pick up and reduction of P, S, and Si. Basic lined coupola may be used in the following instances.

(a) When iron is required with a low sulphur content as in the production of spheroidal graphite cast iron (S.G. Iron).

(b) When molten iron is required for the production of steel in electric furnaces.

With increasing basicity $(CaO\% + MgO\%)/(SiO_2\%)$ the carbon content goes up whereas both silicon and sulphur are reduced. The relationship between carbon and sulphur contents, slag basicity and metal temperature is shown in Fig. 9.2.

The main problems associated with basic refractory are more wear and high cost of lining. The cost may be reduced by use of water cooling.

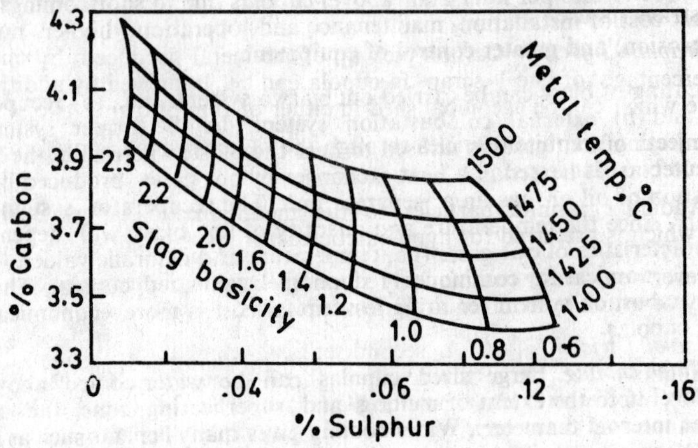

Fig. 9.2 **Relationship between carbon and sulphur content, slag basicity and metal temperature**

(iv) *Use of Liningless Cupola.* A lining-less cupola may be used with increased water cooling of the cupola shell and a carbon refractary in the well. This technique forms a skull of metal at the extremeties of water cooling zone which can be reduced by using a hot air blast and projecting tuyeres. This technique permits the use of high steel scrap percentage due to higher carbon pick up and highest blast temperature.

(v) *Oxygen Enrichment of Air.* Oxygen enrichment of air may be used to increase the temperature of hot blast and melting rate. 1 – 2 pct. O_2 enrichment increases the air blast temperature by about 150°C and by about 10 °C of metal temperature wilth 0.4% C and as such provides identical benefits as using hot blast. Oxygen enrichment may be obtained in two ways, i.e. (a) Oxygen enrichment of blast through tuyeres, and (b) direct injection of O_2 into cupola. The latter method is found to be more effective, but an optimum position of injection (about 300 mm below the tuyeres) is required to produce the best results. The use of oxygen is useful in controlling temperature over the initial melting period, where as during the latter periods use of hot blast is more economical to obtain a specific carbon content and temperature of metal.

(vi) *Injection of Supplementary Fuels in Cupola.* Iniection of supplementary fuels such as oil and natural gas has been found useful in improving combustion conditions and thermal efficiency. However, the calorific values of natural gas and oil are less than that for coke and hence may require an increase in the hot blast temperature to campensate. It has been observed that injection of oil or gas above tuyeres level helps in increasing the melting rate and metal temperature. The main limiting factor in the use of oil and gas is their higher cost. A cokeless cupola has been demonstrated using hydro-carbon injection.

(vii) *Use of Calcium Carbide*. The addition of calcium carbide is helpful in reducing the sulphur content of the metal and raising its temperature. It is found to be more effective with basic slags. The high cost of calcium carbide rêstricts its use.

(viii) *Graphite Injection*. Carbon pick up of the metal produced by melting high percentage of steel scrap in cupola can be improved by addition of graphite which can be accompalished in either of the following two ways;

(a) Injecting graphite particles through the tuyeres which is preferred for hyper eutectic cast irons.

(b) Adding graphite particles to the metal stream tapped from the cupola which is preferred for hypo-eutctic cast irons.

However both these techniques reduce the melt temperature with the recovery of graphite from 27 to 80 per cent.

(ix) *Use of Divided Blast*. A second set of tuyeres at the level of about 90 mm above the main tuyeres, fed from different wind belts having 40 – 50% of the total blast, produces a marked improvement in melting performance due to (a) improved thermal efficiency due to burning of most of the CO to CO_2 and (b) increased heat transfer to the metal.

The use of divided blast results in increased melt temperature with improved carbon pick up for a given coke charge or a given melt temperature can be obtained by reduction in coke charged. At the same time, melting rate is also increased.

9.2.4 Factors Affecting Efficiency of Cupola

The main variables in cupola melting are air blast rate, iron: carbon ratio, metal temperature, and melting rate. These variables are plotted in the form of a net diagram (Fig. 9.3) for cupolas of varying diameters. The blast may vary from 70—125 m^3/sq. m of cupola plan area per minute measured at 0 °C and 760 mm Hg pressure. Thus, knowing two variables, i.e., air blast rate and iron: carbon ratio, remaining two variables, i.e., metal temperature and melting rate get fixed. Use of iron: carbon ratio, instead of iron: coke ratio is preferred since different cokes possess different reactivities.

The net diagram (Fig. 9.3) applies to equilibrium melting conditions and a certain amount of time should be allowed on altering any of the four variables in order to reach this state.

Increasing the air volume at a constant metal: carbon ratio, increases the melting rate, since more coke is burnt and raises the metal temperature as heat losses are less with fast melting rates. On the other hand, keeping the air volume constant and decreasing the iron: carbon ratio, melting rate decreases since there is more coke to be burnt below the iron before it is lowered into the melting zone.

The maximum temperature is obtained when the cupola is operated at the melting rate and the blast rate indicated by the optimum line for any given coke ratio. Therefore, it may be concluded that blast rate and metal:

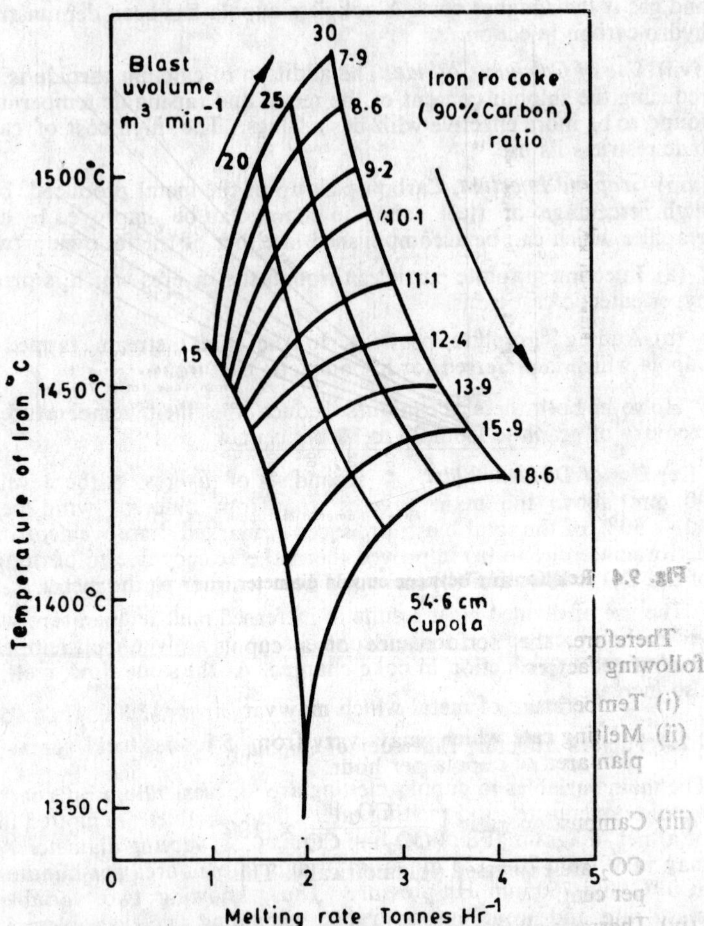

Fig. 9.3 A net diagram showing the effect of air blast rates, iron: carbon ratio metal temperature and melting rate

coke ratio play the most important role in controlling the performance of the cupola. It has also been found that thermal efficiency of a cupola reaches maximum at a particular blast rate which gives the highest metal temperature.

Fig. 9.4 further shows the relationship between the cupola diameter, the melting rate, and the blast rate. These diagrammes offer important informations in cupola control and sometimes even lead to a recommendation in a change of cupola cross sectional area (to achieve a specific melting rate at an increased metal temperature).

The coke size and quality of coke are important considerations in achieving high thermal efficiency. Use of smaller coke size results in lower thermal efficiency, lower melt temperature and less complete combustion of carbon to CO_2.

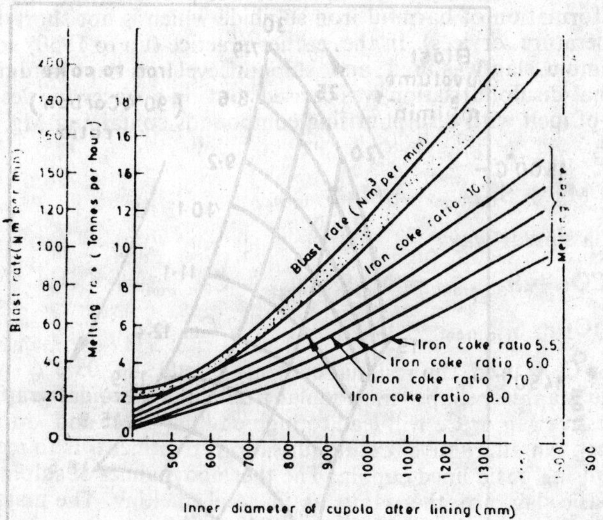

Fig. 9.4 Relationship between cupola diameter, the melting rate and the blast rate

Therefore, the performance of a cupola can be judged in terms of following factors:

(i) Temperature of metal which may vary from 13000 °C to 1550 °C.

(ii) Melting rate which may vary from 5 tonnes to 12 tonnes per sq. m plan area of cupola per hour.

(iii) Cambustion ratio $\left(\frac{CO_2}{CO_2 + CO} \times 100\right)$ in which amounts of CO and CO_2 are expressed volumetrically. This ratio may vary from 45 to 90 per cent.

(iv) Thermal efficiency, i.e., proportion of total heat input absorbed by metal and slag (may vary from 28 per cent to 45 per cent).

After tapping from cupola, melt may be directly poured or held in hot metal receivers having some form of heating (oil, gas, or electric) or relying on the heat from cupola. The use of metal receivers evens out the variations in composition during melting and variations in demand of iron from foundry. Post-cupola treatment and superheating are usually carried out in electrical induction furnace.

9.2.5 Compositional Control in Cupola Melting

Control of cupola melting can be achieved with an acid or basic operation, hot or cold air blast, and use of oxygen, hydrocarbon (gas or oil) and/or graphite injection. The choice of process mainly depends on the melting stock charged and the type of cast iron required. The main constituents of cast iron are sulphur, carbon, phosphorous and silicon which are responsible to the quality of cast iron.

(a) *Sulphur*. Sulphur must be kept as low as possible (less than 0.1%) to lower the formation of harmful iron sulphide which is hot short (fail under high temperature services). In the earlier practice (up to 1950) acid cupola with minimum slag was used and sulphur level was kept minimum in the charge. Final disulphurisation was carried out in a receiving vessel by the treatment of melt with disulphurising compounds containing Mg, Ce, CaC_2 or Na_2CO_3.

i.e. $$Mg + S_{(in\ iron)} = MgS_{(in\ slag)}$$

Ce also in a similar way.

$$Na_2CO_3 + S_{(in\ iron)} = Na_2S_{(in\ slag)} + C_{(in\ iron)} + CO_2$$

$$CaC_2 + S_{(in\ iron)} = CaS_{(in\ slag)} + 2\,C_{(in\ iron)}$$

If Mg or Ce is used, the residual contents should be 0.06% which also ontrol the graphite shape in nodular iron or spheroidal graphite (*S.G.*) ron. Excessive Mg or Ce will lead to high chilling effect and an increase in rocess cost. An alternative desulphurisation treatment is to melt under a asic slag using basic lined cupola. The thermodynamics of sulphur removal nder a basic slag are the same as in steel melting. The major limiting actor in basic cupola is the high cost of basic refractories.

(b) *Carbon*. The carbon content of the iron is controlled by the transfer of carbon from the coke which depends on the temperature, reactivity of he coke, the interfacial conditions between the iron and coke, i.e. the case vith which the two surfaces wet each other. The composition of initial harge is equally important since this and the temperature will determine he equilibrium carbon solubility in iron. In general, a high temperature vill favour carbon pick up in the iron. A basic slag tends to reduce the dherence of slag to coke and provides a more reactive surface for carbon lissolution in the iron.

(c) *Phosphorus*. Like sulphur, phosphorus should also be kept to low concentrations in cast iron to prevent cold shortness (a harmful mechanical lefect). For phosphorus removal, the conditions required are identical to hose required in steel making, i.e., basic processing, low melt temperature and oxidising conditions. As basic processing is quite expensive, usually acid processing is used keeping the phosphorous in the charge as low as possible by dilution with steel scrap and using charge of low phosphorus content.

(d) *Silicon*. Timing of silicon additions (usually as ferro alloys) is important since marked difference in the graphitising behaviour of cast iron can result even when final silicon contents are the same.

Melting under acid slags produces higher silicon iron than basic slag melting since acid slags can hold less SiO_2 whereas SiO_2 reacts with basic slag lowering its activity. Increasing temperature of metal also increases silicon content in the iron, i.e. $Si_{(in\ Fe)} + O_2 = SiO_{2\ (slag)}$ $\Delta G° = -95{,}000 + 199\ TJ\ mole^{-1}$. Thus as temperature, T, increases, ΔG becomes more positive forming the reverse reaction and increasing the silicon in iron.

9.3 OPEN HEARTH FURNACES

The *Open hearth furnace* (Fig. 9.5) is a squat rectangular structure having a shallow eliptical hearth to hold the charge. In foundries, usually small sized furnaces (capacities ranging from 10 tonnes to 100 tonnes per batch) are used for melting steel or producing steel from pig iron direct for casting. The furnace may have basic (magnesia bricks) or acid (silica bricks) lining depending on the pig iron used. The charge mainly consists of pig iron (solid or molten) in varying amounts along with steel scrap. In acid process, it is difficult to control the amount of sulphur and phosphorus in the molten steel and thus a high grade metal charge is needed. In the basic process it is possible to remove phosphorus and to some extent sulphur and thus a lower grade metal charge can be used. Lime is added with the charge as a fluxing agent.

The open hearth process is based on the principle of regenerative heating which gives very high temperatures required for steel melting. An open hearth furnace can be heated by means of gas or oil (atomized) and sometimes powdered coal is used. The furnace consists of gas and air ports at each end, and two pairs of regenerators (one for gas and other for air) with necessary flues and chimneyt At a time, the burners are fired at one end offurnace for 20 minutes and the hot exhaust gases heat up the regenerators (a brick work system beneath the far end of the furnace). After 20 minutes, the air is passed through this heated brickwork and burnt with the fuel at the other end. The hot exhaust gases heat the regenerative systems alternatively, i.e., the cycle of burning and preheating is reversed every twenty minutes. Burning from both ends improve the melting and refining rates. Oxygen enrichment can be used to increase melting and refining rates.

The main advantages of open hearth process are as follows:

(i) Shallow bath of molten metal provides a maximum surface area fora given volume of metal which gives quick transfer of heat and better reaction between metal and slag increasing the refining and melting rate.

(ii) Since iron ore is used as an oxidising agent and the heat source is external, the temperature and composition of the bath are under proper control.

(iii) A variety of raw materials can be used to produce almost all types of steels.

(iv) Due to use of iron ore, the yield is more than that of other methods.

However, the slow melting and refining rates due to the quiescent bath have made open hearth furnace uneconomical with compared to converter or electrical arc steel making.

9.4 REVERBERATORY FURNACE

A reverberatory furnace (Fig. 9.6) is a large wide refractory lined rectangular furnace in which fuel (gas, oil or pulverised coal) is burnt at one end. The heat produced is radiated and reflected down off the sloping roof to the metal charge on the hearth. Charging is carried out through doors on the front of the furnace, whereas the molten metal is tapped at the back of the furnace. The reverberatory furnace has relatively a shallow molten metal bath and thus provides large slag-metal interface area and thus high degree of control on refining reactions.

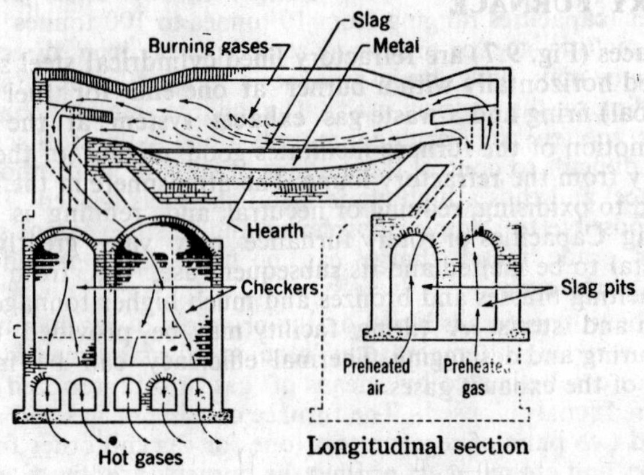

Fig. 9.5 Open hearth furnace

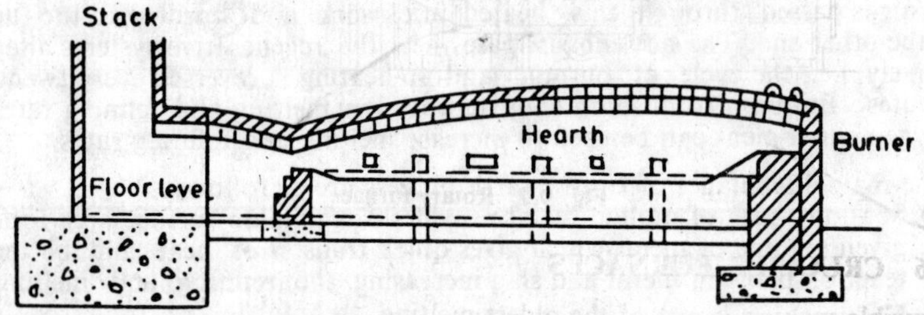

Fig. 9.6 Reverberatory furnace

These furnaces may vary in capacities ranging from few hundred kilograms to 100 tonnes. These furnaces are principally used in foundry holding furnaces, permanent mould dip wells, ingot production and melting of scrap, pig and alloy hardeners and for remelting of ingots. Thc reverberatory furnace may be made in several designs (single hearth or double hearth). The double hearth design is more common in which one is for melting and other for holding. The two hearths may be entirely separate or just divided by a refractory wall. The reverberatory furnace may be either stationary or tilting Lip axis reveraberatory furnace is being used increasingly on account of the economics and less metal losses.

Advantages of reverberatory furnaces over crucible furnaces include the following:

(i) By using sloping hearths, oxides and tramp iron remain on the hearth, while molten metal passes to another hearth.

(ii) Since heat transfer is fast, it is an efficient method of melting scrap.

(iii) Increased thermal efficiency.

(iv) High rate of melting ensures minimum exposure of the molten metal to the combustion gases and thus high quality metal is produced.

9.5 ROTARY FURNACE

Rotary furnaces (Fig. 9.7) are refractory lined cylindrical steel shells. These are positioned horizontally with a burner at one end for fuel (gas, oil, or pulverised coal) firing and a waste gas exhaust system at the other end. Rotational motion of the furnace facilitates good mixing of the melt and heat recovery from the refractory lining. The atmosphere in the furnace can be controlled to oxidising, reduing or necutral, and refining is accompalished with slag. Capacities of rotary furnance may vary greatly depending upon the metal to be melted and its subsequent use, e.g. from 50 kg to 5 tonnes for melting brasses and bronzes and much higher tonnage for melting cast iron and steels. A tilting facility may be provided to facilitate charging, pouring and deslagging. Thermal efficiency can be increased by recuperation of the exhaust gases.

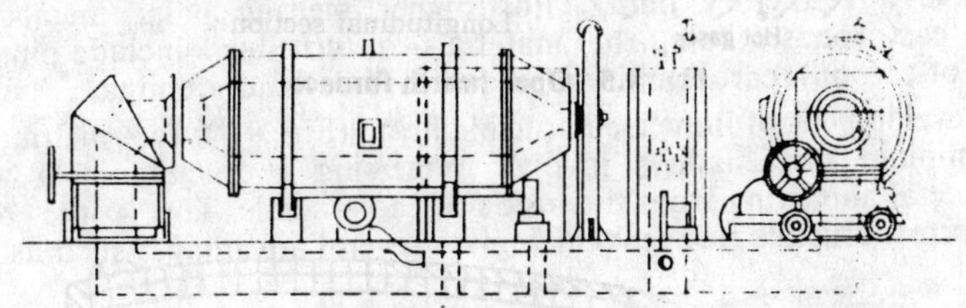

Fig. 9.7 Rotary furnace

9.6 CRUCIBLE FURNACES

Crucible melting is one of the oldest melting techniques. In this case, the charge to be melted is placed in the crucible and heated externally by oil, coke, gas or electricity via conduction through the walls of the crucible. Thus combustion products do not come in direct contact with metal and there is no contamination problem. The low thermal efficiency of the process limits the melting in small batches. However, due to very low capital outlay, crucible melting is still used where small amounts of metals (particularly lower melting point non-ferrous metals) are required. For many melting applications, induction furnace has replaced crucible melting. Crucible furnaces can be classified into following categories depending upon the method of pouring.

(i) *Lift out Furnaces*: In this case, the crucibles are removed from their place and transferred to the mould or die.

(ii) *Bale out Furnaces*: In this case, the metal is ladled out from crucible to the mould. The crucible is stationary pot of refractory or cast iron and is generally used for holding molten metal at a constant temperature prior to casting.

(iii) *Tilting Furnace*: In this case, the crucible or pot is fixed (Fig. 9.8) and metal is discharged by way of tilting the furnace. The metal may be poured with the mould at melting site or metal is transferred in a small portable ladle.

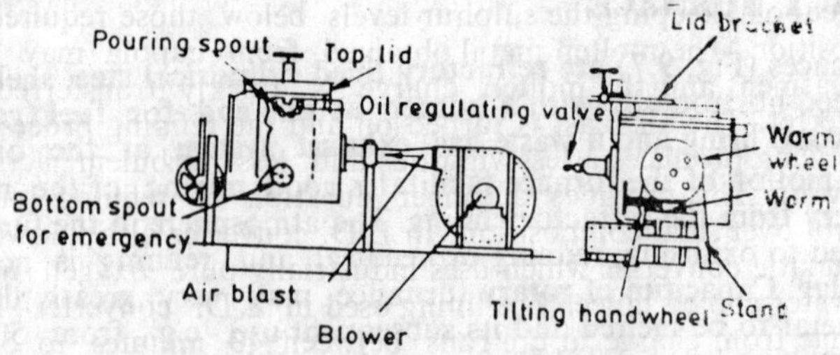

Fig. 9.8 Tilting crucible furnace

Crucible for melting low melting point metals/alloys (aluminium, magnesium, zince, and other alloys) are generally made of cast iron or steel and may have a refractory lining. For higher melting point metals/alloys (copper, cast iron, steel etc.) the materials of crucibles include plumbago (a form of carbon), carborundum (silicon carbide) and alumina.

The development of immersed crucible firing (Fig. 9.9) for melting zinc and aluminium metals/alloys involve immersing a crucible in a well of molten metal and firing from the inside of the crucible. This is the reverse of the normal crucible furnace which gives several advantages such as:

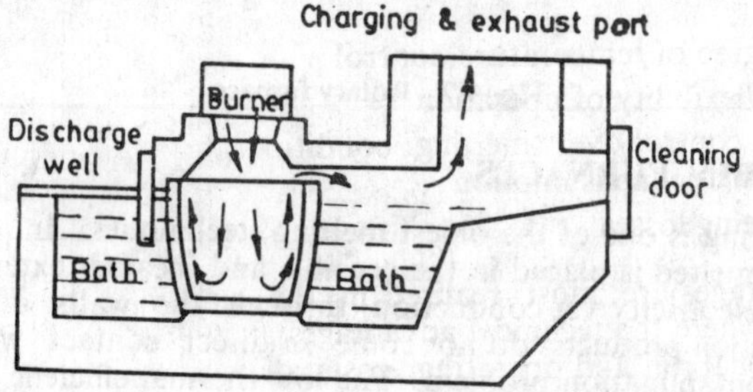

Fig. 9.9 Immersed crucible firing furnace

(a) higher thermal efficiency (up to 40 per cent higher)

(b) reduced metal losses

(c) reduced danger of crucible damage on charging

(d) more uniform distribution of temperature, and

(e) high crucible life

9.7 CONVERTERS

Converters are pear shaped tilting vessels lined with suitable refractories (basic or acid) and are blown with air/or oxygen through liquid iron to produce steel. The original converter blows air through the molten iron via tuyeres at the bottom, producing a low carbon steel with in 15 – 20 minutes. But in foundries the only type of converter used is side blown converter and to lesser extent *LD* converter is used which treats cast iorn. These are called Tropenas converters. The capacities of these converters may vary from 1 tonne to 50 tonnes.

The initial melting is generally carried out by melting iron and scrap and pig iron in cupola, keeping the sulphur levels below those required in the final composition. The molten metal obtained from cupola may be given soda ash treatment and the molten charge is quickly transferred to side blown converter. The air blast is turned on and the refining process begins. The side blowing produces a less vigorous and less turbulent action than bottom blowing. This provides a longer duration of refining but allows better control over final composition. In L.D. converter also known is top-blown pneumatic converter which uses industrially pure oxygen instead of air with a water cooled lance. The lining used in L.D. converter is basic. The total time from charge to tap runs between 40 minutes to one hour depending upon the capacity. In steel industries, converters of large capacity (upto 400 tonnes) are used whereas in foundries only small converters are used. The charge consists of scrap and hot metal. After bringing the converter into vertical position, oxygen is blown in at about 3 m^3/min/tonne of charge. The reaction between the slag and the metal takes place rapidly due to high oxygen input ratio, high temperature, and the constant mixing of slag and metal. The process possesses the various advantages such as great degree of dephosphorisation and partial removal of slag and nitrogen, low capital outlay, flexibility and higher rate of production.

9.8 ELECTRIC FURNACES

Electric furnaces are increasingly important because of the following advantages:

(a) High degree of temperature control
(b) Greater flexibility of operation
(c) Greatest control over melting conditions and pollution due to the absence of fuel consumption
(d) Low melting losses

These advantages are most realised in the case of expensive and hgih quality alloys where the technical advantages and low melting losses outweigh the high capital and operating costs of electric melting plant. Electric furnaces are being increasingly used for both ferrous and non-ferrous metals. The electric furnaces generally fall into three main categories, i.e. arc furnaces, induction furnaces and resistor furnaces.

9.8.1 Arc Furnaces

Arc furnaces are analogous to fuel fired hearth furnaces which are used for refining as well as straight forward melting. These furnaces obtain heat generated from an electric arc within the furnace. Arc furnaces can be subdivided into direct arc and indirect arc furnaces. Both can be lined with acid or basic refractories so that melting can be carried out with acid or basic slags.

Direct arc furnaces: In direct arc furnaces, an electric arc is struck between an electrode (graphite) carrying a high power supply (fairly low voltage and high current) and the metal being melted. A simplified diagram of direct arc furnace is shown in Fig. 9.10. The furnace consists of a circular steel shell lined with refractory (acid or basic) and a top cover or roof containing three holes through which the electrodes can be lowered or raised. After the electrodes are withdrawn, the roof can be swung back to facilitate charging. The distance between electrodes and molten charge is automatically controlled.

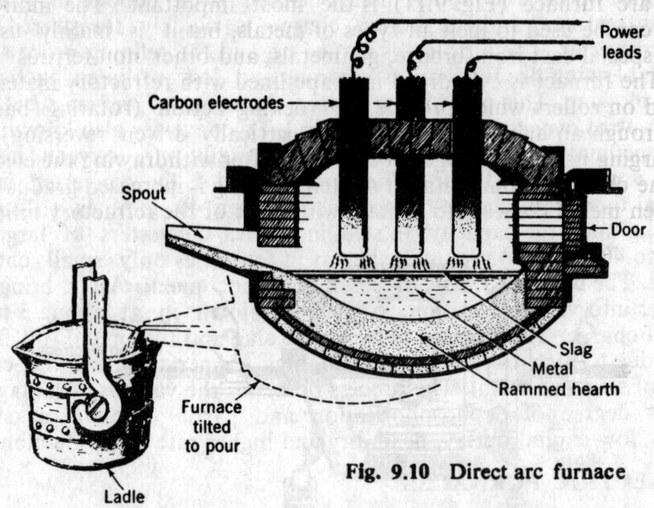

Fig. 9.10 Direct arc furnace

The furnaces have a tilting arrangement and a pouring spout. A charging door is usually provided just opposite the pouring spout. In larger furnaces this door is used for slag removal and oxygen lancing, whereas in small furnaces it is used for making additions to the melt. Electromagnetic stirring for larger refining processes is affected by incorporating a coil underneath the hearth of the furnace.

During melting, relatively higher voltages (180 – 450 V) are used while lower voltages (80-100 V) are used for refining and maintaining temperature. The currents usually range from 20,000 to 80,000 amps. The larger furnaces require higher voltage for melting and refining. The arc furnaces are rated according to their power inputs, i.e. volts × amps. (e.g., 900 KVA furnace). The capacities of direct arc furnaces employed in foundries usually range from 5 to 25 tonnes. A 10-tonne arc furnace may have the following dimensions.

Outside diameter	3 m
Height	2.25 m
Thickness of steel shell	25.0 mm
Diameter inside the refractory lining	2.5 m

The direct arc furnaces are mainly used for steel melting, but can also be employed for special cast iron, copper and nickel alloys. The direct arc furnaces possess the following specific advantages:

(a) The rate of heat application can be closely controlled.
(b) The furnace atmosphere above the molten metal can be controlled, i.e., it can be maintained oxidising, neutral or reducing.
(c) Most alloying elements (Cr, Ni, W, etc.) can be recovered from the scrap with negligible loss.
(d) Pig iron and steel scrap can be used for making steel as for open hearth furnaces.

Indirect Arc Furnace: In indirect arc furnace, heat is generated by arcing between two horizontal carbon electrodes and thus heating is effected by radiation from the arc to the charge. There are several designs, but the rocking arc furnace (Fig. 9.11) is the most impottant. The indirect arc furnace can be used to melt all types of metals, but it is mainly used for melting special cast iron, bronze, gunmetals, and other non-ferrous metals/alloys. The furnace is cylindrical in shape lined with refractory material and mounted on rollers which provide the rocking action (rotating back and forth through an angle of 180°) via an electrically driven reversing switch. The charging is done through an end door after withdrawing the electrodes. After the charge becomes almost molten, rocking is increased gradually until the moten meltal comes into contact with most of the refractory lining.

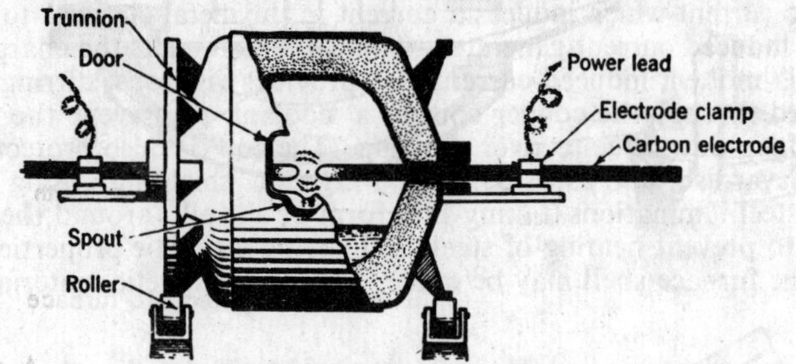

Fig. 9.11 Indirect arc furnace

The indirect arc furnace ensures good mixing of the melt and heat recovery from the refractory. As this furnace operates on a single phase electrical supply, the capacity is limited to about 100 Kg. However, this furnace is ideal for intermittent foundry melting.

9.8.2 Induction Furnaces

In induction furnaces, all the heat is generated in the charge itself. A high voltage electrical source from a primary coil induces a low voltage, high current in the metal (secondary coil) to be melted. The high induced current in the charge metal provides the necessary heat for melting. A crucible of high resistivity is used to prevent induced current. The material used for making crucible should possess good thermal conductivity to heat transfer. The main advantages of using induction furnace are as following:

(a) The melting process is extremely simple
(b) Induction furnaces have high flexibility in melting even small quantity of metal of any composition.
(c) Induction furnaces are ideal for melting and alloying with minimum metal loss.
(d) The control of temperature is obtained easily and quickly.
(e) The metals and alloys melted in induction furnaces are of high quality and free from gases.
(f) Operating costs are lower because of low refractory consumption, lower power consumption and absence of electrodes.

However induction furnaces have the following disadvantages.

(i) The initial cost is high

(ii) Only little refining of the metal is possible as it is difficult to maintain hot fluid reactive slag on the metal surface (slag is not effectively heated by the induced current).

(iii) Since the melting is very fast, bath sampling is difficult.

Induction furnaces are mainly of two types i.e. coreless furnace and channel or cored furnaces.

Corℓless Induction Furnaces. These furnaces (Fig. 9.12) usually have a cylindrical shell with a crucible or a monolithic lining. Around the crucible or lining, a helical coil of flattened copper tubing is wound. This coil carries the electric current which induces a current in the metal charged to the furnace. The induced current generates the heat which melts the charge. Once the metal is molten, induced current also provides vigorous stirring. Water is circulated through the copper coils as a coolant to prevent the furnace from breaking down due to over heating. The coil is also protected with insulating varnish and asbestos tape. Magnetic shielding using packets of silicon steel laminations (taking the form of a yolk around the shell) is employed to prevent heating of steel shell having magnetic properties. Alternatively, the furnace shell may be made from non magnetic materials.

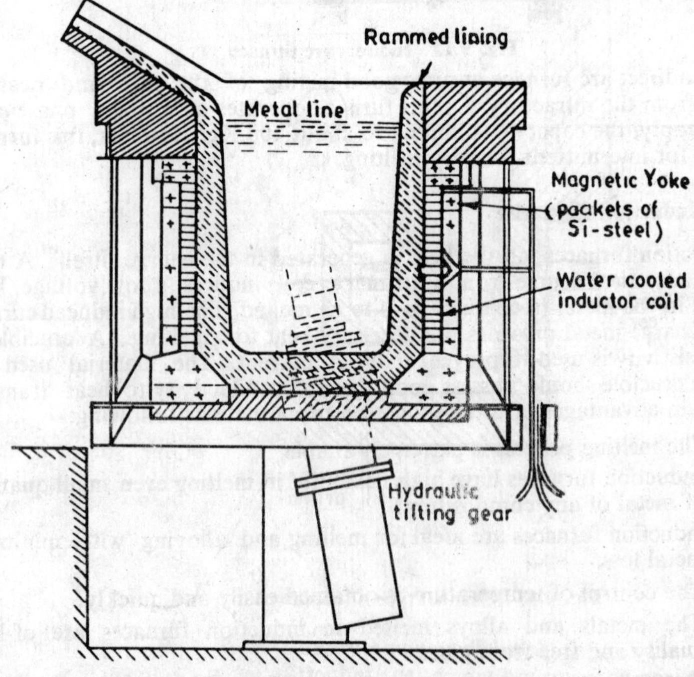

Fig. 9.12 Coreless induction furnace

The furnaces have the tilting facilities and the frequencies vary from 15 cycle per second (low frequency) to as high as 100,000 cycles per second (high frequency). In general, the higher the melting point the higher the frequency required. The furnaces with high frequencies readily melt metal from cold charge. whereas low frequency furnaces require a molten heel of metal to which solid metal can be added.

A high frequency furnace is mainly used for melting of steels, high alloy steels, stainless and magnetic steels. nickel-chromium heat resisting alloys and other high melting point alloys containing expensive alloying elements. A low frequency furnace is employed for melting non-ferrous metals and alloys (aluminium, brass etc).

Line frequeney (60 cycles per second) induction furnace has been developed into an efficient and economical unit for melting cold charge. Stirring action is greater in the line frequency induction furnace than in the higher frequency furnaces. These furnaces quickly melt down additional charges when a liquid heel is present. Line – frequency induction furnace can be employed for the melting, holding, superheating, or duplexing of all ferous and non-ferrous metals.

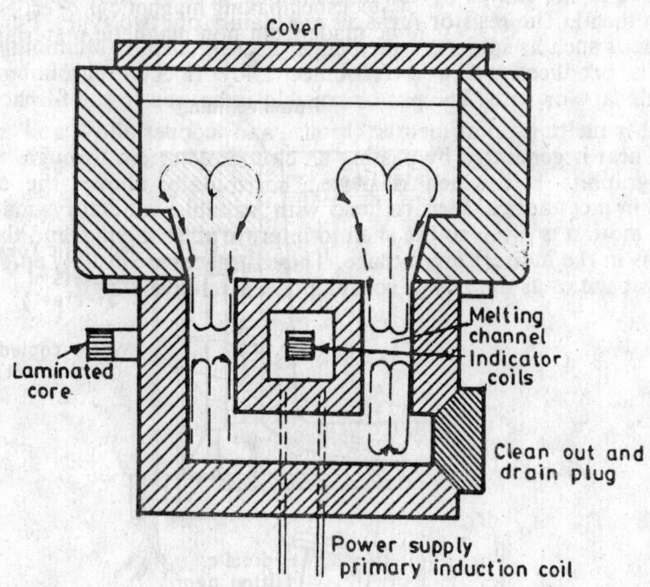

Fig. 9.13 Channel type or cored induction furnace

Channel-Type or Cored Induction Furnace. This type of furnace (Fig. 9.13) has an iron core around which the induction coil is wound. The induction coil is essentially immersed within the bath of molten metal. The current from the induction coil produces a changing magnetic flux in the core which induces a secondary current in the loop of molten metal. The heat generated in the loop of molten metal is circulated into the main well of the furnace which is situated above the channels.

The thermal and electromagnetic effects circulate the metal in the hearth providing uniform temperature and composition of the alloy throughout the bath. For carrying out the melting operation, it is essential to maintain furnace atleast one-third full of molten metal and the solid charge is added above this molten metal. There may be two types of channel furnaces. i.e.,

(i) Core and channel are positioned on the side of the main furnacewell, and

(ii) two separate chambers are provided one for melting and other for holding. The latter type is particularly useful for diecasting operations.

Channel induction furnaces are generally used for melting aluminium, zinc and copper alloys and as a holding furnace for cast iron and steel.

9.8.3 RESISTOR FURNACES

A resistor furnace is based on the principle of heat generation by passing an electrical current through a high resistance material. The generated heat is transferred to the charge by conduction and not by induced current in the molten metal. The resistor furnaces are mainly of two types. For lower melting points such as solders, type metals, bearing metals, aluminium and zinc, heat is produced using a resistance alloy (nickel—chromium alloy or Kanthal) tape or wire. The pot or crucible is housed in the furnace.

For higher melting point metals (bronze and copper alloys and special cast irons) heat is generated by passing an electrical current through a high resistance graphite rod which is placed horizontally above the charge positioned in a cylindrical furnace lined with suitable refractory material. A rocking motion is given to the shell to improve alloy mixing and thermal efficiency as in the indirect arc furnace. These furnaces are usually employed for small capacities as the power consumption is relatively high,

10

Cleaning of Castings

After the solidification of metal and cooling of sand mould, the casting is knocked out by breaking the mould either manually on the pouring floor itself or mechanically. In the latter case, the mould is rapidly jarred by mechanical vibrations so that the sand becomes loose and falls through a grate or screen into a pit or on a belt conveyer arranged below the floor. The casting and moulding boxes are then removed from the grate. The casting removed from the mould as such is not fit for use since it contains unwanted parts such as sprue, risers, gates, etc., and sand particles. Therefore, the cleaning of castings may be considered as the group of operations involved in the removal of adhering sand and core residues, separation of feeder heads and runners and elimination of flash, fins, wires, chaplets and other metal not a part of casting.

The cleaning operations are carried out in the shop known as fettling shop and thus operations are also referred as fettling operations. The various cleaning operations include the following:

(i) Removal of dry sand cores

(ii) Removal of feed heads and runners.

(iii) Removal of unwanted projections such as fins, wires, flash etc.

(iv) Final cleaning and smoothening of casting surfaces.

10.1 REMOVAL OF DRY SAND CORES

The main mass of the moulding material is usually removed at the knock out station, but heaviest castings and castings with confined internal cavities contain residual sand and dry sand cores. This residual sand and cores are loosened with crow bars and pneumatic or hydraulic iron pricks. These devices knock out the cores and also help in cleaning and smoothening of the casting.

10.2 REMOVAL OF FEED HEADS AND GATES

Numerous alternative methods are available for the removal of feeding and gating systems and the choice depends upon the size and shape of the casting and type of the metal. The options extend from simple knocking off through the standard methods of machine cutting by saw, abrasive wheel or parting tool to a highly versatile range of flame cutting processes. All these methods are useful in carrying out the useful functions in the foundry depending upon the type of alloy and dimensions of the casting.

10.2.1 Knocking off or Breaking with Hammer

This method can be applied to the castings having low ductility, particularly where the area of junction with the casting is small. This method is particularly suitable for grey cast iron castings and other brittle metals. The fracture may be facilitated by the use of various types of necks and knock off cores, both to reduce the metal section and to provide a notch eflect. However, one of the slitting processes may also be used to produce initial notch.

10.2.2 Sawing

Although, sawing can be adopted for most types of cast metals, it is most commonly employed for the softer non-ferrous alloys. The various types of saw used include bandsaws, hacksaws and circular saws, but the band saw is the most versatile as it can be used over a wide range of speeds (25 m/s – 0.25 m/s). At still higher speeds (75 m/s) the principle of friction cutting comes into effect in which sufficient heat is generated to bring the temperature at the root of the cut into the plastic range causing the high rate of stock removal. Friction cutting finds its use in ferrous alloy castings including work hardening components not amenable to sawing upto about 25 mm in section thickness.

10.2.3 Abrasive Wheel Cutting

Abrasive cut off machines can be used for all metals but these are specially designed for hard metals which are difficult to saw or shear. On a full sized machine the wheel can be mounted in a fixed position and the casting is pushed against it. or the casting may be clamped to the table and the mounted wheel is swung into the contact position. In general, much use is made of portable machines on which abrasive disc is mounted on a right angled shaft. The wheel thickness may vary from about 3 mm to about 500 mm. For each type of wheel there is an optimum peripherial speed (48–81 m/s) and thus the machine should have the provision for an increase in rotational speed to compensate wear of wheel.

10.2.4 Machining

In some cases heads can be easily removed by machine parting. The method is mainly suitable for large heads and those where a continuous cut can be achieved, e.g. in case of axially positioned and annular heads. In this method, a narrow parting tool is employed and further finishing becomes unnecessary.

10.2.5 Flame or Torch Cutting

Flame cutting is most commonly based on the oxyacetylens system, but propane and other fuel gas can also be used. This technique is widely employed for the removal of heads from steel castings. The principle of oxyacetylene cutting involves

(i) preheating the metal to its ignition temperature by heat evolved due to oxygen-gas reaction

(ii) then the heat is derived mainly from the combustion of the metal itself with excess oxygen supplied through the torch and

(iii) Cutting occurs by a cambination of melting and oxidation to slag. Depending on the metal thickness, a suitable size of torch nozzle is used with necessary oxygen pressure so that cutting reaction and slit move along.

The conventional cutting torch can cut carbon and low alloy steels to considerable depths, bnt alloys having high oxidation resistance (stainless steels and heat resisting steels) are difficult to cut since the progressive action of the flame is impeded by stable oxide films.

10.2.6 Powder Cutting

This technique can be successfully used in cutting oxidation resistant alloys. In this case a stream of preheated iron powder is dispersed through the torch into the fiame, providing additional heat of combustion which assists local melting and increases the fluidity of the oxide products.

10.2.7 Electric Arc Cutting

This is an important process in cutting the heads by local melting of the cast metal with the help of an arc. Either metal or carbon arcs can be used for melting. The process is suitable for oxidation resistant alloys. Compressed air is blown to remove molten metal clean of the cutting zone. Care is required in the high temperature, cutting processes to prevent undesirable metallurgical changes due to local heating, e.g., in high carbon and alloy steels, martensitic transformation may lead to cracking. However, this danger can be minimised by preheating as it is done in welding.

10.3 REMOVAL OF UNWANTED PROJECTIONS AND FINS

Either after, before or during the surface cleaning, the castings are subjected to dressing for the removal of excess metal which includes flash, fins, pads and the stumps of feeder heads and gates, chaplets, wires, etc. which are not a part of the casting's final dimensions. The principal techniques employed in dressing of castings include chipping and grinding but hot fettling techniques with specially designed torches are also used.

10.3.1 Chipping

Though chipping of fins, riser pads, wires, etc. can be carried out by hand hammers and chisels, the principal tool is the pneumatic chipping hammers or chisels operating at the pressure of 500-700 KN/m^2. A variety of hammer and chisel sizes are used for different casting alloys. Light grey and white iron castings can be chipped conveniently by hand tools, whereas heavier castings and casting more difficult to trim, are chipped conveniently with pneumatic tools.

10.3.2 Grinding

The grinding operations to remove excess metal are carried out by three principal types of grinders, i.e. swing frame, pedestal or bench stand and portable grinders. In addition, specialised machines such as disk grinders, belts and cut offs are also used. The swing frame grinder is suitable for the removal of stumps of feeder heads and often heavy duty grinding on large castings. In this case, the grinder is mounted on a swing frame, the casting is positioned under the grinder and then grinding is carried out. Pedestal or bench grinders using wheels upto about 750 mm diameter are suitable for small castings which can be handled individually. In pedestal grinders mechanised feeding and positioning fixtures may be employed to speed up grinding, when large number of same type castings are involved.

Portable grinder is the most universal tool for general dressing. This uses a wide variety of abrasive wheels (made of either aluminium oxide or silicon carbide) for manual operations on castings of all sizes. These grinders are electrically powered. The commonest design of portable grinder uses an axially mounted wheel but angled shafts with flexible drives. Though many shapes of wheel are available, plain wheels are most common, e.g 150 × 25 × 16 mm size. Delicate shaping and polishing are accompalished by small high speed pneumatic or electric machines with mounted points and tungeston carbide burrs.

Abrasive belt machines are the further development in grinding of castings. A very high rate of metal removal can be achieved with these machines. Other methods of surface abrasion include filing and wire brushing which are usually employed for the softer non-ferrous metals in which fettling problems are less severe.

10.3.3 Hot Fettling Processes

These methods have the particular advantage of avoiding dust generation and are adopted for ferrous alloys. Hot fettling processes include powder washing process, air-carbon arc process and plasma torches. In the first method, iron powder is dispersed through a specially designed torch into an oxyacetylane flame, which is played across the surface of the casting to remove excess metal. In the air-carbon arc process, an arc is applied to obtain local melting of excess metal which is blown away by a stream of compressed air as in the cutting process (section 10.2). Plasma torches can be used for cutting as well as removal of metal.

10.4 CLEANING AND SMOOTHENING OF CASTING

In general, surface cleaning operations are carried out after the removal of gates. fins, projections etc. However, surface cleaning may also facilitate gate removal, e.g,, elimination of sand favours sawing and torch cutting particularly in case of nonferrous alloys and in these instances, surface cleaning may be carried out before the removal of feeder heads and gates. Some times surface cleaning operation may be repeated after heat treatment to remove adhering scale on the castings. The various methods available for surface cleaning arc blasting (sand and shot), tumbling, hydraulic cleaning or hydroblasting and chemical cleaning.

10.4.1 Blast Cleaning

Blast cleaning is accomplished by streams of dry abrasive particles to impinge on the castings. The abrasive particles are energised either by compressed air or high speed centrifugal impeller. Depending upon the system of particles and energiser used, the blast cleaning may be either sand blasting or shot blasting.

In sand blasting, a high velocity stream of compressed air along with nonmetallic abrasive particles such as sand, alumina, slag, silicon carbide, etc. is directed with the help of a blast gun against the casting surface. The blasting operation is usually carried out in special chambers. The discharged sand drops through a perforated floor from where it is conveyed to the moulding section for reuse. Suitable mechanical means are provided to handle the castings. The sand blasting method can be used for both fragile and large sized castings. This method also imparts good finish to the castings.

In case of shot blasting, instead of compressed air, centrifugal force is extended by means of high speed centrifugal impeller (Fig. 10.1) to impinge abrasive shots on the castings. The abrasives used in this case may be steel shots or chilled iron grits. The latter is cheaper but it is subject to loss by fragmentation.

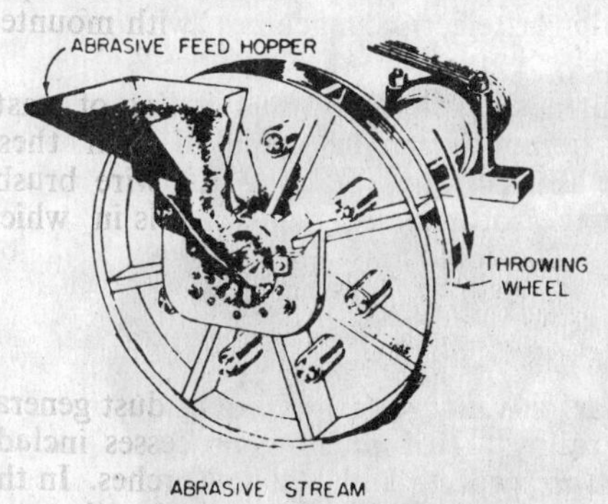

Fig. 10.1 High speed centrifugal impellor for shot blasting

A number of designs of shot blast equipment are employed Heavy castings are usually treated in a fixed chamber and the nozzle is manipulated either by operator or remotely controlled from outside. Very small castings are blast cleaned in the glove box cabinet units. Castings of a wide range of intermediate sizes are processed automatically in two ways, i.e.,

(i) castings are supported on a turnable table or suspended from a pendulum conveyor with in a fixed chamber and the shot stream is traversed to obtain complete coverage of the exposed surfaces or

(ii) the eastings can be tumbled in the shot stream through continuous rotation of chamber itself about the horlzontal axis. Automatic blast plants give high production rates but involve high operative costs due to high abrasive wear of plants. Provision is made for dust extraction and for cleaning and recirculation of the abrasive.

Similar to the blast cleaning operation is shot peening in which angular grits are avoided in favour of round shots or prerounded cut wires which ensure smooth indentations rather than pitting of the surface. Shot peening is employed to improve fatigue resistance by cold working the surface layers of castings.

10.4.2 Tumbling

In this case castings are tumbled in a large steel shell or barrel (1 – 12 m^3 capacities) which is closed at its ends by steel or cast iron lids The barrel is supported on horizontal trunions and rotated at a speed of 25—50 rpm (Fig. 10.2). Small pices of waste cast iron (known as stars) may also be charged along with castings to help the cleaning and polishing operations. During tumbling, continuous pressing action takes place which not only makes the castings cleaned and polished but also eliminates sharp edges and fins. The internal stresses in the castings are also relieved. Care is taken to ensure the proper packing of castings in the barrel, since very tight packing will prevent the relative motion of the castings, whereas the loose packing will cause breakage. This process is suitable for small sized castings which are not fragile in nature. This can not be used for heavy castings and small castings of fragile nature.

Fig. 10.2 Tumbling barrel

10.4.3 Hydraulic Cleaning

In this system the castings are exposed to high pressure water jets (1.6 MN/m^2 pressure) which may also incorporate sand as an abrasive. This can also be adopted in tumbling barrel. The barrel is arranged at an angle of about 30° and the castings are tumbled with a high velocity (about 100 m/sec.) stream of water and sand. This method offers several advantages such as (i) total elimination of dust, (ii) separated sand can be classified conveniently for reuse, (iii) more efficient cleaning and polishing, and (iv) reduced cleaning time.

Arrangements for applying the jets are analogous to those used in blasting, i.e., the nozzle is handled directly by the operator or remote controlled from outside the work chamber. Hydraulic cleaning is more suitable for non-ferrous castings, since ferrous castings tend to get corroded in the presence of water.

10.4.4 Chemical Cleaning

There has been very little application of chemical cleaning compared to blast cleaning. Chemical cleaning of ferrous castings can be carried out in following two ways.

(i) A bath of molten caustic soda is used in which additional reagents are introduced to react with and breakdown the surface oxide layer and thus loosening adhering sand.

(ii) Electrolytic method using caustic soda bath in which oxide is reduced by nascent sodium liberated at the casting used as cathode in the cell.

Non-ferrous castings can also be cleaned by scratch brushing for which wire brushes are used.

11

Casting Defects, Their Causes and Remedies

The melting, casting and solidification of metals involve many intricate operations and processes. Consequently, perfect control is impossible and a variety of casting defects may arise if some step in the manufacturing process is improperly controlled. It has to be appreciated that casting defects are usually not an accident. These can be minimised and controlled by having a sound grasp regarding their origin and fundamental causes in the context of quality specifications imposed by the design and service requirements. Casting defects lead to rejections thereby considerably reducing the total output of castings and also increase the cost of production. These may be of the following types.

1. Major defects resulting in complete rejection of the castings and total loss because the defects can not be rectified.
2. Defects that may be rectified but the cost of salvaging attempt may be uneconomical.
3. Minor defects which are remedied economically without adversely affecting the properties and performance of the casting.

The relation between the assumptions made by the designer and the castings actually produced are largely dependent on the design of the casting, materials used, foundry practices followed and the nature of the alloys—their melting characteristics and solidification behaviour. It is therefore, imperative to maintain close control over the raw materials used, moulding and core making practices followed, adoption of proper melting techniques and procedures, judicious metallurgical control at different stages to minimise the occurrance of casting defects.

A defect may result from a single clearly defined cause or may result from a combination of factors which are somewhat obscure. All foundrymen are familiar with persistant defects which defy explaination and finally disappear without clarification of its original cause by exercising close control and standardisation on all aspects of production techniques.

Although the logical classification of casting defects is difficult because of a wide range of contributing causes, a rough classification may be made by grouping the defects under certain broad types of origins as follows:

1. Faults arising in pouring
2. Inclusions and sand defects
3. Gas porosity
4. Shrinkage defects
5. Contraction defects occurring mainly after solidificatlon
6. Dimensional errors
7. Segregation and compositional errors
8. Other defects.

11.1 FAULTS ARISING IN POURING

When the liquid metal does not fill the mould cavity satisfactorily, misruns or short runs castings are obtained in which prematured solidification of metal occurs and some part of the casting isimproperly filled. A less severe manifestation called cold lap arises owing to imperfect fusion of two streans of metal in the mould cavity. When the discontinuity extends completely through a casting member in which streams of metal have conveyed from different directions, the more serious defect called cold shut results. The reasons for the above defects may be poor fluidity of metal caused by low pouring temperature or improper alloy composition etc, slow and intermittent pouring, improper gating system, too thin sections and small wall thickness, excessive chills from the mould face arising from heavy chilling or from too high a moisture content in green sand or as a result of back pressure of metal in a badly vented mould cavity. Thus several aspects of the casting methods can contribute to the above defects which result from an inadequate rate of mould filling relative to the freezing rate of the casting. Alloys showing poor fluidity especially those which carry strong oxide film are particularly susceptible to defects arising from improper flow.

11.2 INCLUSIONS AND SAND DEFECTS

Non metallic inclusions in the castings may be considered in two main groups i.e.

(i) Indigeneous inclusions and
(ii) Exogeneous inclusions

(i) *Indigeneous inclusions or Endogenous inclusions.* Indigenous or endognous inclusions which are normally dispersed throughout the casting and to a large extent are characteristic of the alloy and melting practice. These are formed by reactions involving common impurities such as oxygen, nitrogen and sulphur together with more reactive metallic constituents of the alloy. Oxidation reaction is deliberately promoted for refining in steel making and other metallurgical processes and final oxygen is controlled by the addition of deoxidisers before pouring. The products of deoxidation are eliminated by gravity separation or fluxing and slagging which results in coelesence into large globules for rapid separation. In case of light alloys, owing to relatively small differences in the densities of the metal and the inclusions, gravity separation is not very effective. In such cases it is important to disturb the metal surface as little as possible during melting and also treatment with suitable fluxes which absorb and dissolve suspended nonmetallics. Filtration through a porous bed of granular material before casting also gives very large inclusions contents in light alloys. Melting in vacuum or inert atmosphere also reduces the inclusion content.

Indegenous inclusions represent the general condition of cast metal rather than of a particular casting and these can be minimised by maintaining soluble impurity content at low level by suitable selection of charge melting and refining technique. To some extent, these are inevitable because of shifts in melt equilibria with fall in temperature and precipitation and segregation of impurity elements can occure during freezing.

(ii) *Exogeneous Inclusons.* These result from entrainment of non-metallics during melting and pouring, e.g. dross, slag and flux residues, refractory fragments from furnace and ladle lining and inclusions originating in the mould itself consisting of moulding material dislodged during closing or pouring. Such inclusions can be regarded in specific defects and can be controlled and minimised by paying proper attention during melting, pouring, and mould making. Slag, dross and refractory fragments can best be prevented by retention in the furnace and by careful skimming at the pouring stage along with the use of well maintained tea pot or bottom pouring laddle.

Avoiding defects arising from moulding materials involves careful selection, proper ramming during mould making, maintenance of mould properties and characteristics as per the alloy and its overall casting characteristics, proper design of the gating system to prevent impingement of metal against mould and core surfaces and incorporate smooth curves rather than sharp corners and abrupt changes in direction. Exogenous inclusions tend to be concentrated in certain regions of the casting such as at the upper surfaces and adjacent to the ingates and are revealed on machining or after scaling in heat treatment.

The formation of erosion scab is associated with major inclusions and appears as a rough projection formed on the casting where sand has been washed from the mould wall. Even though the scab itself may be removed by fettling, its presence suggests the possibility of wide spread inclusions within the casting. Closing a mould over an ill fitting core or badly finished joint surface may cause shearing of a section from the edge of the mould cavity, thereby producing on the large inclusion and a corresponding projection on the casting called *mould crush.* To avoid this defect the pattern equipment has to be carefully maintained to eliminate distortion and the need for manual patching. Partial or complete spalling of a section of a mould face and penetration of molten metal behind the surface layer of sand results in *expansion scabs* seen as an irregular metal scab which can be removed to expose the underlyings and entrained. In case of insufficient expansion to bring about metal penetration, surface fissure or line defect know as *rat tail* occurs owing to bulging and shear cracking of the mould face.

Expansion scabs and *rat tail* are associated with intense local heating produced by the flow of molten metal over the surface during pouring and are influenced to a considerable extent by the constituents, grading, moisture content, ramming density of the sand and the pouring condition in the casting. Expansion defects are associated mainly with silicious moulding material which shows high expansion on heating to relatively low temperatures owing to the allotropic modifications of quartz at 575 °C, leading to sharp change in volume which produces a tendency to cracking and spalling. In case of persistent scabbing, a change to nonsilicious moulding materials is considered desirable because of their lower rates of expansion. Addition of wood floor, coal dust and other organic materials which provide a cushion to the sand grains by softening or decomposing is also adopted. Cereal binders such as stareh and dextrin also produce a similar effect and frequently 0.5 pct. addition is preferred to clay bonded sand for reducing the scabbing tendency.

(iii) *Sand Adherence and Surface Rooughness.* These defects get aggrevated in the production of castings involving higher temperature causing loss of appearance and increased dressing cost. In the worst case the mould material may become impregnated with metal to form a solid mass in a confined pocket which may be difficult to remove and the casting itself may have to be scrapped. It has been established that sand adhesion is primarily caused by molten metal penetration in voids in the moulding material and both metal and mould variables are involved.

The factors which influence surface roughness and sand adherence include the pressure head of liquid metal, the design, orientation and feeding techniques associated with a particular casting, the degree of superheat, and the initial porosity and finish of moulding material. The penetration by molten metal occures at a critical pressure directly proportional to the surface tension of the metal and is also affected by alloy composition. Other factors such as formation of liquid oxide and slag which in certain cases cause wetting of the sand grains thereby reducing the pressure needed for metal penetration. Carbonaceous dressings containing volatile matter improve casting surfaces by generating a back pressure which opposes penetration.

11.3 GAS DEFECTS

These take the form of internal blow holes, surface pin holes, air locks, intergranular cavities, embrittlement and cracking caused by retention of gases in solution in solid state and the occurrence of gas-metal compound as solid inclusions already discussed earlier (under indigenous inclusions). The distribution of shrinkage cavities and segregation in casting is also affected by the gases. The gas defects may be broadly considered under the following two heads:

(i) Physical entrapment of gases on pouring.
(ii) Gases resulting from precipitation on cooling.

11.3.1 Entrapment Defects

These defects arise (a) When air is trapped within the casting owing to excessive turbulance or aspiration during pouring combined with molten metal with inadequate superheat or the failure of the mould to exhaust air displaced by the liquid metal being poured at a faster rate. The permeabiliy of the mould and the venting system being unable to exhaust the gases leads to a cushioning effect of the entrapped air which delays further flow of the molten metal into the recess, thereby producing an air locked defect in upward facing projections. Confined air can also produce internal cavity by expanding sufficiently to blow back through metal and in extreme cases metal may be ejected from the mould. The importance of effective venting of the mould cavity and permeability to sustain high rate of gas flow can not be over emphasized to minimise the defects arising from retention of mould gases.

Excessive moisture in green sands, volatile organic compounds in moulding mixtures, underbaked or damped cores can cause gas cavity specially when associated with low permeability and insufficient venting. Improperly dried mould coatings, porous chills, oxidiced metal shots in reclaimed sand, rust on chills and chaplets can cause surface blow holes and pin holes. These can be effectively reduced by precautions designed to reduce gas forming substances in the mould and cores and insuring adequate venting in the mould.

11.3.2 Precipitation Defects – Gases Evolved by the Metal

These defects are caused by precipitation of gases which may have been dissolved during melting or may arise because of interaction between liquid metal and mould surface or reactions involving elements already in solution.

The evolved gases may include hydrogen which shows appreciable solibility in most casting alloys. Aluminium alloys are pyrticularly susceptible to gas porosity from hydrogen. A solubility of 0.6 cm^3 per 100 gm just above the melting point falls by a factor of about 20 to approximately 0.03 cm^3 per 100 gm. during freezing. This represents about 1.54 pct, by volume of the metal by gas evolution from saturated solution. The behaviour of nitrogen is similar in iron and steel although practieally it is insoluble in most non-ferrous casting alloys. In a way it can be used as a scavenging gas. Oxygen is not usually precipitated directly as it tends to form stable oxide with many metals. However, it can lead to reaction porosity owing to the evolution of mainly carbon monoxide and steam formed in combination with other solutes, particularly when appreciable residual oxygen exists in the metal.

The compound gases may be formed by reactions between elements such as carbon, hydrogen, and oxygen which may be present at the outset or may get absorbed by reaction between the molten metal and its surroundings. Such reactions include those between carbon and oxygen in iron and nickel alloys, between sulphur and oxygen in copper and between hydrogen and oxygen where either or both elements have been absorbed by the molten metal. These reactions occur by a change in temperature or by concentration due to differential freezing or surface absorption, thereby disturbing the relationship between solute elements. The main precautions against compound gas porosity is to minimise the final oxygen content of the melt. The procedure usually involves strong oxidising conditions to eliminate hydrogen, carbon and other impurities during melting and controlling residual oxygen at the final stage by the addition of appropriate deoxidiser before pouring to form stable oxides. The presence of excess residual deoxidant in the molten metal prevents surface pin holing caused by reaction between carbon and local oxygen concentration produced by surface oxidation.

11.3.3 Gas Absorption in Melting

Metallic constituents of the furnace charge and furnace refractory, fluxes, slag making additions and furnace atmospheres may lead undesirable gas absorption. Finely divided material such as swarf and turnings contaminated with hygroscopic corrosion products and lubricants can result in high hydrogen content in the molten metal. Similarly metals from previous melting cycles or from electrolytic sources are liable to contain high hydrogen resulting from the reducing conditions employed during smelting. Refractories such as furnace spouts, tundishes, ladles, fluxes and slag making additions if not properly preheated before contact with liquid metal may lead to gas absorption through dissociation of steam at the contact surface. Humidity and gases present in the atmosphere may also cause gas absorption which may be aggrevated by conditions peculiar to the melting furnace e.g. by dissociation of gas molecules in electric are melting.

11.3.4 Metal Mould Reaction

Reactions between the metal and mould result in the formation of gases leading to porosity on the surface and such reactions ore prevalent in light alloys containing magnesium and in phosphor bronze. Such reactions are caused by direct contact between an alloying constituent with high oxygen affinity and free or combined water in the mould. Aluminium-magnesium alloys are susceptible to such unsoundness and pinholing may also occur in magnesium treated spheroidal graphite cast iron which contains only minor amounts of magnesium. The metal mould reaction can be supressed by forming a continuous surface film by introducing an alloying element with a still higher oxygen affinity than the offending element. An addition of 0.004 pct. beryllium can eliminate metal mould reaction in light sections of Al – 10% Mg alloy casting and addition of small quantities of aluminium to phosphor bronzes is effective in eliminating metal mould reaction. The addition of inhibitors to the moulding sand is commonly used to prevent metal mould reaction in light alloys by replacing the oxides with another compound film. Boric acid and ammonium bifluoride are used as inhibitors and may be appalied as paint coatings. The strong metal mould reaction tendency in magnesium based castings in curbed by using sands containing approximately 1 pct. sulphur along with boric acid.

Reaction pin holing in the case of magnesium treated iron castings can be prevented by the elimination of free moisture with the aid of clay free moulding dressings and by the use of coal dust to provide a reducing atmosphere. Metal mould reaction may cause pin holing in steel castings by the reaction of iron with free moisture in green sand to produce hydrogen which diffuses into metal. This combines with dissolved oxygen producing water vapour which leads to pin holing. Effective deoxidation with aluminium can prevent pinholing from this and other causes.

11.3.5 Influence of Gas in Solid State

Apart from the cavities formed during freezing, gas remaining in solid solution or as stable gas—metel compound can result in lowered meehanical properties or even embrittlement causing cracking of the alloy owing to the reduction in solid solubility. On further cooling, the dissolved gas diffuses in existing voids and imperfections. This may generate extremely high internal pressure with consequent increased susceptibility to fracture at much lower level of applied stress. Such stress to induce cracking may arise from the structural change or resistance contraction or in service itself. The embrittlement caused by residual hydrogen in steel is a typical case of damage caused in solid metal by dissolved gases. Conventionally such a susceptibility was eliminated by prolonged high diffusion treatment in the solid state. However, it is now possible to produce gas frèe metal by precautions and vacum treatment during melting and casting.

11.3.6 Preventive Measures for Gas Defects

These may be boradly discussed as follows:

(i) Low gas content in the molten metal when poured can be achieved by carefully designed melting techniques and precautions such as

(a) Using selected grade raw materials from known sources with low gas content. Contaminated materials are processed separately for the preliminary removal of gas before they are used with the main charge.

(b) Preheating the charge material to evaporate surface moisture and volatilise oils, paints or other organic contaminants as well as water of crystallisation from hydrated corrosion product.

(c) Treatment of turnings and borings from machine shop in special degreasing plant incorporating a centrifuge followed by treatment with organic solvents. Such a treatment is justified in case of expensive alloys.

(d) Employing fast melting techniques and the use of compact charges to reduce the time of gas metal contact in melting. Such a practice also minimises melting losses.

(e) In copper and nickel alloys as well as steels, maintenance of oxidising condition is beneficial for the removal of dissolved hydrogen. However, this is not suitable for removing hydrogen from melts containing appreciable amount of aluminium, magnesium or similarly reactive elements. Protective fluxes are employed to minimise gas metal contact and undue absorption of hydrogen.

(f) Final alloy additions should be completely dry and gas free and the temperature of the molten metal should be maintained at the lowest level to ensure adequate fluidity during pouring and casting. The holding time should be reduced to minimum and furnace spouts, ladles, and furnace tools should be preheated to eliminate moisture and the danger of gas absorption at the final stage.

Vacuum melting can be considered for the production of gas free metal by excluding atmospheric contamination, extracting dissolved gas to the influence of reduced pressure, eliminating elements which precipitate compound gases during freezing and preferentially evaporoting elements with high vapour pressure. Inspite of the clear technical advantages in vacuum melting, it has found limited application in the production of castings because of the constraints of higher cost and scale of operation. Its use is confined for the production of high duty precision castings in special alloys where quality requirements justify a substantial premium on the selling price.

11.3.7 Degassing of Molten Metal

By careful material selection and observing proper precautions during melting, the absorplion of gas can be reduced, but it is difficult to avoid the presence of gases in melt under routine foundry conditions. Making use of the equilibrium existing between melt and gas atmosphere, following types of degassing treatment before pouring are practised:

(a) *Gas Scavenging*

The flushing gas is bubbled through the melt, thereby providing a large gas metal interface and general agitation. In the case of aluminium alloys, argon, nitrogen or chlorine can be employed to reduce hydrogen absortion. Calcium carbonate which generates CO_2 is used for gas scavenging in case of copper base alloys. During steel making, the oxygen injected for the removal of carbon and other elements also reduces the hydrogen and nitrogen contents of the melt. The CO boil in steel making is also valued for its effect in eliminating dissolved gases.

(b) *Oxidation, deoxidatiod and precipitation of compounds.*

During the melting of iron, copper and nickel alloys, the liquid metal is exposed to oxidizing conditions which increase the concentration of dissolved oxygen in the melt. This is beneficial for the removal of hydrogen by direct chemical reaction to form H_2O molecules. Deoxidation before pouring is accompalished by the addition of suitable elements which form stable oxides. Stabilisation of the impurity by precipitation as a compound can be employed for reducing nitrogen particularly in nickel base alloys. Formation of stable nitrides of lithum, titanium and zirconium effectively suppresses nitrogen pinholing in such alloys.

(c) *Vacuum degassing*

General processes of vacuum degassing have been developed all of which depend on decreased gas solubility in molten metal by reduction of the external pressure. This is in accordance with Sievert's law which can be represented as following

$$s = K\sqrt{(p_1 - p_2)}$$

where s = Amount of gas dissolved in an alloy $cm^3/100$ gm.

K = solubility constant dependent on temperature

p_1 = partial pressure of gas in the surrounding atmosphere.

and p_2 = partial pressure of gas in the melt.

The rate of degassing depends partly on the geometry of the system, i.e. the surface area and the mass of the metal and partly upon the extent of the agitaation and stirring which help in the transport of the dissolved gas to the nearest surface. A number of processes have been developed for use on an industrial scale which include.

(a) static bath treatment (b) induction degassing (c) fractional degassing and (d) stream droplet degassing as illustrated in Fig. 11.1.

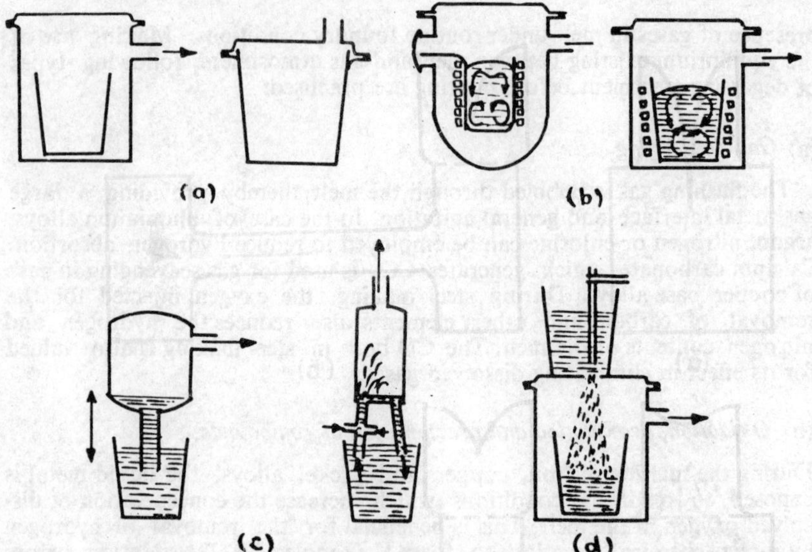

Fig. 11.1 Processet used in degassing of molten metal: (a) static bath treatment (b) induction degassing, (c) fractional degassing, and (d) stream droplet degassing.

The static bath treatment involves enclosure of the molten metal ladle in a chamber which is sealed and evacuated by induction stirring which helps in bubble formation. Fractional degassing involves treatment of fraction of the molten metal which is raised through a section nozzle into a vacuum chamber by alternate raising and lowering of the chamber. In stream droplet degassing, a ladle is tapped through a sealed annulus into a second ladle kept in a previously evacuated chamber. Exposure of the falling stream to vacuum results in gas evolution owing to the formation of small droplets.

In special cases however, advantage is designed from limited gas evolution on freeing, e.g., in some copper alloys with long freezing range, gas evolution obtained from controlled mould reaction can produce a wide dispersion of micrososity rather than a more localised void concentration, this being beneficial for pressure tightness. In case of die casting of aluminium alloys, controlled gas evolution is employed to cause slight expansion of the casting. This helps to offset the linear contraction which may lead to hot tearing under restraint.

11.4 SHRINKAGE DEFECTS

These defects arise mainly because of inadequate gating and risering technique where liquid and solidification contratiom is not adequately compensated. These may manifest in the form of: (a) primary pipe (b) secondary cavities (c) discrete porosity (d) sinks and (e) puncture.

The actual form of defect forming depends on design factor, cooling conditions and the mechanism of freezing of the alloy. Various forms of shrinkage defects are shown in Fig. 11.2.

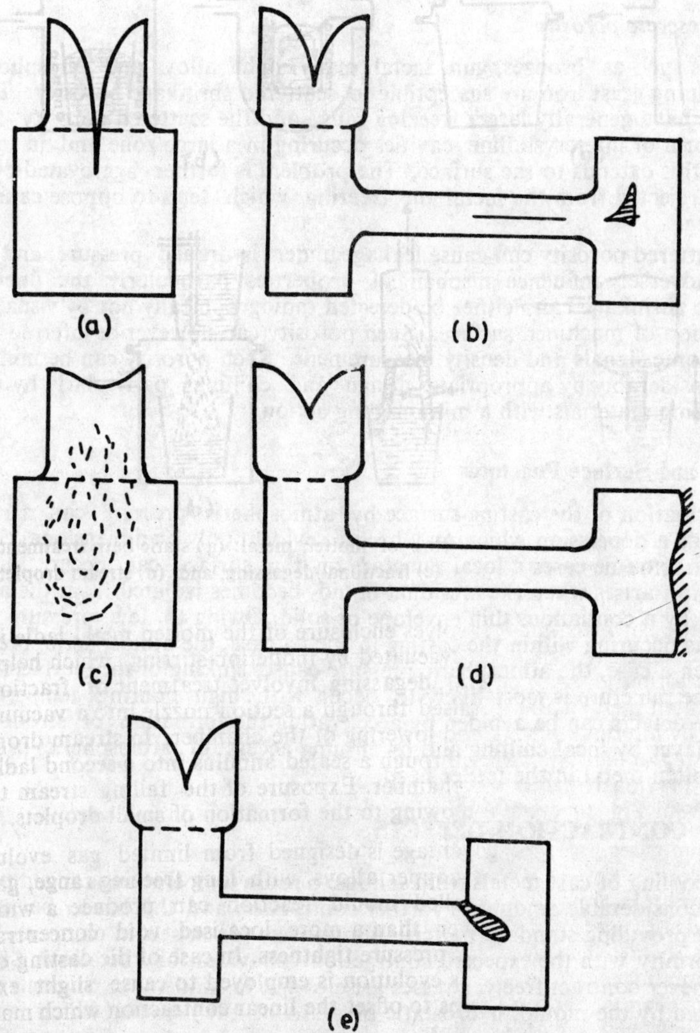

Fig. 11.2 Various forms of shrinkage defects: (a) shrinkage cavily, (b) discrete porosity and (c) contraction defects.

(a) *Shrinkage cavities*

Inadequate feeding of the casting can lead to the formation of primary shrinkage cavity or pipe depending upon various factors mentioned above which restricts directional solidification throughout freezing. However, secondary shrinkage is only internal and occurs in locations remote from the feeder head such as central zones of parallel wall sections, in the absence of provision for direct feed or selective chilling.

(b) *Descrete porosity*

Alloys such as bronzes, gun metal, many light alloys and phosphorous containing cast iron are susceptible to scattered shrinkage porosity. These alloys have generally larger freezing range and the scattered porosity takes the form of intercrystalline cavities occuring in a large zone and in many cases this extends to the surface. The problem is further aggrevated by the gases rejected from the metal on freezing which tend to oppose capillary feeding.

Scattered porosity can cause leakage under hydraulic pressure and can also adversely influence mechanical properties particularly the ductility. Micro shrinkage can neither be detected radiographically nor by visual examination of machined surfaces. Such porosity can however be inferred from ultrasonic signals and density measurement. Such porosity can be minimised considerably by appropriate design and chillings particularly by using moulding materials with a mild chilling action.

Sinks and Surface Punctures

Deformation of the casting surface by atmospheric pressure can manifest itself as a depression which may become evident only on dimensional checking. In extreme cases a local puncture on the surface can occur. Such a situation arises when the residual liquid becomes isolated from the atmos phere by a continuous thin envelope of solid. Owing to low pressure conditions occuring within the casting in such a case, the atmospheric pressure in such a case, the atmospheric pressure causes sinking of the surface. The surface puncture is most likely to occur with high pouring temperature. These defects can be avoided by accelerating the formation of strong surface layer by local chilling and by ensuing access of atmospheric pressure to molten metal in the feeder head.

11.5 CONTRACTION DEFECTS

The cooling of cast metals from solidus to room temperature is accompanied by a considerable amount of further contraction and this has to be adjusted by providing standard pattern allowances (discussed in chapter—2) in conformity with the expected contraction behaviour of the alloy. As castings never contract freely, stresses arise owing to hinderance to contraction offerred by the mould, hydrostatic pressure of the residual liquid and by other parts of the casting itself due to differential cooling. The contraction defects may manifest hot tears.

(i) *Hot Tears*: This is the outcome of contraction resistance at an early stage during cooling when the metal is unable to withstand elastic strain due to the pressure of liquid films. Hot tears can cause partial or complete fracture and are often located at changes in sections where stress concentration occurs. Alloys with very short freezing range are comparatively free from tearing tendency whereas alloys with a comparatively large temperature range of freezing are more volunerable. Susceptibility to hot tearing is also influenced by segregation of alloying elements and impurities. Microsegregation or coring aggrevates the critical period of weakness of hot tearing because of prolonged retention of intergranular liquid film with consequent increase in the accumlated strain, e.g. the serious lowering resistance to hot tearing caused by high sulphur and phosphorous contents in steel.

Design and production factors have significant inflnence on hot tearing owing to effects upon temperature distribution which governs the distribution of mechanical properties and resistance to concentration. To minimise susceptibility to hot tears, a number of specific precautions have to be observed in the casting design namely section uniformity, gradual changes in sections and providing adequate radii at all corners to reduce stress concentration. Webs and brackets are provided as reinforcements across critical regions to accelerate cooling through the critical temperature range and strength volunerable zones. Such reinforcements may be removed in fettling or may be retained as permanent feature of the design. For sharing the stress concentration in members volumerable to hot tearing corrugated surfaces may be provided.

(ii) *Cracking, distortion and residual stress*: After cooling below the tearing temperature, resistance to cooling of the casting for most alloys and rates of cooling can be accommodated by continuous elastic deformation. During this stage fracture can occur only if the cooling rate is so rapid as to cause the ultimate stress of the alloy to be exceeded. Usually during such a phase of cooling, dimensional errors can occur through failure so contract to rule and such minor errors on critical dimensions are usually corrected satisfactorily by appropriate machining allowances.

The casting may retain a high level of residual stress specially where sections are of widely varying thickness. These residual stresses are propational to Young's Modulus, cofficient of expansion and the temperature difference between the members. In case of poor ductility of the alloy, the residual stresses can cause fracture in extreme cases during the late stage of cooling, even if a fracture does not occur. The residual stresses make the casting volunerable with respect to both strength and dimensional stability and a small super imposed stress such as rough handling or local heating during head removal, grinding or welding can lead to cracking/fracturing. Castings with residual stress are dimensionally unstable and many change either spontaneously with time or if heating occurs in service. Residual stresses can be removed by heat treatment and the temperature required for the rapid stress relief generally lies in the region 0.3-0.4 T_m where T_m is the melting temperature of the alloy expressed in K. It is helpful sometimes to withdraw castings from the mould while still in the plastic region and charge it immediately into a hot furnace where cooling can be controlled and equilised throughout the critical period. Stress relief of castings has been attempted by ageing or weathering for long periods before use. While ageing at an ambient temperature, brings about some dimensional changes, its effectiveness is uncertain and probably accounts for the relief of only a small proportion of residual stress.

11.6 DIMENSIONAL ERRORS

Pattern making, moulding and casting or fettling, if not properly carried out, gives rise to dimensional faults. Major defects are due to misalignment of mould parts and cores such as cross jointing and displaced cores, wornout pins and pniholes which produce lateralship along the joint line. Dimen sional errors can be considerably reduced by a high standard of box main tenance and jig checking at regular intervals for pin centre dimensions and for pin and pin hole clearances by proper attention to core print design and clearances, defects from misplaced or ill fitting cores can be greatly minimised.

Distortion of mould involving enlargement of mould cavity during pattern withdrawal, manual patching, low green strength and soft ramming, lack of rigidity in the assembled mould, inadequately reinforced cores or badly rammed mould parts, core shifts, swelling, and growth across the joint line permit movement owing to pressure and buoyancy force on casting. High pressure moulding and hardening of moulds and core in contact with the pattern has reduced the incidence of such distortion considerably.

11.7 COMPOSITIONAL ERRORS AND SEGREGATION

These defects can arise: (a) from loss of reactive elements during melting and (b) Compositional differences arising during solidification and persisting in cast structure.

Alloys with strong segregation tendency usually have long freezing range, gentle liquidus slope, and low solid solubility. Such segregation may be present as intergranular or dendritic segregation (microsegregation) following the form of the grain or substructure. These can affect tensile strength, ductility, impact and fatigue resistance and other mechanical properties which are affected by intercrystalline conditions which differ from the matrix.

Segregation on a macroscopic scale is produced by various mechanisms during freezing of the alloy. Normal type of segregation may result from progressive concentration of the impurities towards the centre of the casting. Differences in density of the precipitated phases as compared with the parent liquid can cause gravity segregation where as nonmetallic inclusions generally rise to form local concentration. Susceptibility to gravity segregation is markedly enhanced in heavy sections because solid phases can be suspended in the liquid for prolonged time. Another form of macrosegregation called *inverse segregation* occurs by flow of solute rich liquid through interdendritic channels from deeper regions within the casting. Thus a high concentration of a solute appears in the outer zone in place of central region of the casting. Coarse columnar dendritic structure is particularly susceptible to inverse segregation because of the relativety straight capillary feed passages. Local differences in composition get aggrevated since the outside zone also contains the purest solid in the first crystals to freez.

Homogenisation treatment consisting of reheating the casting and keeping it at high temperature for a prolonged period helps to remove concentration gradient by diffusion. The spacing of the composional variation and the highest possible temperature of homogenisation consistant with other metallurgical requirements help to increase the diffusion considerably. However, because of the large distances involved, practically there is no possibility of eliminating macrosegregation by heat treatment.

11.8 SIGNFICANCE OF DEFECTS

The significance of the casting defects hasto be evaluated in the context of the function of the casting and its appearance, the former being of overiding importance. The characteristics of importance are:

(i) Mechanical properties
(ii) Hydraulic soundness and
(iii) Surface finish.

(i) *Mechanical properties*

For a proper appreciation of the influence of defects on mechanical properties, the size, shape and position of the defect have to be assessed in the context of the stress pattern in the casting. The following generalised discussion will highlight the relevance and importance of the different factors involved.

(a) Freealom from surface defects and proper surface finish is particularly important under fatigue condition because generally the skin of the casting contains regions of highest stress.

(b) A flaw on an extensive flat surface is generally regarded less harmfull than one occurring at change of section where an appreciable applied stress concentration is anticipated. Defects like centre line porosity or segregation located close to a neutral axis of stress are considered less severe. However, their detection and radical repair is more difficult because of inaccessible location.

(c) Any defect introduces a stress concentration, however, such a risk increases manifold with irregularity of the defect, e.g., a fine crack as com pared to a spherical defect such as a blow hole or slag inclusions.

(d) The presence of a large number of small defects can influence the mechanical properties depending upon metallurgy of individual alloy and the location and the distribution of the defect. Decision regarding using the casting or discarding it can be taken in the contex of specific service conditions.

(ii) *Hydraulic soundness*

Castings like hydraulic cylinder, valves and pipe works are tested under pressure to demonstrate their capacity for the retention of fluids under service conditions. Presence of very fine porosity can lead to seepage under high pressure and many copper and nickel base alloys are prone to inter granular porosity and resulting leakage because of their long freezing ranges. Proper control of feeding and gas content can largely eliminate such defects.

(iii) *Surface Finish*

This is important to ensure proper appearance as also for contings which operate under corrosive conditions where surface pits or inclusions can form nuclei for corrosive attack. Surface imperfections become more marked after honing, polishing, enamling, plating and anodising.

11.9 SALVAGE AND RECTIFICATION

Although salvaging of castings can be undertaken by restoring the pro perties and service performance to a standard equivalent tc that: (a) no defects were present or (b) to improve the appearance in such cases where the defects do not impair the performance. However, such a consideration should not bring about an indiscriminate relaxation of standard in the different stages of production. In general, rectification processes in crease cost, consume time, and undermine the confidence of the user. In mass production of castings such as light castings, a few defective castings are scrapped and replaced because of the high cost and inconvenience of salvage operation. However, in case of heavy castings or when production of few castings of a particular type is involved, it becomes more difficult to eliminate defects completely and the cost of rejecting the casting may involve high financial loss. In such cases salvaging the castings and making them serviceable is justified.

Depending upon the objective in view regarding the salvaging of the casting i.e., restoration of properties and for improvement in appearance, different techniques are used which include welding, brazing and soldering, caulking and impregnation, patching and plugging, filling compounds etc.

Welding

Rectification of defects by welding if properly carried out and followed by appropriate post welded heat treatment can give properties fully equivalent to those of originally cast metal. By welding a true metallurgical union between filler metal and the parent casting is achieved giving continuity of crystal structure across the original interface. This is usually not possible with other methods of rectification including soldering and brazing where the union is only superficial.

Rectification by welding involves the observance of the following steps and precautions.

(i) For proper matching of the weld with the parent casting, for good corrosion resistance, appearance as well as mechanical properties, it is advisable to use a filler metal of approximately the same composition as that of the casting. A difference in composition of the weld can accelerate corrosion and minor difference in hardness or colour gets revealed on machined surfaces.

(ii) It is necessary to remove completely the defective material from the casting because any attempt to weld over porous metal can result in gas evolution and porous weld deposit. Cracks if not completely removed can get further aggrevated owing to thermal stresses during welding. It is therefore imperative to remove the defect completely by grinding or machining or by gouging with arc, gas flame, or chisel before executing the repair successfully irrespective of the processes to be used.

Metal Arc Welding

This process is widely used for casting repairs such as large size cracks, blow holes and other imperfection. Consumable electroder provide the source of filler metal and the arc welding technique is now applied to most types of alloys. Coated electrodes produce denser, stronger, tougher and more ductile weld. Welding rod coatings often contain slag forming constituents, deoxidising agents, and alloying elements in addition to arc stabiliser which promote better ionisation in the arc causing it to be more stable. This increases the melting rate as also the depth of penetration of the weld. Slag formers such as titanium oxide, calcium fluoride and silican dioxide, protect the metal deposit from the air and also act as heat insulater, thereby, reducing the rate of cooling, permitting the escape of gases and reducing the cooling stresses. Deoxidisers such as ferromaganese give denser and stronger weld by reacting with oxygen from the weld metal. Different alloying elements such as molybdenum and chromium present as ferro alloys impart added strength to the joints by mixing with the weld metal.

Other variations of arc welding used for repairing of castings include tungsten inert gas welding (TI.G. process) metal inert gas welding (M.I.G. process), carbon arc welding and atomic hydrogen welding. Carbon arc welding is particulaly suitable for use with all ordinary foundry alloys except those of magnesium and it is extensively used for welding copper base and aluminium alloys. In this process the eletric arc is struck between the carbon rod and the casting and a separate filler rod may be fed into the arc at the desired rate. TIG process uses a non consumable type of tungeston electrode together with an inert gas shield such as helium or argon to prevent oxidation. The process is widely used for welding of magnesium alloys, aluminium alloys as also stainless steels and alloys of eopper and nickel. In MIG process, a consumable type electrode such as electrode wire is automatically fed and protection from oxidation is provided by inert gases such as argon, or helium. MIG welding is economical for salvaging large castings where its high production rate can be fully utilized. Atomic hydrogen welding is suitable for repair welding of metal moulds and alloy steel dies as also welding of thin castings in stainless steels, aluminium alloys etc. The method consists of passing a continuous stream of hydrogen through an arc produced between two tungsten electrodes. The hydrogen gas dissociated to atomic form under the heat of the are subsequently reunites, there by releasing an enromous amount of heat necessary for melting and joining the base metal, and also providing a shield against oxygen and nitrogen from the atmosphere. Filler metal is fed separately from a wire and a very homogeneous smooth and strong joint is produced.

Gas Welding

For the repair welding of most commercial metals, a mixture of oxygen and acetylane is used. This method is popular and widely used because of the lesser cost and portability of the equipment and the possible flexibility in maintaining an oxidising, reducing or nentral flame. Untike arc welding, it is possible to use broad flame to preheat the area ahead the section being welded. By using the proper type of flame and technique almost all cast metals and alloys except magnesium can be gas welded. A neutral flame is preferred for most alloys and a filler alloys rod with suitable fluxes is fed into the hot zone. In view of the larger heat affected zone produced in the final structure as compared to more intense and localised electric are weld ing there is greater tendency for distortion in gas welding.

Thermit Welding

The process consists of igniting a mixture of iron oxide and aluminium powder with appropriate additives to produce high temperature required for melting the metal produced by the reduction of iron oxide with alumi nium owing to the exothermic reaction. A typieal arrangement for such a process to accompolish welding is shown in Fig. 11.3. Thermit welding is in a way more of a casting process and is employed for repairing large and heavy steel castings such as steel mill rolls and gears as also for joining relatively simple castings to produce excessively large and intricate parts.

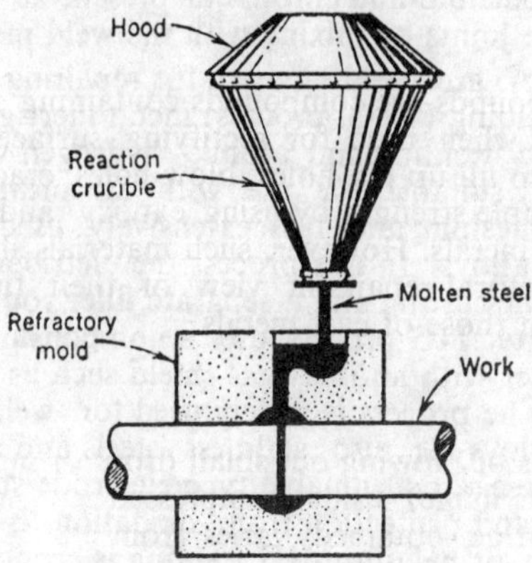

Fig. 11.3 Thermit welding.

Brazing and Soldering

Brazing is employed for such parts and components that tend to get distorted or cracked when welding at high temperature. In such cases the entire process is carried out below the melting point of the casting using a low melting alloy as filler. There is little penetration into the parent metal and the hardness and colour of the filler alloy are usually different. This method is used for making castings water tight and to repair pipes and pipe fittings and other thin plate tight castings, for filling crevices, porosity, fine cracks etc., Copper ond nickel base alloys and silver solders with melting temperatures above 430°C are employed as fillers for brazing. Use of brazing flux is commonly based on borax and barates and fluorides of the alkali metals is necessary to remove tenacious oxide film for satisfactory brazing. These fluxes being highly corrosive, are carefully removed before the casting is put into service.

Soldering is carried out by employing lower melting point alloys based on tin, lead, cadmium and zinc which melt at much lower temperature. The process is confined to filling of minor surface cavities and other imperfections, when high strength is not required, such as making porous area in copper base alloys pressure tight.

Patches and Plugs

In some cases, the defective zone of the casting is removed by machining or drilling and then inverting a plug for final sealing of the a perture by tapping and threading and finally finishing by pinning or tack welding. Apart from rectification of defective castings, this method is used for the final sealing of temporary core print apertures in the original casting.

Impregnation

To fill pin holes blow holes, cracks, etc. resin impregnation under pressure is undertaken while the casting is kept under vacuum. The compound is finally cured usually by thermal treatment to produce an inert and stable filling. This method is widely accepted to make pressure tight castings in ferrous and nonferrous metals, particularly for those alloys specially prove to microporosity.

Filling Gompound

Non-metallic compounds are compounds containing metallic powders and epoxy plastic fillers when used for rectifying surface lemishes to improve appearance and also fill up pin holes, blow holes, cracks, etc. It is possible to develop appreciable strength by using expoxy and acrylic resins which strongly adhere to metals. However, such materials should not be regarded as suitable for structural repairs in view of their fundamentally different characteristics from those of cast metals.

Metal Spraying

The process consists of blowing out small drops of molten metal (melted by an electric are orgas flame) using compressed air with the help of spray gun. The metal coating obtained varies from 0.3 to 0.8 mm. in thickness, each pass giving a coat of metal about 0.03 mm thick. The method is suitable for building up undersize castings and all types of metals can be sprayed. As the bond obtained is of mechanical type with negligible diffusion, the strength of the sprayed metal is inferior to that obtained by welding or brazing. This technique is also used to improve the corrosion resistancc of iron and steel castings by spraying anticorrosive metal layer.

Straightening

Rectification of deformed or warped castings can be carried out by straightening in a press by applying pressure. Since the procedure involves plastic deformation, this operation is possible only in case of metals possessing reasonable ductility. Generally a hydraulic press is used with appropriate press fixtures and packing to achieve the correct final dimensions.

12

Inspection and Quality Control

Inspection comprises all those operatlons which check the quality of the castings. The inspection of casting serves the following basic purpose:

1. Inspection ensures that the castings conform to design and other requirements and thus separates defective castings from the non-defective ones.

2. Inspection provides information required for quality control in the foundry and thus inspection locate defects and flaws at different stages of castings production and use of raw materials. Therefore, the purpose of inspection is also to prevent the occurrence of defects in castings to be produced in future.

Although production and technical staff are concerned in many aspects of quality control, inspection department has a special responsibility for the maintanance of standards laid in contracts and commercial relationship. Thus systematic collection and interpretation of quality data is generally channelled through an inspection department which is a vital element in an integrated quality control system extending throughout the manufacturing sequence. Close contact and mutual confidence between inspection and production staff are needed for the two way flow of information required for the maintenance of standards. A concept of organisation for inspection and quality control as suggested by Quilter is shown in Fig. 12.1. A modern quality control organisation is concerned with a great deal more than the routine testing in relation to specified limits. Much data is obtained by the use of highly sophisticated techniques and more detailed treatment is required in interpreting the results. Adverse trends should be detected before quality deterioration leads to a high proportion of rejects.

12.1 THE COST OF IN PROCESS INSPECTION

The cost of inprocess inspection should be justified by savings in subsequent manufacturing operations or in the quality of the product. Since inspection involves expenses, the purchaser is required to specify the kinds and itensity of inspection. The inspection of castings before shipment usually is the responsibility of the foundry. Unless other inspection methods are specified usually a visual inspection is given to the standards of that foundry. In case additional inspection proceures are require-d, detailed information on these

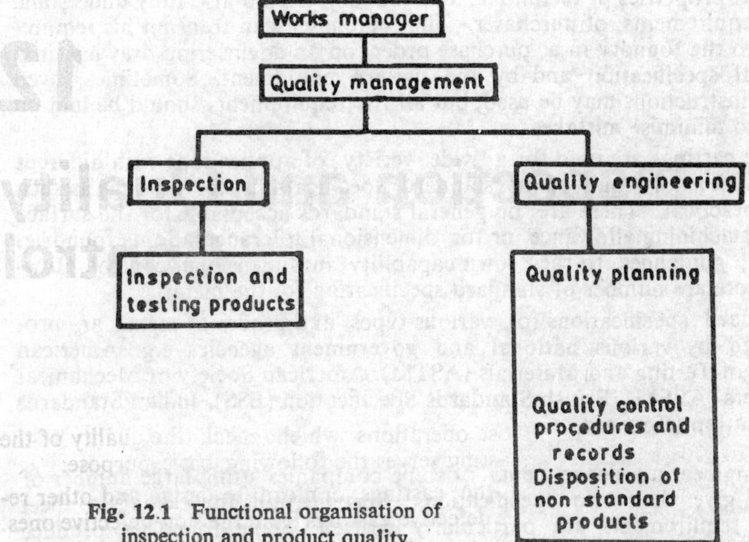

Fig. 12.1 Functional organisation of inspection and product quality.

should be conveyed to the foundry, since the foundry's inspectors do not know the use of casting and what is critical to its application. Unnecessary tight requirements also increase the cost of castings.

There is also a concern for duplicat'cı of inspection, i.e. inspection by the foundry as well as inspection by ..ıc purchaser. The inspection at the foundry provides the opportunity to correct process variables in the production schedule. However, inspection by the purchaser is mainly related to the serviceability of castings. Due to the different advantages of two inspections, both shipping (foundry) and receiving (purchaser) inspections are carried out.

12.2 SAMPLING PLANS FOR INSPECTION

The castings produced in small or moderate quantities are usually all inspected by the foundry before despatching, whereas for castings made in large quantities, a sampling plan of inspecting a portion of the castings may be preferable, Sampling plans employed should be mathematically designed so that the size of the sample and the frequency of sampling provide the degree of dependability required for each application. The selection of the sample castings should be entirely random as the selection of convenient samples and previewing will make the plan unreliable. Good supervision and auditing is required for successful sampling inspection. The sampling plans are based on the mathematics of probability and in any plan there is a risk of rejecting lots of good castings or of accepting castings of below standard. However, this risk may also occur with the routine inspection of all the castings.

12.3 SPECIFICATIONS

In addition to the quantity of castings and delivry date, the purchaser is required to supply the necessary informations regarding the characteristics of castings such as on surface finish requirements, dimensional tolerances required properties of metal, etc. The foundry should also fully understand these requirements of purchaser. The purchaser can transmit his requirements to the foundry in a purchase order on an engineering drawing, in a material specification and by the pattern equipment. Sometimes, even verbal instructions may be used, but all the requirements should be laid on paper to minimise mistakes.

Since castings are used for a wide veriety of applications with different combinations of requirements, standard specifications for castings as such are impractical. There are no general standards acceptable for the surface finish, machining allowance, or for dimensional tolerances. Some foundries establish guidelines to their own capability in dimensional control. However, there are number of standard specification for the metal.

Standard specifications for various types and grades of metals are promulgated by various national and government agencies e.g. American Society of Testing and Materials (ASTM), American Society of Mechanical Engineers (ASEM), British Standards Specification (BSS), Indian Standards Specification (ISS), etc.

The engineering departments of some companies using large number of castings, give their own specifications. Specifications for castings to meet specific requirements are particularly written. Some foundries produce castings under a trade name and grads for which the properties and characteristics are provided by the foundry.

The various properties of the castings relating to their serviceability must be established by the purchaser and appreciated by the foundry to avoid misunderstanding. The thickness and shape of the castings influence their properties as these are influenced by the manner in which they solidify and cool. Further, the rate of cooling is determined by the individual design of a casting, the process of casting, the way the molten metal enters the cavity, and pouring temperature. Thus, it is necessary the specifications be clear to both purchaser and producer and that they agree on the referred properties as being in the cast part or in separately cast test specimens. There are standard specifications for both the cases and as and when necessary special requirements may be added. In general, foundrymen also advise their customers on these characteristics of castings.

12.3.1 TYPES OF SPECIFICATIONS

There are several methods of specifying castings so as to obtain a satisfactory product. Specifications can be written to designate the properties of the metal that are specifically needed, i.e., composition of metal, stregth, hardness, finish, accuracy or special characteristics such as resistance to heat or corrosion. There are many casting applications in which the ordinary properties of the designated type of metal/alloy are quite adequate.

When mechanical properties of the castings are important, the most usual procedure is to utilize a standard test bar which is poured separately but along with the specified lot of castings. This test bar should possess the minimum required tensile properties. The relation between properties in the casting and the test bar will depend on their relative cooling rates. The properties of iron and steels are particularly sensitive to the section thickness of casting and thus a series of standard test bars of increasing size should be cast related to various casting thickness.

A commonly used method of specifying iron and steels for castings produced in larger quantities is based on the microstructure and hardness of the metal in the castings themselves at a specific location. There is a good relation between hardness and tensile properties but this relation is not used as part of specifications.

For some applications, the castings can be tested in the manner in which they will be used, e.g. pressure components can be 100 per cent hydraulically proof tested as this test would be nondestructive to satisfactory units. Similarly, load carrying components can be tested by measuring their deflection under load and rotating parts can be spin tested.

Some cast metals can be specified by chemical analysis alone but this is rarely satisfactory to specify iron and steel castings. Limits of the content of some elements are usually included in specifications as these elements can influence properties or characteristics of the metals which are not easily measureable. For example, the thermal shock resistance of gray iron is influenced by its total carbon and corrosion resistance is influenced by alloy content of iron. However, different foundries may obtain the required properties with metals that differ in chemical analysis because their facilities and the teehniques are different.

Some castings may be required to meet the specific requirements which may include radiographic soundness, stress relief heat treatment, rust pre venting coatings, etc.

12.3.2 Standard Specifications

Standard specifications of most of the castings are laid by various agencies. When metal properties are to be specified, the standard specifications provide the purchaser with a maximum number of sources and possible economic advantages because the use of familar specifications saves time and minimises the possibility of mistakes. Even with unusual requirements of castings. it is preferable to use an oppropriate standard specification as a base and these add the unusual requirements to it as and when possible. However, the additional requirements should be clearly stated and given with the number of standard specification on the purchase order. Standard specification societies publish a summary of commonly used standard specifiction for various types of castings. The summary usually includes specification number, uses, grade or class, yield strength, tensile strength, elongation in tension. chemical analysis, microstructure, hardness, heat treatment, specific requirements and typical applications.

12.3.3 Standards and their Implications

The basic requirement of inspection is the establishment of standards as required by the designer. In a broad sense, the design requires definition of the casting in terms of physical shape and dimensions, properties, composition, and general quality. Inspection is based on permissible deviation from the ideal standards. Some factors such as dimensions, composition and mechanical properties can be defined quantitatively so that numerical tolerance limits can be prescribed whereas in other cases such as soundness and surface finish, subjeetive judgements are required. The latter quantities are closely associated with commercial policy and may be set at higher level than actualy needed to meet the formal specifications. However, standards must not be set at an unrealistic level in relation to the basic capability of the particular casting technique. Further economic implications of high standards must not be over looked.

12.4 QUALITY ASPECTS

In general, the quality aspects for a casting include surface finish, dimensional accuracy, soundness and metal properties.

1. *Surface Finish* The surface condition of castings is influenced by a number of factors such as kind of pattern used, process of moulding, quality of sand and other additives, cleaning process etc. At one time the kind of moulding sand used was most important to casting finish, but now the abarsive cleaning process may be the most important factor. Smoothness of castings can be evaluated only by visual comparison and there is no other satisfactory way.

2. *Dimensional Accuracy* The basic dimensions of a casting are established mainly by the pattern equipment but variation from this base may be caused by several other factors in the casting process i.e., expansion of the mould cavity due to heat and pressure of the molten metal, contraction of the metal after solidifying and cooling, and heat treatment. To compensate all these effects, the pattern is purposely made of oversize. Distortion or warpage of casting may occur due to different cooling rates of different parts. Sometimes it is necessary to distort or fake the pattern to obtain a true casting. Once a sample casting is approved by the customer, castings are not generally checked for their dimensional accurancy. However, when producing large number of castings, critical dimensions are measured frequently to check for possible drift especially due to pattern wear.

3. *Soundness* Because of the different ways in which various metals/alloys solidify, chances of unsoundness may be more or less depending on the metal to be cast. The metals and alloys showing low shrinkage exhibit less possibility of unsoundness. Often a cavity in the casting will be evident to visual inspection because of an opening or depression at the surface of the casting. Internal soundness can be checked by radiography and ultra sonic scanning. Sample castingss may be sawed in half to check for both wall thickness and internal soundness.

4. *Metal Properties* The standard specifications for various metals and alloys provide the means for evaluating their properties.

12.5 INSPECTION PROCEDURES

Inspection procedures may be classified into following categories:

(i) Visual inspection for surface finish and defects.
(ii) Shape and dimensional checking.
(iii) Testing for chemical composition.
(iv) Testing for mechanical properties.

(v) Destructive testing of castings.

(vi) Nondestructive testing four soundness.

12.5.1 Visual Inspection for Surface Finish and Defects

Visual examination of individual casting ensures that none of its features has been omitted by short running, moulding errors, fettling mistakes and errors in other processes of casting. Suıface defects observed at this stage include surface cracks, tears, blowholes, metal penetration, rat tails and buckles, swells, shifts, surface roughness, shrinkage, etc. Visual examination may be carricd out with naked eye or using a magnifying glass.

Sometimes numerical standard may be specified which may be evaluated by visual comparison with a series of standard surfaces graded arbitarily or assessed quantitatively by other means. Alternatively, a styles measuring instrument may be used to obtain representations of surface profiles. However, such instruments are usually suitable for distinguishing between relatively smooth surfaces such as those obtained in die casting and other precision casting methods.

Surface roughness may be expressed as a number (in microns), which is an arithmatical mean of the heights of the peaks and depths of the valleys on a casting surface above and below a mean line within a specified sampling length. The surface roughness can be assessed according to IS: 3073, 1967. The approximate roughness values for different types of castings are as follows:

Casting process	*Roughness value*
Ordinary sand casting	25 – 50
Good quality sand casting	5 – 25
Permanent mould casting	0.8 – 6.3
Die casting	0.8 – 3.2
Shell moulding	1.6 – 6.3
Investment casting	0.8 – 3.2
Ceramic moulding	0.2 – 1.6

Presently, a range of optical devices is available to examine the areas inaccessible to the human eye, e.g., the bore of a cast cylinder block by using Boroscopes (miniature combined periscope and telescope), light guides or miniature television cameras. A simple and useful method of revealing surface defects in acid etching. The etching removes all oxides from the surface of the casting and reveals its crystalline structure and surface defects in sharp relief. The etchant used depends on the type of material, surface condition, and type of defect to be reavealed.

Precise measurements of finished surfaces can be carried out by the use of electrical type of direct reading surface measuring instruments or profilometers such as *Talysurf*. Visual inspection is the simplest one and is carried out to ensure a casting that looks good. Visual inspection is quite satisfactory for many castings such as manhole covers, drains, counter-balance weights etc.

12.5.2 Shape and Dimensional Checking

Dimensional inspection is generally required for all types of castings. In some cases it may not be so critical but in other cases it may be vital. Castings produced by die casting, investment casting. shell moulding, etc., and castings requiring adequate machining need close check for the dimensions. All under sized castings requiring machining are rejected as their dimensions will be further reduced after machining. Initially, when the castings are made from a new pattern, first few castings are carefully checked for the dimensions to make sure that the pattern and coreboxes are of correct size. If there is any deviation from the blue print, the pattern equipments are rectified. In mass production, once the castings are found to be within the tolerances, spot checks of castings, together with regular checks of the pattern and dies being used may be adequate. However, in case of the jobbing type of foundry, each casting may have to be thoroughly checked for dimensional variations.

Dimensional examination is carried out against the drawings with the help of various measuring instruments such as gauges, jigs, templates, special fixture, etc. Tolerances on cast dimensions are usually included on the drawing. In some cases marking out for subsepuent machining is also included with dimensional checking. If castings are to be machined without individual marking out, specific care is required in respect of dimensions incorporating datum surfaces. In cases where the weight of the casting is critical, a specific weight check is also included. Whcn dimensional errors detected, their true significance should be determined. Various methods/instruments used for dimensional checking are discussed below:

(a) *Measuring instruments* such as rule, verniers, micrometers, scribing block, combination set, straight edge, squares, spirit level, and dial indicator are commonly used for checking the dimensions. More advanced instruments such as auto collimater, comparator, ultrasonic instruments for wall thickness, projection instruments, etc. are employed for high precision castings or machined castings.

(b) *Templates and contour gauges* are used for the checking of profiles, curves and intricate shape. Templates having required profile can be easily prepared in mild steel or brass sheet. These save time in measurements and make the entire job easy.

(c) *Limit gauges* are usually employed for tolerance dimensions on castings produced on a repetitive basis. The various types of limit gauges includes plug, ring, snap, plate, etc., the use of which depends on the shape of the parameter to be checked. The limit gauges should be periodically checked for their correctness.

(d) *Special fixtures* are employed where dimensions can not be easily checked by other instruments, e.g. relative dimensions, centre—to—centre distances, angularity of surfaces etc. Inpection fixtures are specially designed for specific jobs.

12.5.3 Testing for Chemical Composition

Testing for chemical composition provides an actual evaluation of the properties and characteristics of the cast metal. Chemical composition is usually determined by tests carried out on specimens cast separately from the same heat of metal. The samples must be drawn at the same time as the castings are poured, since losses of volatile or easily oxidised elements may cause change in composition and thereby resulting also in difference in mechanical properties. When pouring operation is long or successive ladles are drawn from a single heat, frequent sampling should be carried out at the casting point. Analysis for chemical composition may also be carried on discarded heads and runners. The chemical composition is usually assessed in relation to the specified limits for the major alloying elements and expected impurities, tramp elements and gas content. In the case of iron and steel castings, the percentages of C, Si, S, Mn and P are invarially determined. The presence of alloying elements (Cr, Ni, Cu, W, V, etc.) may have to be determined depending upon the composition of alloys. For determination of chemical composition, wet analysis methods are the most accurate and reliable, but take a long time, However, wet methods are unsuitable for determining the composition of the molten metal before pouring to adjust the composition of the final melt. The use of spectrographic, X-ray fluorescence and other physical techniques enables simultaneous and rapid determination of a comprehensive range of elements. The instrumental methods permit more extensive bath sample analysis and thereby provide better control in melting. The composition of last cast should be verified before the further processing of castings to minimise loss of production time.

In small foundries not able to afford costly instruments, the elements of critical importance are checked by either cheap instruments designed specifically for a limited number of elements or quick analytical methods. One quick test developed called carbon equivalent (C.E.) is given by total carbon % + 1/3 silicon %. The instrument works on the principle that C.E. is directly related to the liquidus arrest temperature of the cast iron. The instrument is equipped to hold a small reservior or molten cast iron over a chromel-alumel thermocouple and the temperature of metal as it solidifies, is recorded on a chart from which the liquidus arrest temperature can be easily determined. Following conversion table is used to determine the corresponding C.E.

Liquidus °C	*Temperature of*	*C.E.*
1232.2	2250	3.60
1221.1	2230	3.70
1210.1	2210	3.80
1198.9	2190	3.90
1187.8	2170	4.00
1176.7	2150	4.10
1165.6	2130	4.20
1154.5	2111	4.30

Similarly an apparatus known as Strohlein thermoelectrometer is available for quick determination of silicon in cast irons. It measures the thermal e.m.f. produced on heating a junction of cast iron in which silicon is to be determined and another metal such as copper. Silicon possesses a special position in the thermoelectric series and thus silcon can be determined by means of emperical caliberation curves prepared by using samples of known silicon content.

Spectroscopic analysis is becoming more popular in foundries for quick determination of various alloying and trace elements. Various types of spectroscopic analysers are available. The selection of a particular type depends on the requirements, i.e. elements to be analysed, desired accuracy, frequency of tests and the type of cast metal.

12.5.4 Testing of Mechanical Properties

Specimens for mechanical testing are generally prepared by machining of specially designed cast test bars which may be poured separately from the same heat or as directly attached to the heavy casting itself. With certain exceptions, the test carried out on specimens are representative of the particular heat of metal and not of any individual casting. The measured properties will reflect influence of dissolved gas, inclusions, and melt treatments, but may not necessarily be those which are obtained from test pieces cut from actual castings, since there may be difference in section, structure, and degree of feeding. Standardisation of test bar conditions permits the use of test results to ensure perfect soundness of the test specimens. Thus in casting of test pieces, yield is disregasded and most test pieces are over fed.

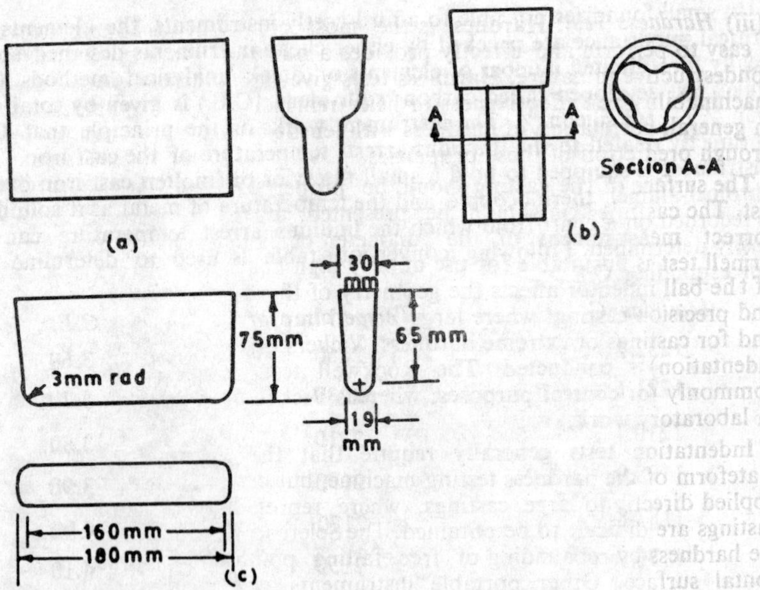

Fig. 12.2 Typical test bar designs: (a) for general purpose keel block, (b) Clover leaf test bar, and (c) chill cast test bar.

Typical test bar designs are shown in Fig. 12.2 However, in some cases more benefit can be derived by having the cast surfaces of test bars in close proximity to the castings. Further to avoid major differences in cooling conditions, all test bars are generally cast in moulds of similar material to that used for the actual casting. The pouring temperature should also be standardised as far as possible.

Casting may be evaluated on the basis of a wide range of mechanical tests for which detailed procedures are given in numerous text books and standards. Various mechanical tests are as following:

(i) *Test for Tensile properties* Tensile properties are directly determined by testing an especially shaped sample made from separately cast test bars in a tensile testing machine (Universal Type of Testing Machine or Hounsfield Tensometer). However, to establish the actual tensile properties of the metal in a particular section of a casting, a tensile test bar can be machined from the actual casting but it may be remembered that this test only provides data on the metal that is tested. Tensile properties can be estimated by load testing of the entire casting and also from hardness, microstructure, sonic or eddy current data. The tensile results include proof or yield stress, tensile strength, modulus of elasticity, elongation and reduction of area. Casting specifications generally include one or more of these properties.

(ii) *Microstructure* Microstructure of metals and alloys establishes the type and possible heat treatment in case of iron and steels. Microstructure is identified by viewing a polished and etched surface of the metal under a metallographic microscope. The microstructure can be studied either of a metal piece sectioned out of the casting or on the casting surface itself but the latter involves difficult techniques.

(iii) *Hardness Test*: Hardness is the most commonly used test because it is easy to perform and directly provides a numerical value and the test is nondestructive in nature. Hardness tests give the guide to wear resistance, machinability and effectiveness of heat treatment applied to actual castings. In general, correlation of hardness with tensile strength and ductility enables a rough prediction of these properties.

The surface of the casting should be properly prepared for a hardness test. The casting skin should be presented to the indentor to enable the correct measurement of the diameter of impression with microscope. Brinell test is unsuitable for use at very high hardness surfaces as distortion of the ball indentor affects the geometry of the impression. For high quality and precision castings where large Brinell impressions can not be tolerated, and for castings of extreme hardness, Vickers or Rockwell test (gives small indentation) is conducted. The Rockwell test is quick and is used more commonly for control purposes, whereas Vickers pyramid test finds its use in laboratory work.

Indentation tests generally require that the casting be placed on the plateform of the hardness testing machine, but certain hardness tests can be applied directly to large castings, where representative samples of heavy castings are difficult to be obtained. The Scleroscope can be used to measure the hardness by rebounding of free falling pointed indentor on any horizontal surface. Other portable instruments give readings which can be correlated with standard identation tests. However, these additional tests are used for control purpose rather than for formal inspection. Hardness testers should be routinely checked for accuracy with a calibrated hardness test block.

Sometimes file hardness test is very useful because it evaluates the hardness of an edge. Particularly iron castings not heat treated can have hard edges which can be destructive to cutting tools. In the file test, a sharp medium cut mill file is held against the edge of the casting and moved just enough to determine whether it is cutting or not. A tendency to slide without cutting under moderate pressure indicates a hard edge.

(iv) *Impact test* Though the impact test values do not have direct significance for quantitative design, the important values are included in specifications of some castings. Both Charpy and Izod tests can be used at room temperature (for general castings) or subzero temperature (for castings for cryogenic and other low temperature applications). In the case of cast iron castings, notched bar impact testing is hardly of any significance in discriminating between the quality of different irons, but an un-notched plain round specimen is held in stepped grips and the fracture energy is determined in the normal Izord Test.

(v) *Bend, transverse and compression tests*; These tests are conducted to evaluate the ductility and resistance to shock of the cast metal. All these tests can be conducted on the Universal type of testing machine. In the bend test, a bar specimen supported at its end is bent round a former through which the load is applied. In the transverse test the specimen is loaded centrally and supported at its ends to obtain the amount of deflection and breaking load from which transverse rupture value is derived. This test is mainly used for grey cast iron.

(vi) *Fatigue test* This is carried out where an appraisal of the life of the casting service is to be known. Though frequently the castings are used under fatigue stresses or for high temperature works, fatigue tests and stress rupture or creep tests are seldom used for routine inspection owing to a very lengthy time involved in tests.

(vii) *Other tests*: Various special tests may be required to establish the response of castings to service conditions. The various tests include for weld decay as applied to austenitic stainless steel, ballistic tests on plates, weldability tests and other tests based on analogies to the conditions to be encountered by the casting.

12.5.5 Destructive Tests on Castings

In general, variation of properties within a single casting results from inevitable differences in microstructure, and composition produced by local differences in solidification conditians. The normal test blocks (separate cast bars or block attachedto the casting) are designed to evaluate the metal under ideal conditions of feeding, but tests on sectioned castings can provide a more direct indication of properties within a particular design. Systematic sectioning of casting also aids in direct investigation of structure and internal soundness by metallographic techniques. This also facilitates the accurate interpretation of nondestructive test and radio-graphic results,

In one form of the destructive testing, direct loading of representative casting is carried out to failure in special fixtures simulating service stresses. The deformation, breaking load and position and nature of fracture provide guide to the quality and behaviour of the design under highly stressed conditions. One simpler form of test consists of fracturing the casting by compression in a press.

Destructive tests can be carried out on rejects from other causes, but in some cases it may be necessary to sample a proportion of castings from regular production. Such techniques can prove to be a vital aid to the interpretation of test results and inspection find.ngs derived by more conventional methods.

12.5.6 Non Destructive Testing of Castings

Non destructive tests are conducted to examine the internal soundness of the castings without recourse to sectioning or fracturing. The tests may be carried out for any sub-surface or internal defects, surface defect not detected by visual examination, pressure tightness, and overall soundness which may be required in service. These tests not only assist in maintaining high quality standards, but provide a valuable aid in the development of foundry techniques. These factors have greatly improved the position of the casting in engineering. However, inview of increased sensitivity of inspection by non-destructive testlng, the techniques should be adopted and interpreted with discretion to avoid unnecessary rejections. Various non destructive tests carried out can be classified under following categories:

1. Tests for defects in the surface zone.
2. Tests for detecting internal defects.
3. Other tests.

Tests for Defects in the Surface Zone

Cracks and other flaws in the surface of the casting can be detected by a number of methods which include chemical etching, penetrant method, and magnetic flaw detection. All the methods require clean and relatively smooth surface for effective detection.

(a) *Chemical etching method.* Very fine surface cracks, laps and similar defects may be covered by (i) thin layer of metal resulting from superficial plastic deformation in shot blasting, fettling, and machining of casting or (ii) oxide scale formed in heat treatment. In such cases, usual penetrant methods are no effective. Such surface defects can be revealed by a simple chemical etching method in which controlled corrosion attack takes place. The method employs acid or acid salt mixtures developed for particular alloys. The casting is immersed in the etching bath or swabbed down with etching solution. After etching, excess acid is neutralised, the castings are rinsed and dried. This process removes all oxides and superficial metal from the surface of the casting and reveals crystalline structure along with defects. Many etchants are available and the type of etchment used will depend on the type of material, surface condition, and type of defect to be revealed.

castings which can not be readily pressure tested. Penetrant solution can be applied to one face and the opposite face may be examined for sign of seepage. The penetrant test consists of cleaning of casting, treatment of casting with penetrant, removal of excess penetrant, development, and inspection of penetrant indication.

Cleaning. The casting to be tested should be clean and throughly free from contaminants such as loose scale, greasc oil, rust, etc. Cleaning may be accompalished by using one or more of the following methods.

(i) Vapour degreasing by using trichloroethylene.
(ii) Solvent degreasing.
(iii) Pickling in acid or alkaline solutions.
(iv) Steam cleaning.
(v) Cleaning by ultrasonic vibrations given to the casting suspended in a tank of liquid.

Carry over of residues from cleaning solutions, particularly chlorine from degreaser and acid or alkali from pickling solutions should be avoided as these will react and contaminate the penetrant. The physical methods of cleaning such as shot blasting, brushing, etc. should also not be used as these may cause the closing of cracks and holes at the surface by peening action.

Treatment of casting with penetrant. This consists of allowing penetrant to seep into the surface discontinuities by capillary action for a suitable interval of time (5 minutes to 3 hours). The penetrants employed fall into two categories, i e. (i) brilliant dye penetrants which are solutions of brilliant red dye and reveal red indications on white back ground and penetrants (ii) Flucrescents which are solutions containing fluorescense and emit brilliant green fluorescence when viewed in darkened room with an ultraviolet (often known as black light). Brilliant dye penetrants do not require black light and are sensitive to fine defects and very suitable for the inspection of light alloy castings having smooth surface. However, these are difficult to be used on rough surfaces and dangerous when used in open tanks.

The choice of penetrant is influenced by various factors such as. (a) sizc, shape and quantity of castings, (b) analysis and surface condition, (c) size, shape and position of expected defects, (d) compatibility of the penetrant with the casting material, (e) sensitivity required and (f) availability of equipment and services. There are no hard and fast rules for the selection of best penetrant, but can only be found by experience,

The penetrant can be applied to casting by brush or aerosol spray or by complete immersion of casting in a tank of penetrant for a recommended period of time, the surplns penetrant must be removed.

Removal of excess penetrant. The penetrants from either category may be solvent removable, water washable or water emulsifiable (removal of penetrant is possible only after an emulsifier is added). In the case of solvent removable penetrants, operation can be performed by using an aerosol, whereas for water washable penetrant, water hand spray is used with a pressure of 2 — 2.5 Kg/cm^2 at a temperature of 45°C. When using emulsifiable penetrant (paraffin oils), an emulsifier is added to the penetrant before water washing. The emulsifier breaks up the continuous oily film into an aggregate of small droplets which can be washed with water spray. Washing of the casting should be carried out from bottom upwards and so that water running down does not remove penetrant from the defect. Further, violent wash should be avoided as it will also tend to remove penetrant from the defects.

Development. After the excess penetrant is removed, the casting may be treated either for dry development or wet development. In case of dry development, the casting is dried either by normal evaporation or by forced air circulation (temperature of air should be less than 80 °C). After drying, a developer in the form of dry light coloured powder is uniformly applied to the surface by tumbling the casting in the powder or by dusting the powder over the casting or by using an air gun to blow the powder on to its surface.

For wet development, excessive moisture is removed from the casting after washing. Wet developer is applied in the form of a powder suspended in water or in a volatile solvent (aerosol). For the correct observations, uniformity of developer and thickness of the applied coating are very important (too thick coatings may mask very fine defects). The developer should be vigorously agitated just before its use.

The castings are then allowed to stand for development. The time taken for development may vary from 15 minutes to 36 hours and is known as 'bleeding time'. The penetrant inside the defect will seep out by reverse capillary action (blotting paper action).

Examination of penetrant indication. The examination of penetrant flaw indicators must be carried out under adequate illumination. The area to be examined should be equally illuminated without any shadow or glare. The interpretation of the indicators is the most difficult aspect of penetrant flaw detectors. Due to the blotting paper action and porous nature of developer, usually an enlarged outline of defect is obtained. On the other hand if excessive developer is applied to the surface, only a very small amount of penetrant may seep out and thus indicating an extremely small defect. Penetrant methods for crack detection are most suitable for non-ferrous metals and austenitic steels.

(c) *Magnetic Flaw Detection*. This method is highly effective and sensitive for revealing linear surface discontinuities (cracks and similar defects) on or just beneath the casting surface in case of magnetic materials (iron, steel and their alloy). The magnetic flaw detection is based on the fact that the magnetic permeability of ferromagnetic materials is greater than that of air. If magnetic lines of an applied field are allowed to pass through a casting of ferromagnetic material containing a defect, they will tend to pass more readily through the material than the surrounding air (Fig. 12.3). As can be seen from Fig. 12.3 distortion of magnetic lines will be caused in the neighbourhood of the defect. Further, it can be noted from Fig. 12.3 that if the defects are positioned across the flux path at 90° to it, maximum distortion of magnetic flux will be observed, whereas minimum distortion of magnetic flux occurs when the defects are inline with the flux. In the latter case, the defects may not be revealed and thus two or more tests will be required to ensure 100 pct. detection of defects.

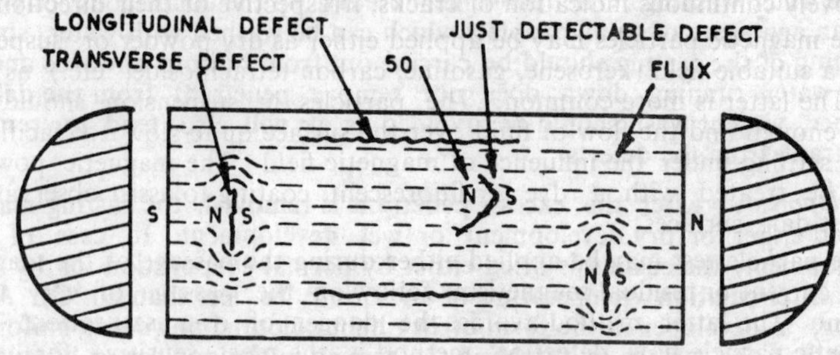

Fig. 12.3 Magnetic flux flow around flaws in a plate.

The resulting disturbance in magnetic flux can be detected by its effect on magnetic particles sprinkled on the surface of the casting. In the case of sound casting, the magnetic particles will be distributed uniformly all over the surface, whereas in a defective casting, the magnetic particles will be attracted to the region of defect. The depth of defect may also be judged by studying the concentration of the particles. However, for an accurate estimation of defects, a considerable experience is needed.

Although permanent magnets can be employed for small scale testing, commercial crack detection equipment employs a.c. or d.c. current (low voltage and high current) to produce the required magnetic field. An alternating current is employed for high surface sensitivity, whereas the direct current is preferred for detecting the cracks beneath the surface. The current can be applied in a number of ways as shown in Fig. 12.4.

In contact method (Fig. 12.4(a)) heavy current is passed directly through the casting by placing it between two contacts in the form of solid copper clamps or flexible gauge pads. Another contact current flow method known as prod technique is illustrated in Fig. 12.4 (b) which is excellent for selected areas of large castings. In this method the flux is set at right angles to the line joining the prods. The shape of the prod chosen depends on the contour of casting. The material of prod may be lead, copper, aluminium or steel depending upon the compatibility with the casting material (prod material should not contaminate the casting.

For the castings of annular shape, magnetic field can be obtained by using an axially disposed conductor bar threaded through the orifice (Fig. 12.4 (c)). Another method of magnetising consists of energising an external electromagnet (Fig. 12.4(d)) with the holes situated at the ends of the section. A permanent magnet can also be used in the similar manner. The required magnetic field can also be produced by using a flexible braid conductor, which can be coiled round any metal section to induce an axial field on passing the current (Fig. 12.4 (e)). This technique is specifically suitable to castings of irregular shapes.

Small castings can be tested directly on magnetic crack detectors. The critical regions of large castings can be tested by the use of coils or contact probes carried on flexible extension cables. For examining the casting in more than one direction, either casting is rotated or successive magnetic fields are applied in two directions at right angles. The sinusoidal variation of the a.c. induced field superimposed on the permanent field of electromagnet results a field of continuously changing direction which gives an effectively continuous indication of cracks, irrespective of their directions.

The magnetic particles may be applied either as dry powder or suspended in a suitable field (kerosene, gasoline, carbon tetrachloside, etc.) as an ink. The latter is more common. The particles in suspension should be large enough and the flow of fluid over the surface quite slow to facilitate rapid settling under the influence of magnetic field. The magnetic powder may be treated with a dye or fluorescent coating to assist observation against dark surfaces.

The particle test may be applied either during the passage of the magnetising current or residual magnetism following the passage of very high current. The latter method avoids the danger of over saturation. The magnetic particle fiow detection method is the most sensitive for crack detection in ferrous castings. A round hole at the surface or subsurface disconformity is not readily indicated because the magnetic lines can go around these defects and stay within the metal so that the magnetic particles are not attracted. This type of disconformity can be revealed by using a very strong magnifying force and careful technique.

12.6 TESTS FOR DETECTING INTERNAL DEFECTS

12.6.1 Radiography

This is a means of obtaining a two dimensional shadow picture of variations in the density of shapes in metal by using X-ray or gamma ray radiation which are electromagnetic oscillations and share the same properties. A radiograph is a photographic record obtained by passing X-rays or gamma rays through a casting on to a film. The amount of radiation reaching the film depends upon, (a) nature and thickness of material, (b) the number, position, and size of defects, and (c) quality of radiation used.

X-rays are generated from X-ray tube which consists of a highly evacuated glass tube containing an electrically heated filament (cathode) and a target comprising a thin plate of tungesten (anode) embedded in a block of copper. X-rays are generated by application of high voltage between the cathode and anode causing electrons travelling at high speed from cathode to collide with anode.

Gamma rays arise from the spontaneous and continuous disintigration of atomic nuclei within radioactive substances. The radioactive source itself governs the energy of gamma radiation and can not be controlled. As the rate of disintegration cannot be changed, the intensity is uncontrollable. For economy, it is desirable to use radioactive isotopes having slow decay rate. An adequate quantity of radiation is required to keep exposure time short and thus the source must have high activity and be concentrated into a minimum volume.

Natural radioactive sources can not be obtained with dimensions small enough to give required intensity and thus artificial radioactive sources are more suitable. The most suitable radioactive isotopes used for industrial radiography include cobalt 60, indium 192, thalium 170 and caesium 137. It is necessary to select a source (compromising for penetrating power und decrease in contrast) which gives sufficient sensitivity to reveal the defect with the shortest practicable exposure.

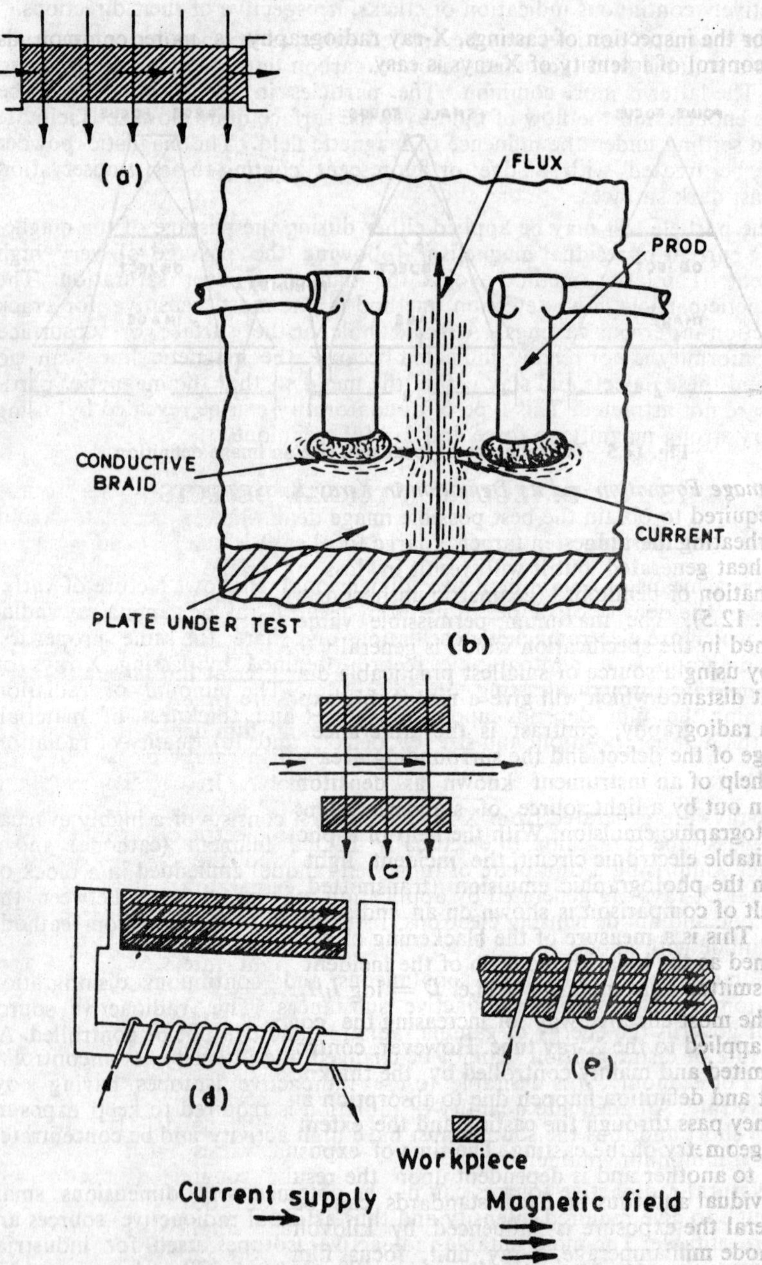

Fig. 12.4 Various techniques of current application for magnetisation. (a) contact method, (b) with prods, (c) threading bar, (d) electro-magnet and (e) coiled conductor.

For the inspection of castings, X-ray radiography is more common as the control of intensity of X-rays is easy.

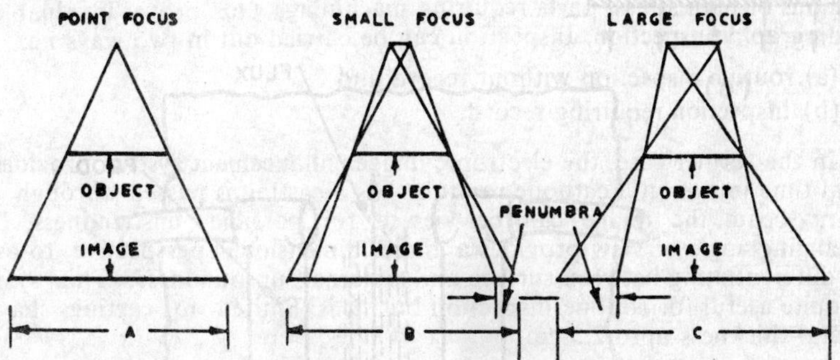

Fig. 12.5 The influence of focal spot size on image definition.

Image Formation and its Definition in X-ray Radiography. A point source is required to obtain the best possible image definition. In order to avoid overheating the tungesten target, a large focal spot is necessary to dissipate the heat generated, but it will result in loss of image definition by the formation of penumbra (geometric unsharpness) at the edge of the image (Fig. 12.5). The maximum permissible value of penumbra is usuably defined in the specification which is generally 0.2 mm. This value is achieved by using a source of smallest practicable diameter at the largest convenient distance which wlll give a reasonable exposure time.

In radiography, contrast is the difference in film density between the image of the defect and the surrounding area which can be measured with the help of an instrument known as densitometer. In this incident light given out by a light source of standard intensity is passed through the photographic emulsion. With the help of a photo electric cell connected to a suitable electronic circuit, the incident light and the coming out light from the photographic emulsion (transmitted light) are compared. The result of comparison is shown on an indicator calibrated directly in density. This is a measure of the blackening of a photographic film (D) and is defined as the logarithmic ratio of the incident light intensity (I_i) to the transmitted light intensity (I_t) i.e. $D = \log I_i/I_t$.

The most effective way of increasing the contrast is by decreasing the *KV* applied to the X-ray tube. However, control of contrast by this method is limited and mainly controlled by the thickness of casting. Loss of contrast and definition happen due to absorption and scattering of the X-rays as they pass through the casting and the extent depends on the material and geometry of the casting. The time of exposure varies from one X-ray unit to another and is dependent upon the results sought and openion of individual and thus no fixed standards can be laid down. However, in general the exposure is influenced by kilovoltage applied to x-ray tube, cathode milliamperage, X-ray unit, focus film distance, film speed and developer used, material and thickness of casting, and the use of screens and filters.

Radiography of Castings. The radiography of castings presents many problems due to the complex shapes and variations of each section. Before a casting is radiographed, it should be fettled and the gates and risers be removed. Knowledge of some facts such as method of casting, position of chills and chaplets, changes in section, likely defects and their position, regions of high stress, parts requiring machining, etc. prove invaluable in radiography inspection. Inspection can be carried out in two ways i.e.

(a) routine inspection without record and

(b) inspection requiring record.

In the former case, the electronic image enhancement system provides a real time image on a cathode ray tube. As a casting is passed through the X-ray beam, the image can be viewed for possible unsoundness. The changing angle of view provides a three dimensional perspective to assist in differentiating between surface and internal unsoundness. This system is quite useful for routine inspection but it is limited to castings having metal thickness upto 2.5 cm.

Castings having heavier section and requiring inspection record are radiographed on film. The proper interpretation of radiographic films requires reference radiographs and considerable skill.

For taking a radiograph of a casting following operations are involved;

(i) *Selection of area to be radiographed.* Castings of complex shape present difficulty in adequately covering all parts. Some areas containing defects may be missed if attention is not paid in positioning the casting and angling the beam. Variation in thickness of castings causes some parts of the film under exposed while other over exposed. This problem can be minimised by

(a) using blocking media

(b) using two or more films with different speeds and

(c) using filters between the source and object to reduce contrast.

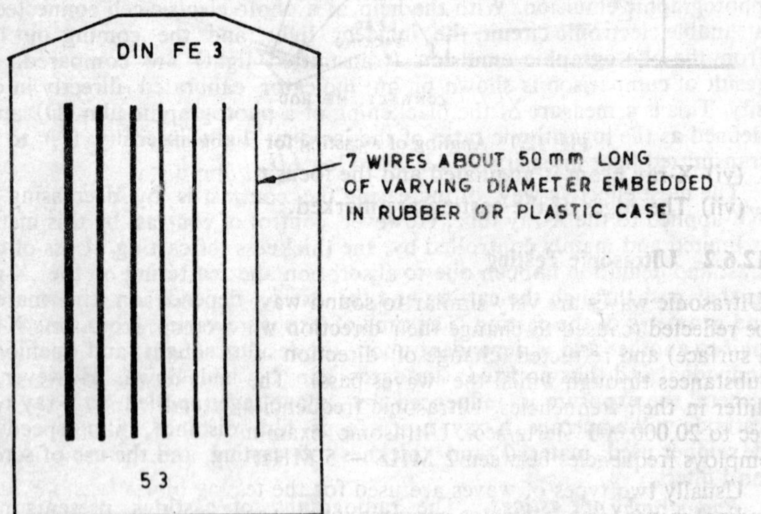

Fig. 12.6 An image quality indicator.

Minimum number of shots are used which give satisfactory coverage and adequate flow detection together with the minimum use of film and total exposure time.

(ii) *Marking off.* The area selected for radiography are identified, punched or etched letters or numerals to facilitate the correlation of radiograph with the casting.

(iii) *Image quality assessment.* Assessment of quality is made by the device known as image quality indicater (I Q I). One form of this consists of a series of fine wires of varyfing thickness sealed into a plastic envalope which incorporates identifying symbols as shown in Fig. 12.6. The IQI is placed on the specimen surface remote from the film. The material under it should be of known and constant taickness. After the film has been exposed and developed the sensitivity is equal to the thickness of the thinnest wire visible on the radiograph divided by the specimen thickness

(iv) Exposure in decided from exposure chart.

(v) Film is placed in cassette and the cassette is placed in correct position (Fig. 12.7).

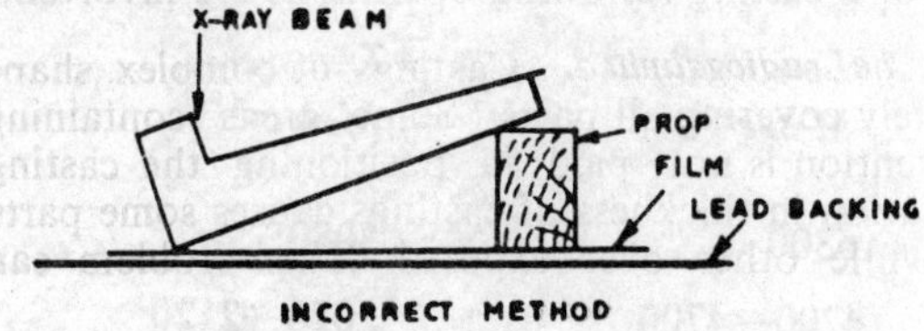

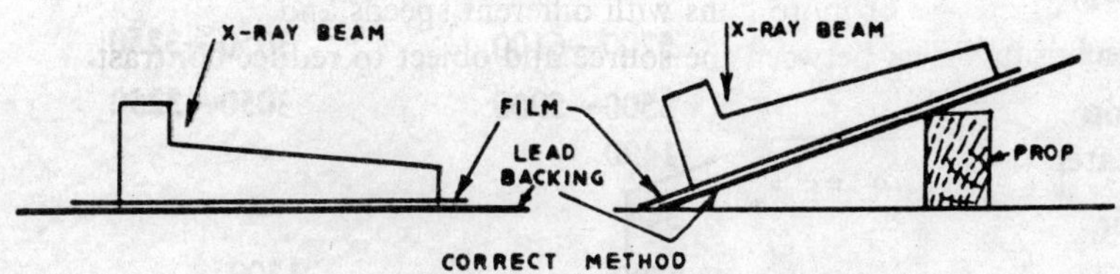

Fig. 12.7 Angling of a casting for radiography.

(vi) X-ray beam is angulated and the focus to film distance is checked.

(vii) The edges of the casting are marked.

12.6.2 Ultrasonic Testing

Ultrasonic waves are very similar to sound waves in characteristic, i.e.. can be reflected (caused to change their direction when coming in contact with a surface) and refracted (change of direction with change of density of substances through which the waves pass). The sound waves essentially differ in their frequencies. Ultrasonic frequencies extend from 20 M Hertz/sec to 20,000,000 Hertz/sec. Ultrasonic examination of castings usually employs frequencies between 2 MHz – 5 MHz.

Usually two types of waves are used for the testing of castings, i e. longitudinal waves (also known as compressional waves) and transverse waves (also known as shear waves). Longitudinal waves are much easier to generate than shear waves and can be propagated in liquids, gases and solids, whereas transverse waves can only be propagated in solids. The velocity of longitudinal waves can be calculated from the formula

$$V = \frac{\sqrt{K + 4/3\,G}}{\rho}$$

Whereas velocity of transverse waves can be calculated from the formula

$$V = \frac{G}{\rho}$$

where K = Bulk modulus

G = Shear modulus

ρ = density

Ultrasonic velocities (frequency X wave length) for some casting materials and other media are given below:

Metal/Alloy	*Longitudinal* $M\ sec^{-1}$	*Transverse* $M\ sec^{-1}$
Aluminium	6300	3080
Brass	4200—4700	2020—2120
Copper	4700	2300
Steel	5700—6100	3050—3250
Iron	3500—5900	3050—3250
Water	1490	—
Air	344	—
Perspex	2700	1300

The ultrasonic waves are successfully employed to test internal soundness and wall thickness. In this test an electronic instrument known as Ultrasonic Testing Oscilloscope (Fig. 12.8) is employed. The instrument provides an electrical pulse at an ultrasonic frequency which is converted tomechanical vibrations and directed through a casting by a transducer (transmitter+receiver) coupled to the surface of casting. The echo of the pulse from opposite surface is picked up by the same transducer and re-converted to an electrical signal. The time delay between the original pulse and its echo is measured and empirical calibration indicates the thickness of the metal through which the ultrasound traveled. The receiver amplifier supplies positive charges which are applied to the plate at the instant that the ultrasonic energy is transmitted and that at which it is received. These pulse of charge cause the spot of light on the screen to be deflected upwards, so that on either end of the horizontal line there are two peaks which indicate the difference in time from when the ultrasonic waves started to travel through the specimen to when they were reflected back from its bottom surface.

A flaw in the specimen gives rise to a reflecting surface closer to the probe than the bottom of the specimen. Ultrasonic waves from the transmttter are reflected off this to the receiver in a shorter duration, and therefore give an impulse to the receiver amplifier when the spot of light is still travelling across the screen. This gives rise to a peak past of the way

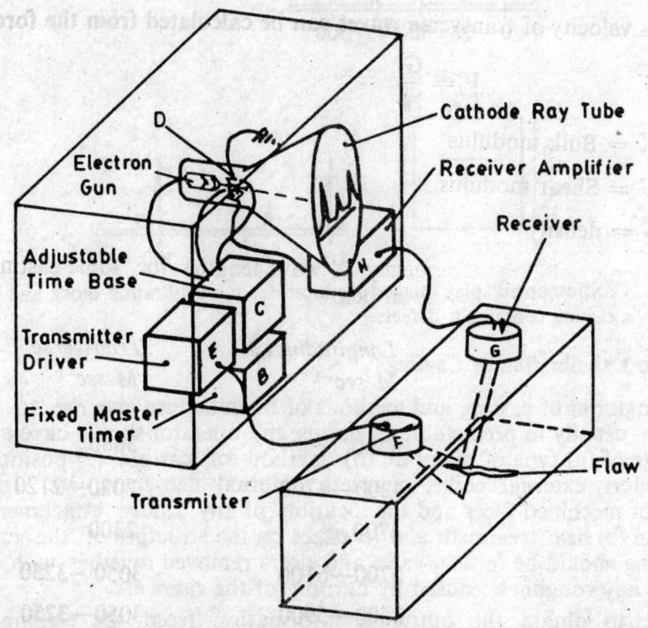

Fig. 12.8 Schematic representation of an ultrasonic testing oscilloscope.

across the horizontal line. The position of the peak indicates the depth while its height gives an indication of the size of the flaw.

Before commencing the ultrasonic testing, the time base must be calibrated with the help of a calibration block. Fig. 12.9 represents the oscilloscope display images from a calibration block and those from a casting containing a defect.

The ultrasonic waves are successfully employed to test internal soundness and wall thickness. Ultrasonic method is very sensitive and can even indicate very small non metallic inclusions. The complicated shape of some castings does not provide any opposing parallel surface which will reflect pulse. However, multisurface reflections have been successfully employed to inspect the castings but it requires special equipment and high engineering skill. The speed and accuracy of ultrasonic testing can be increased by using special test equipment which scans the casting while it is submarged in water.

Ultrasonic thickness gauges are now small battery operated having a digital read out. These are calibrated for the type of metal being tested and the two surfaces must be reasonably parallel.

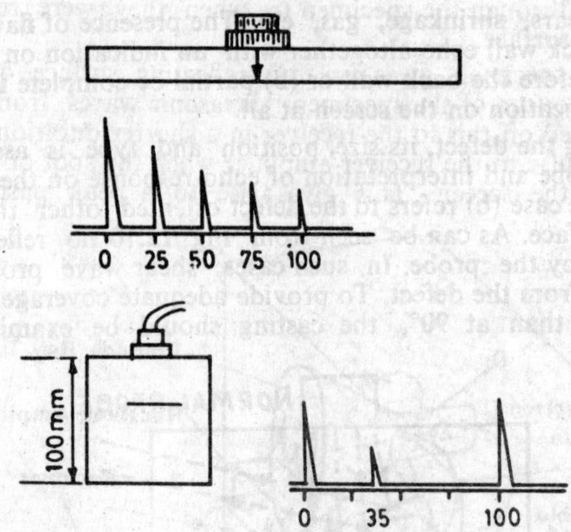

Fig. 12.9 Oscilloscope display images produced from a calibration block and from a casting containg a defect.

Ultrasonic Examination of Castings

Since material of casting and methods of manufacture give rise to specific defects, usually in predictable positions, the operator should have a good knowledge of (a) type of material, (b) method of casting (c) position of risers, feeders, external chills, chaplets, internal denseners, etc.. (d) the position of mechined faces and the location of any future attachments or welds, and (e) heat treatment and its effect on the structure of the casting. The casting should be fettled, gates and risers removed together with loose scale and any roughness caused by burning of the risers etc.

In order to obtain the optimum information from the examination, probe selection is made considering the following factors:

(a) *Metallurgical structure*. High frequency waves produce more scatter than low frequency waves.

(b) *Penetration*. Low frequency waves penetrate more than high frequency waves.

(c) *Resolution*. High frequency waves provide greater accuracy and have superior resolution to low freqnency waves.

(d) *Accuracy*. The beam spread decreases with increase in frequency.

The initial examination is carried out using compressional wave probes (single or twin crystal type). Twin crystal probes are advantageous when searching for near surface defects. Using a full scale drawing, the casting is inspected for wall thickness and core shift. This initial examination gives vital information concerning the ultrasonic response of the material. If the screen is full of spurious lines (commonly known as grass), and a poor back wall echo is received then the test frequency is lowered. In general the test frequency should be between 2 MHz and 6 MHz. The casting is then examined using still a compressional wave probe for significant flaw such as hot tears, shrinkage, gas, etc. The presence of flaw is revealed by (a) loss in back wall echo altogether with an indication on the screen at some distance before the back wall or (b) partial or complete loss of back-wall with no indication on the screen at all.

After detecting the defect, its size, position and type is assessed by the movement of probe and interpretation of echo response on the cathode ray oscilloscope. The case (b) refers to the defect oriented other than at 90° to the scanning surface. As can be seen from Fig. 12.10 no reflected energy will be received by the probe. In such cases, shear wave probe is used to receive a signal from the defect. To provide adequate coverage for defects oriented other than at 90°, the casting should be examined with a

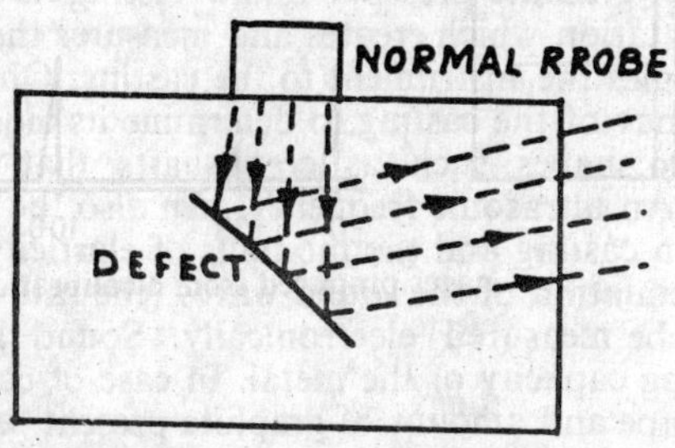

Fig. 12.10 Reflected energy not accepted by the prob.

45°, 60° and 70° angle probes (Fig. 12.11), Manipulation of the shear wave probe together with screen interpretation are helpful in assessing the type and shape of the defect.

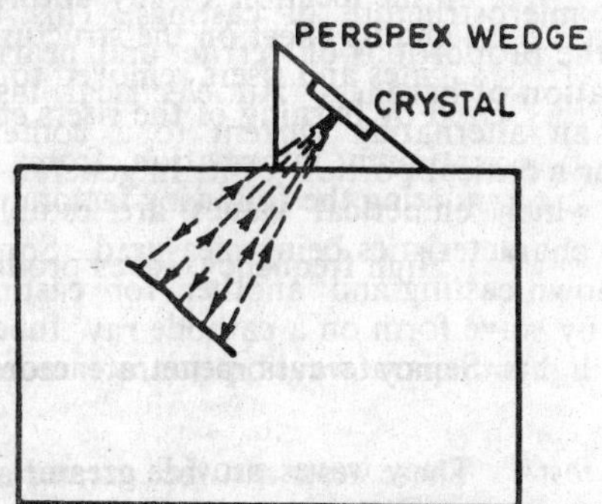

Fig. 12.11 Reflected energy accepted by the probe.

12.6 3 Other Tests

(a) *Leak testing*. This test is applied to castings which function as containers for liquids or gases under pressure. Leak testing can be carried out as a routine procedure on unmachined castings if the openings in the casting can be satisfactorily sealed by some means (rubber plug etc) otherwise the test may be done on the machined castings. In unusual circumstances leak testing may be performed after assembly.

The most common procedure for leak test is to pressurize the casting with air and then to submerge it in a tank of water. Air bubbles indicate a leak. The air pressure used should be sufficient for the degree of perfection desired. However, for safety, the pressure should never be that high which may cause fracture. A drop in pressure over a certain duration also indi eates a leak. A very sensitive test utilizes a gaseous hologen compound (fluorine refrigerent gas) under pressure in the casting and a halogen sensing cell is used to indicate the leak. This test is highly sensitive and is used where the castings are required to meet the highest standard of soundness. This test can only be performed when the casting is sealed with machined fittings.

(b) *Sonic testing.* This test utilizes the velocity of sound to establish the modulus of elasticity of the metal. In iron, modulus is influenced by the shape and amount of graphite present. Sonic testing is carried out with electronic instrumentation which creates and measures the vibration and a transducer that couples the instrument to the casting. One approach utilizes the resonant frequency of the casting to determine its modulus. This method is more applicable to shapes such as crankshafts that have a dominant mode of vibration. An ultrasonic frequency can also be used to measure the sound velocity in casting and the modulus of elasticity can be calculated from this. The attenuation of the sound waves (the rate at which they are absorbed), can also be measured electronically. Sound attenuation is the result of the damping capacity of the metal. In case of cast irons, this can be related to the shape and amount of graphite present and thus indicates the range of other properties.

(c) *Eddy current testing.* This test utilizes an electromagnetic field to measure a combination of the magnetic induction and electrical resistance and thus is very useful for iron and steel castings. The test results can be correlated with the microstructure of castings. This test is particularly useful in indicating the proportions of ferrite and pearlite in the matrix and thus is an indication of hardness. An electrical instrument is used to supply and measure an alternating current to a conferming coil placed around the casting or a critical portion of it. In general, these tests are of comparative type in which emperical values are established with sample castings that vary in characteristics being measured. Some systems employ two coils, one for known casting and another for casting under test. The observations can be by wave form on a cathode ray tube or instrumented to a set of go no-go lights. Separate calibration is needed for each design of casting.

(d) *Service load tests.* These tests can be carried out on the entire casting to evaluate properties. Castings required to carry a structural load can be tested by applying a load in a fixture and measuring the deflection. Pressure containing parts can be hydraulically tested to destruction or proof load. Rotating parts can be spin tested. These tests check the sound ness properties of metal, and appropriatness of the design of casting.

13

Use of Computers in Founders

13.1 INTRODUCTION

A computer is an electronic data processing machine acting on the basis of specific logical instructions fed into it by a human operator. The first generation of computer (1949-1955) contained vacuum tubes, the second generation (1956-1965) used transistors, the third generation (1966-75) are made from integrated circuits, fourth generation (1975-present) are made from very large integrated circuits and fifth generation of computers is expected to be delivered in 1988. It is expected that these computers will use a large number of distributed processors. In foundry third and fourth generations are widely used as large number of variables have to be processed. In future the fifth generation will also find it's application in foundry owing to its high processing capacity.

Various variables or data can be either in the form of numerals or words which are basic facts required to arrive at conclusions or solution to problems. Once the 'Logical Instructions' are fed into the computer along with the data, it takes over to process and delivers the results. The instructions and data together form the input whereas the results given by the computer form the output. Between these two, operations are carried-on in the unit known as central processing unit (CPU), which is the heart of the computer. The CPU consists of three main parts, i.e. the '*store*', the '*control*' and the '*mill*'. The *store* or storage unit holds the input material, processes it, and then stores the processed results until an out put is called for. The *control* unit directs the flow of information and instructions, with their interpretation and issue signals to the rest of the machine. The *mill* is a arithmatic logic unit (ALU), and performs the actual computing and arithmetical operations on the stored number. These three mechanical and electronic components of the computer are referred to as 'Hardware'.

Instructions given to the computer using specific languages are called programs. The computer works solely on the basis of these programs, which are described as 'software'. Software is of the two main kinds, i.e., applications programs, which are useful for specific jobs for the user, and system programs, which control the performance of the computer. Systems saftware is a part of the computer which translates application programs into binary digits, and 'tell' the computer how to run these programs.

Programs and data can be fed into the computer through various input media such as punched cards, magnetic discs and tapes or a key board. The key board is connected directly to the CPU and is commonly used with a T.V. screen which display input as well as output.

Data fed into the computer can be indefinitely held in the storage unit, which is also called the memory. Different computers have different memory capacities, which can be expanded by using auxiliary storage devices. A block diagram of a computer system is shown in Fig. 13.1.

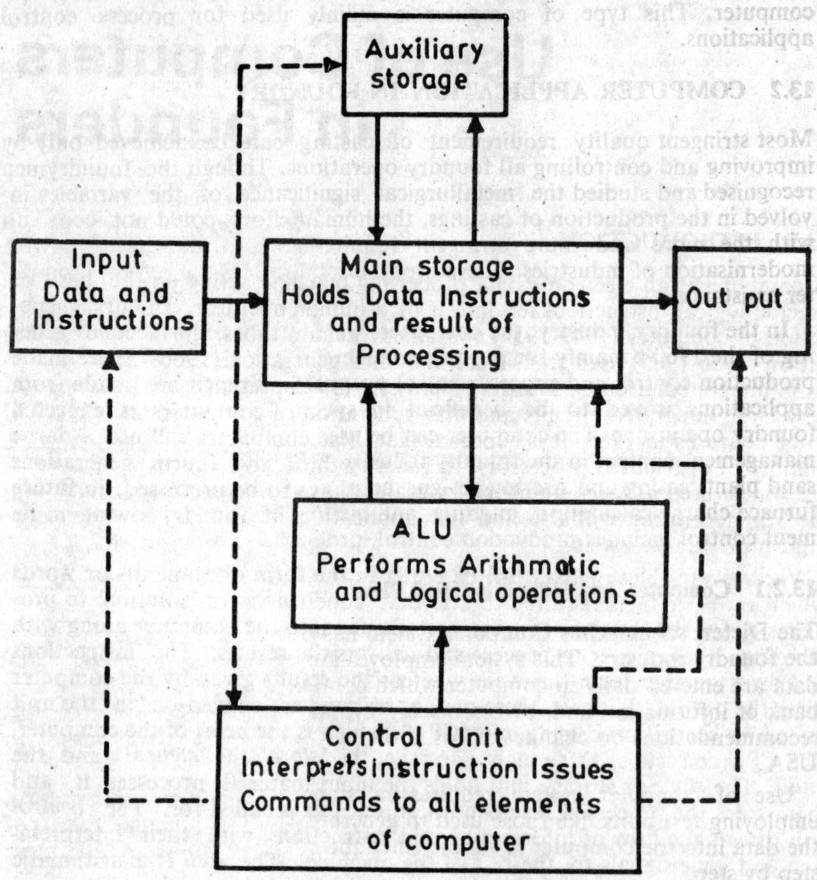

Fig. 13.1 A block diagram of a computer system.

Computers may be classified as micro, mini, super-mini, midi and maxi-based on the CPU speed, word length, memory capacity, the software supporting the system and the cost.

A computer can be associated with the foundry process in the following two ways:

(i) *Off-line Computer*. In this, variables of process are recorded and then fed to the computer from time to time or whenever required. The change in operating variables suggested by the computer is then carried out as and when possible. It is cheap and less complicated. These computers are mainly used for management and commercial applications.

(ii) *On Line Computer*. In this, all variables are continuously recorded by a computer which compares and if there is any deviation, it sends out signals for immediate correction. This is also known as '*In line*' computer. It is costly and complicated but more effective as compared to the *off line* computer. This type of computer is mainly used for process control applications.

13.2 COMPUTER APPLICATION IN FOUNDRY

Most stringent quality requirement of casting can be achieved only by improving and controlling all foundry operations. Though the foundrymen recognised and studied the metallurgical significance of the variables involved in the production of castings, the human efforts could not cope up with the more and more stringent requirements of the castings with modernisation of industries. Therefore, the foundrymen looked for computer assistance.

In the foundry industry, the computers were first applied in the beginning of the 1960's mainly for payroll, commercial purpose and later on for production control and process control purposes. More recently computer applications proved to be beneficial in specific departments of typical foundry operations. The computer can be used in process control as well as management control in the foundry industry. The process control includes sand plant, gating and risering design, core and casting design, moulding, furnace charge calculation, melting and quality control where as management control includes production control, production schedules etc.

13.2.1 Computer Application in Sand Plant

The Dietert schumacher Computer system has been used for sand plant in the foundry industry. This system employes a time sharing computer. Test data are entered daily in computer which compares the data with a vast bank of information and prints out an evaluation of the sand mix, and recommendations on changes. This is being used in Cleveland foundry in USA.

Use of computer system requires a standardized sand laboratory, employing test units like those used to generate the source data. Entry of the data into the computer is simple since the computer requests each item step by step.

Data required for the daily computer program are moisture content, compactability, permeability specimen whight, green compression strength, splitting strength. methylene blue test for total clay and dispersed clay and ball jolt toughness. Values required on a weekly basis are clay content, grain fineness and volatiles.

The green compression strength, compactability, moisture, permeability and weight data can be obtained from the conventional test equipments but some new data, i.e. splitting strength, ball jolt toughness and methylene blue test require specified test procedures.

For splitting strength a standard A.F.S. 5.08 cm (2 inch) diameter specimen is laid on its side and loaded diameterically so that it splits in half with a nominal 25.80 cm^2 (4 sq. inch) area of rupture. The test value is taken from the tensile scale, divided by 10 and expressed as splitting strength in Kg per cm^2.

In ball jolt toughness test, an unsupported 5.08 cm (2 inch) diameter test specimen is subjected to a load by a steel ball. The supporting plate form is jolted unit and the specimen splits under the weight of the ball. The ball jolt toughness is measured by the number of jolts before the specimen fractures.

Total clay and dispersed clay is measured by titrating the clay with methylene blue.

The computer is programmed to compare the input data with an exhaustive bank of sand data, prints out three reports on the conditions of the sand mix and recommends how to improve the mix, if necessary. After input feeding following three types of reports can be obtained:

(i) *Short Control Form Report.* It typically gives instructions to the sand technician about addition of moisture, clay, sea coal. cellulose and even new sand. Additions are listed for three different sand to metal ratios that the foundry might be using at the time.

(ii) *Management Form Report.* In addition to the information included in the short control form, it provides additional background information of value to the manager, i.e., material consumption, manhours, schedules etc.

(iii) *Complete Form Report.* It contains all the information available on the sand system and is intended for close scrutiny by technical management. With this report the plant metallurgist can make decision about the sand system. This report can be generated on a daily basis or as needed.

Fig. 13 2 represents a typical sand plant control system. It also represents how the three reports fit into the overall system. The computer itself requires only about 3 seconds of operating time to prepare the information to be printed out in the three reports, but the mechanical speed of the terminal slows down the process so that the short form, the management form and the complete form report take 2 minutes, 5 minutes and 6 minutes respectively.

At present, the program is designed to provide a once-a-day check of a foundry's sand and on line control remains to be developed.

Computer controlled sand plant offers the following advantages:

(i) It permits operations at rate 10% higher than the conventional method.

(ii) The sand quality is improved, which directly affects the product quality and scrap level.

(iii) Control of the operation and detection of defects lead to locating of breakdown and thus more preventive maintenance can be achieved.

(iv) More important logic equations can be taken into account.

(v) The accuracy of computer control is greater than that of any other method.

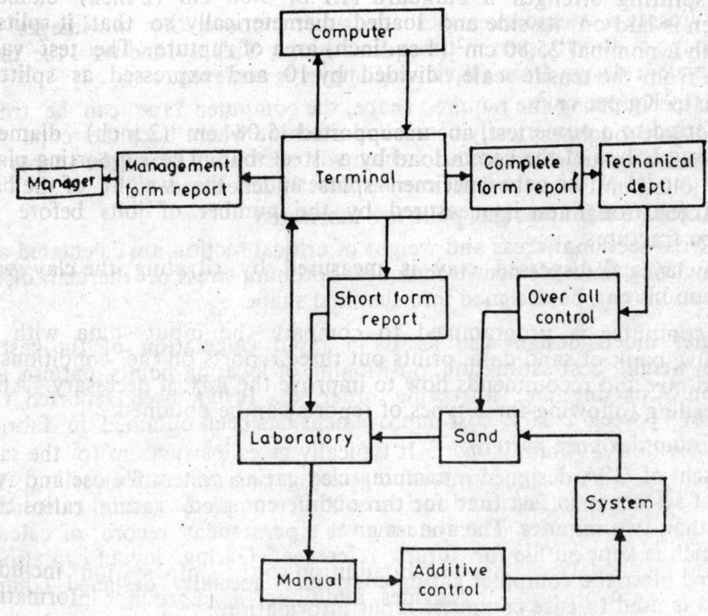

Fig. 13 2 Typical sand plant control system.

(vi) The computer aids management for cost control.

13.2.2 Computer Application in Casting and Core Design

Computer aided design (CAD) and camputer aided manufacturing (CAM) system are developed for manufacturing the complex shapes required with coremaking and casting. The Cambridge University's Computer Laboratory has developed a system which enables a simple pipe bend casting to be designed and the patterns produced within two days. In addition to the benefits of such a short lead time, other important benefits are as following:

(i) Better shape definition and the production of patterns by numerically controlled machining which result in improved dimensional control. In consequence, less metal needs to be removed during fettling operation with corresponding improvements in profitability and yield.

(ii) More reliable implementation of casting methods leads to better yields through reduced feeder sizes and scrap production.

Computer aided design is based on the following:

(i) Critical design elements of a core are first drawn to scale according to print and entered into the computer.

(ii) Different smoothing connections are then applied between the patches so as to arrive at an approximately acceptable shape.

(iii) The entire shape is then reproduced by the computer on an X-Y plotter to form a stereo pair.

(iv) By using a stereo magnifier (to super nnpose the two images) the shape is directly seen in three dimensions. This is done until the most efficient design from all viewing angles is achieved.

(v) After finalizing the required shape, the computer tape can be transferred to a numerically controlled machine where both the cope and drag halves of the corebox (which will accurately make the shape) or a model of the core itself can be directly machined.

(vi) Accurate left and right pairs are fabricated.

(vii) Cross-sectional areas and weight of critical section are calculated and by using finite element techniques, optimum stress or thermal distributions can be designed into the final shape.

Computer aided design also assists in rapid calculation of the casting volume, weight, cost estimation, positioning of locating points, pattern and selection of parting line, a freezing order etc. It has been reported that saving of 7½ week at 20% cost improvement has been obtained in fabrication of complex core pattern.

Kotschi of USA designed a computerized gating system for casting. The speed of system is so fast that one can obtain complete gating calculation in less than two minutes. The system gives a permanent record of calculation which is kept on file for future reference. Gating layout can also be improved after the computer gating analysis. A specially designed computer card is used to ease computer input information.

The system requires the input data: (a) metal code, (b) casting weight, (c) pouring temperature, (d) average thickness, (e) alloy density, (f) sprue height. (g) pattern depth, (h) pattern height, (i) gating ratio, (j) number of runners (k) H/W ratio of gate, (l) number of gates, (m) part number, and (n) customer name.

The output displayed by the computer on T.V. screen are: (a) fluidity, (b) effective sprue height, (c) sprue diameter, (d) pouring time in seconds, (e) cross-sectional area and dimensions of choke, (f) cross-sectional area and dimensions of runners, and (g) cross-sectional area and dimensions of gates.

This system has good response to the non-pressurized system (i.e. gating ratios of 1 : 2 : 2 and 1 : 3 : 3). In case of the pressurized system some modifications are required.

By using this computer program, the danger of runout and misrun defects of casting are almost eliminated as this allows the metal to enter the casting in less time and thereby owing to adequate layout of the gating system, the amount of scrap is reduced.

Computer program can be designed for the determination of the most cost effective feeding system. Simplification and speeding up calculation represent only a part of the total benefits of computerisation, the real benefits include (a) consistent results (b) permanent printed records of calculations, and (c) computer data storage with the facility for almost instant data – retrieval. The program requires the input data (a) the weight of the casting is to be fed from one feeder, (b) inscribed sphere, (c) largest dimension, (d) shrinkage factor, (e) shape, (f) number of non-cooling surfaces, (g) binder type, (h) ingate, (i) type of feeder (side or top), and (j) safety margin The program contains following facilities.

(i) *Fecder size calculation* : This is the main function or the program, occupying over 50 % of the total program content. The optimum height/diameter ratio of the feeder can vary according to the type of feeding aids used. The program offers several (usually three) height/diameter ratios for sand feeders (1/1, 1.5/1 and 2/1 and sleeved feeders (0.7/1, 1/1 and 2/1). The computer prints out the different feeder dimensions for each type of feeder (i.e. sand feeder, sleeved feeder) for different height/diameter ratio, the amount of anti-piping compound to be added to each feeder, the minimum neck modulus connecting the feeder to casting, and yield of each feeder.

(ii *Feeding distance* : The program facilitaties the calculation of different feeding distances which help to determine the number of feeder and position of the feeders. With the help of these distances, the position and number of the risers is adjusted in such a manner that the system yields a sound casting.

(iii) *Cost analysis* : Using the co st data supplied by the computer or by the operator comparision can be obtained between the costs involved in susing different feeding systems. A typical example of comparision (sand feeder and sleeve feeder) by computer is shown in Table—13.1.

(iv) *Wcight calculation* ; Operator has choice to select a best shape out of various options, such as block, clinder, annulus, truncated cone, triangular prism, sphere, spherical cap etc. The best choice can be found out with the help of computer as it facilitates to calculete the weight, volume, surface area and modulus of all these shapes with constant weight of casting.

(v) *Side neck calculation* : The program can be made to calculate the optimum size of the feeder necks using modulus values obtained from feeder section of the program.

(vi) *Print customer details* : In addition to above mentioned facilities, computer gives a copy of customer name with casting details.

(vii) *Study of effect of one input change*: Often the operator may wish to ascertain the effect of changing only one of the data inputs (such as ingate position) or compare a side feeder with a top feeder and to enable rapid assessment of input change, the program contains a facility for changing any of the input without the need for going through the whole input routine.

13.2.3 Computer Application in Moulding

Mould design can be facilitated by computer in the following way.

(a) After receiving an engineering drawing of a casting, the numerically controlled machine programers quickly duplicate the part on the graphics terminal.

(b) The mould design is then generated by defining the mould geometry around the existing part geometry with all the allowances.

(c) Core setting requirements can be computed automatically permitting proper availability of cores, manpower, fixtures and chaplets at the mould line.

(d) Automated mould system cycles and operations can be maintained and altered and therefore the control on jolt duration, squeeze pressure, moulding rates and sand quantity can be obtained.

(e) Where in the mould inoculation is used, correct amount of materials can be calculated by computer.

(f) Programmable controllers can be used to assure correct placement of materials.

13.2.4 Computer Application in Melting

In melting section microprocessers are used to identify the least cost charge and alloying additions. As this involves linear programming (L.P.) problem a computer giving quick solution on the shop floor itself is required. By using computer, quality and analysis consistency can be tracked and corrected to compensate for deviations from accepted norms. Variances in scrap metal and chemical analysis can be recognized and thus alloy additions adjusted.

Fluid metal requirements can be adjusted automatically to match variations in moulding operations. Melts rate, chemical consistencies and temperature can be monitored automatically and altered by minicomputers and programmable controllers. In melting section, computer can be employed for (a) charge calculation and (b) furnace control.

(a) *Charge calculation*: Linear programming is potentially of great value in furnace charge calculation. Prior to advent of cheap, ready available electronic data processing, the volume of data, and the extent of calculations involved in linear programming prevented industry from utilizing the technique on realistic sized problems. Hand solution with simplex method can be used for a system containing only 3 or 4 elements. The solution to more complex problems can be obtained quickly only with the help of computer.

Package programs for linear programming have been developed which contains 'instructions' on the method of data processing. With these packages, the data is input to the computer in a specific form and run with the program which is usually stored on magnetic discs or tapes. Large problems involving 40 or 50 materials and as many elements can be handled by most of these packages. Some programs handle considerably more variables.

Following advantages have been claimed for computerized linear programming in the area of charge calculation.

(i) Reduction in alloying costs due to the most effective use of the elements present in raw materials.

(ii) Increased accuracy of charge calculations resulting in fewer 'Off specification' castings.

(iii) Reduction in time required for the calculation of charge.

(iv) Indication of stock materials uneconomic in use.

(v) The availability of advice for purchase department upon the suitability of materials offered to them.

(b) *Furnace control*; The computer special system staff of Digital Equipment Corporation evolved a furnace control system which yields saving in power demand while maintaining the same production level. The system monitors total foundry power usage and controls operation of the electric furnace to reduce power demand in the following three ways.

(a) Averaging the demand of each furnace over two or more demand intervals.

(b) Delaying activation of melt power at the start of a melt cycle.

(c) Turning furnace melt power on and off during a melt cycle.

In electric arc furnace the computer pulls the furnace electrods as and when actual furnace demand rises to a critical level (calculated by the computer according to the value of the self imposed demand limit setting). All important process signals and data on charging variables are recorded continuously.

The computer controls and measure (a) melt down of scrap charged (b) refining of the steel (c) alloy additions necessary (d) oxidation losses and (e) energy input and losses. Amount of exhaust gas and cooling water requirement also are part of the computer printout. Following melt down, metal temperoture is recorded continuously which helps in achieving better production rates. Heat logs are written automatically and close control of all operations is achieved. A 15 % saving of melting costs has been reported by the use of computer in electric arc furnace. In induction furnace, use of computer helps in automatic control in input of power.

In the case of cupola, the applications consist of control of several operating parameters in a manner which optimizes the cupola operation at minimum cost. The computer surveys the quality and the quantity of the metai produced and maintains these by acting on the control variables of wind and charge. The benefits obtained are (a) reduction of coke and oxygen comsumption (b) higher and better controlled yields, and (c) imposed uniformity of analysis.

In gas and oil fired furnaces, the application of microprocessors improves combustion condition with saving upto the extent of about 35 pct. The closed loop controller monitors air and fuel flow rates, furnace pressure, stack gas temperature, and oxygen content, and automatically adjusts servo-assisted valves and dampers in order to optimise the furnace performance.

Various other furnaces which are rarely used in foundry industry (i.e. L.D. convertor, open hearth etc.) can be computerized to give best performonce.

Various benefits claimed by the computer application in melting can be summerised as follows:

(i) Increase in productivity.

(ii) Less energy consumption.

(iii) Better chemical composition control of liquid metal.

(iv) Maximum utilisation of furnace capacity.

(v) Better working conditions for labour.

(vi) Minimum break-down.

13.2.5 Computer Application in Quality Control

Computer assistance in quality control appears to be particularly attractive proposition. A computerized statistical control system with graphics displays and print outs can monitor incoming materials used in the metal casting process. It also permits quality control departments to exert much better control over raw material consistency and quality. On demand the computer allows quick conparisons of vendor product quality to pre-established purchasing specifications.

Fortran IV pragoams are developed especially for this analysis, in which one month's data are processed in less than 1/2 hr. on a computer. The programs produce summary information as well as the specific values repuired for the establishment. The computer adds to the capability to contrl by providing accurate information to management in a short of time, enabling intime correction. The additional advantage resulting from the low processing time is the cost saving over hand calculaion methods.

Pertinent information for the statistical analysis is put into the proper form for input to the computer. The computer processing produces reports that first are analysed by staff personel. At this point, control chart updates would be made and out of control situations identified. Results then would be translated to appropriate levels of management for any corrective actions deemed necessary.

Control charts are setup to monitor rejection rates for each part classification. It enables a guide for identification of the particular part causing problems. Pinpointing the cause or at least the general area of the problems again led to a quick solution.

13.2.6 Computer Application In Management

In foundry management computer can assist in the following areas.

(i) Production scheduling.
(ii) Production control.

(i) *Production schedulimg*: After examining the current order book for jobs required within a specified time period, prodution facility limitations are examined in turn to ascertain which jobs might be loaded into the schedule. The jobs with shortest lead time (or perhaps with some other characteristics upon which priority grading could be given) would be loaded if they satisfy production constraints. The production time required for these is added on to a 'clock' mechanism with in the computer, and the computer would then repeat its reach for jobs which it could load for the next time period.

(ii) *Production control*: Manual production control system still predominates in the industry but in many ways is inadequate as the information provided is very basic and often only of historic interest. Computer offers a ready solution to improved production concrol. From the customer's orders which are classified monthly, the computer gives plan week by week according to the production facilities available. It also calculates the loading of the different shops. It forecasts all the raw materials required for the castings.

The computer analyses the foundry operation and provides various reports, i.e. performance reports, scrap reports and stock availability report which are very useful for the cost control. It has been reported that through the use of a computer for production control, it had cut inventory by 30%

achieved 5% improvement in efficiency from both direct and indirect labour, were able to handle a higher volume of orders without any additional staff and most strikingly scrap had reduced by 66%.

Main advantages of computer claimed in production control are (a) increased flexibility, (b) fewer overmakes (c) less scrap (d) increased productivity through better loading and use of resources and (e) reduced debtors and other intangibles,

13.2.7 Miscellaneous Applications

There are some other fields where computers can assist the foundry industry which include (i) fettling (ii) spectrometer (iii) purchasing, shipping and receiving.

(i) *Fettling*: The computer can play a big role in controlling the shape of a casting with respect to a reference casting as it memorizes all the geometric co-ordinates. The role of the computer in this area of operation is perhaps that played in the numerical command system, i.e. the tendency being towards the mini-computers controlling only a few machines for reasons of security, independence and flexiability

(ii) *Spectrometer with computer*: Connection of a computer to a spectrometer coverts analogue measurements to real time values indicating the composition of the casting and print out the results. The computer using a stored program based on metallurgical theory, determines the amount of elements which need to be added to meet specifications. These are outputs, which ar communicated to a plant operator who adds the requisite quantities. In an automated system, the output will be used to operate feeders for addition of elements. Load cells will measure the weights added and stop the feeders when appropriate quantities are charged into the furnace.

(iv) *Purchasing, Shipping and Receiving*: Computer can be used for inventory control of parts to be shipped, raw materials storage and inventory. They can promote least cost purchasing of raw materials by calculating price and quantity limits of raw materials.

13.3 COMPUTER APPLICATION IN INDIAN FOUNDRIES

Due to heavy investment involved and lack of required knowledge, so far only limited use of computer has been made in India. Some known examples are given below:

(i) In Alcobex Metals Pvt. Ltd., Jodhpur, computer is attached with spectrometer to give the chemical analysis results.

(ii) In Forge and Foundry Division of HAL, Bangalore, computer is used to control the electric are furnaee.

(iii) The BHEL Ranipur and Semi-conductor complex Ltd., Chandigarh are going to instal computer in the foundry shop for process control.

14

Production of Cast Iron Castings

Iron—carbon alloys with more than 2 per cent carbon are classified as cast irons, although the percentage of carbon generally varies between 2.5 to 3.6 pct. Besides carbon, other impurity elements such as silicon, Mn, S and P are always present in cast iron which influence its properties. Typical composition of cast iron may be as follows:

$$C = 2.8 - 3.6$$
$$Si = 1 - 3$$
$$Mn = 0.4 \text{ to } 1.0$$
$$P = 0.1 - 1.0$$
$$S = 0.05 - 0.35$$

From the above it is obvious that as compared to steel, cast iron contains greater amounts of C, Si, P, and S to bring about a significant change in the microstructure. In the absence of plasticity, components in cast iron are produced by casting and machining. Owing to its cheapness as compared to other alloys and also because of the following desirable attributes, cast iron is popular for various uses.

1. The melting range of cast iron is considerably lower (1150 – 1250 °C) and casting properties are comparatively superior as compared to steel.
2. Cast iron exhibits better corrosion resistance in acidic and other common chemical media.
3. Various types and grades of cast iron with a wide range of properties are available which enlarges the area of their applications.
4. Grey cast iron possesses better machining and damping capacity as compared to all other ferrous materials. It also has good wearing properties.
5. Chill cast white iron possesses high hardness and good wear resistance which has proved useful for many applications.
6. Malleable iron and nodular iron have considerably enlarged the scope and uses of cast iron for different engineering applications owing to desirable combination of toughness, impact resistance ane improved

TABLE 14.1 Range of Compositions for Typical Unalloyed Cast Iron

Type of Iron	percent element				
	Carbon	Silicon	Manganese	Sulphur	Phosphorus
1. Grey Cast Iron.	2.5 — 4	1.0 — 3.0	0.2 — 1.0	0.2 — 0.25	0.02 — 1.0
2. White cast Iron	1.8 — 3.6	0.5 — 1.9	0.25 — 0.8	0.06 — 0.2	0.06 — 0.2
3. Malleable Iron (Cast White)	2.2 — 2.9	0.9 — 1.9	0.15 — 1.2	0.02 — 0.2	0.02 — 0.2
4. Nodular (ductile) Cast Iron	3.0 — 4.0	1.8 — 2.8	0.1 — 1.0	0.01 — 0.03	0.01 — 0.1
5. Compacted Graphite Iron	2.5 — 4 0	1.0 — 3.0	0.2 — 1.0	0.01 — 0.03	0.01 — 0.1

plasticity. These cast irons have substituted steels and many non-ferrous metals for specific applications.

7. The properties of cast irons can be modified to develop special characteristics by the addition of different alloying elements.

Because of the above reasons the production of different types of cast iron is greater than all other cast metals and a very large number of uses have been developed for different engineering applications.

14.1 CLASSIFICATION OF CAST IRONS

Chemical composition of different cast irons is given in Table 14.1. Based on chemical composition, structure and properties, cast irons can be classified as follows:

(i) *Grey Cast Ircn* (Fig. 14.1 (a))

Owing to the grey or black appearance of the fracture, these cast irons have been given the above nomenclature. Carbon is generally present in the form of graphite which appears as flakes under the miroscope. The shape and distribution of graphite flakes determines the properties of grey cast iron. This class of cast irons are soft, brittle and easily machinable, their hardness varying between 180 to 240 BHN. These are widely used owing to their low cost and easy melting. The presence of graphite flakes imparts excellent damping capacity, thereby making grey iron very propular for the base of various types of machines.

Ingot moulds, automobile cylinders, water carrying pipes and numerous comparatively less expensive castings are made in grey cast iron.

(ii) *White Cast Irons* (Fig. 14.1 (b))

The fracture of these irons is comparatively white owing to the presence of almost all carbon in the cambined form as cementite. White cast irons are hard (400—500 BHN), almost unmachinable and brittle which require the use of grinding to bring about minor adjustment in shape and size. Main uses of white cast iron include chilled rolls for rolling mills, and chilled balls used for grinding in cement and mineral dressing industry where the wear resistance is improved by judicious control of chemical composition and chilling. While cast iron is also the starting material for the production of malleable iron components where the desired rosette (nodular) structure of graphite is obtained by suitable annealing treatment.

(iii) *Mottled Cast Iron* (Fig. 14.1 (c))

These cast irons show a structure which is a combination of grey and white cast iron structure, i.e. a part of the carbon is present in the combined form as cementite and the rest is present in the form of graphite flakes. For this reason, the fracture is mixed with white and grey portions.

(iv) *Malleable Cast Iron* (Fig. 14.1 (d))

This is obtained by annealing treatment of appropriate white cast iron. In the microstructure, carbon is present as rounded rosettes of graphite

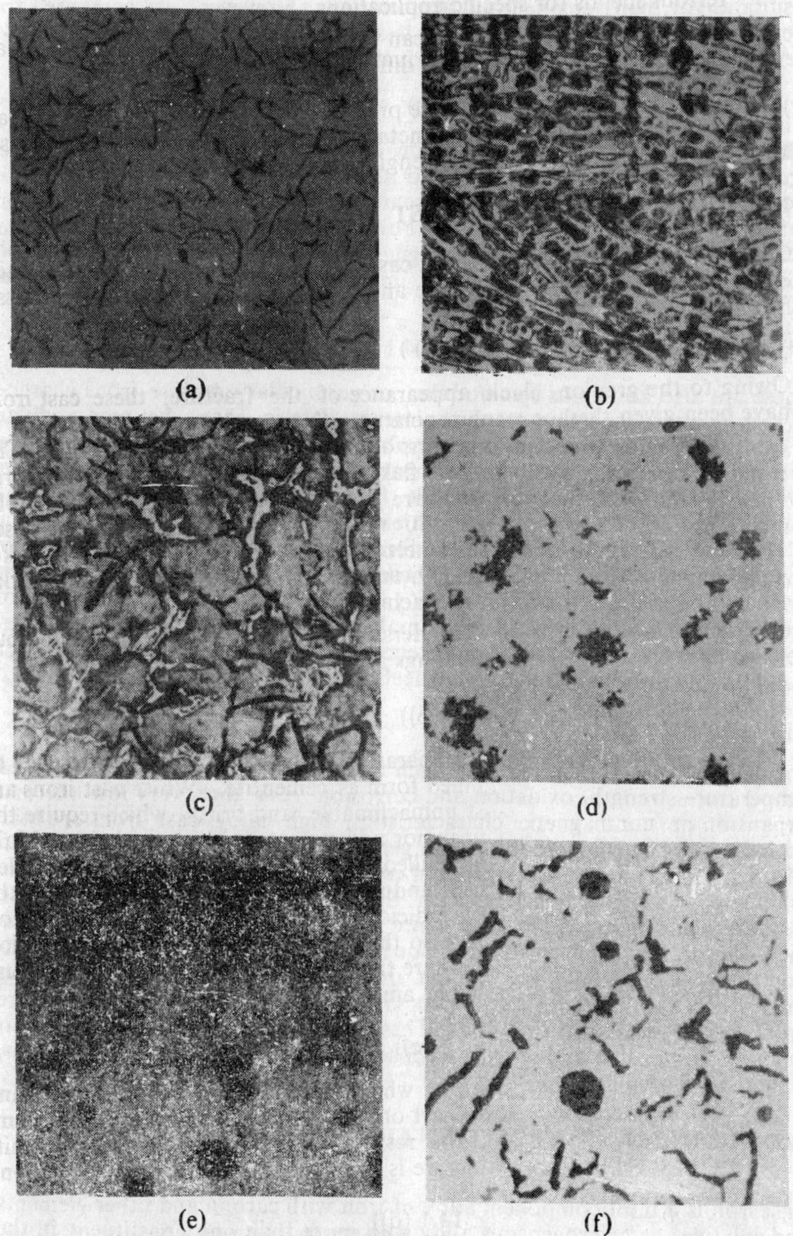

Fig. 14.1 Structres of different cast irons: (a) grey cast iron, (b) white cast iron, (c) mottled cast iron, (d) malleable cast iron, (e) nodular cast iron and (f) compacted graphite iron.

ductility and other properties are modified and improved, thereby making it suitable for many engineering applications. However, as compared to grey and white cast irons, it is much less brittle and can be bent by hammering, but it is not malleable in normal sense.

(v) *Nodular Cast Iron* (Fig. 14.1 (e))

This is also known as ductile iron or spheroidal iron. As compared to ordinary cast iron it is not brittle and its properties are similar to malleable iron. In the structure, carbon is present mostly as nodules of graphite or partly in the combined form. The above structure is obtained by suitable treatment of molten metal with magnesium or cerium. Development of nodular cost iron constitutes a very important metallurgical achievement of the present century.

(vi) *Compacted Graphite Iron* (Fig. 14.1 (f))

These are also known as seminodular or vermicular graphite irons characterised by interconnecting graphite intermediate in shape between nodular graphite of ductile iron and flake graphite of grey iron. These irons have become commercially available comparatively recently and the production method is similar to that for ductile irons requiring close metallurgical control. In addition to the rare earth elements addition, controlled addition of titanium is made to minimise the formation of nodular gtaphite, thereby giving compacted graphite shape. These irons retain the castability of grey irons and have strength and ductility values intermediate between those of grey iron and ductile iron, thereby making them suitable for such applications as heavy duty brake drums, eccentric gears, valve bodies, exhaust manifolds, hydraulic cylinders, flywheels and a variety of housings, etc.

(vii) *Alloy Cast Irons*

For meeting special requirements such as extreme wear resistance, elevated temperature strength, oxidation and corrosion resistance, very low thermal expansion or nonmagnetic characteristics, high alloy cast irons are used with 3 to 40 pct. alloy additions. In general alloying elements like Ni, Cr, Mo, Cu etc. are added to improve these properties. Except malleable irons where heavy alloying interferes with the metallurgy of malleable process, high alloy additions are made to white irons, grey irons, and ductile irons, the addition requiring special technique of melting rather than simple addition to the ladle containing molten iron. In view of the special melting, casting and quality control techniques involved, high alloy iron castings are produced by specialised foundries. These irons are usually specified by their chemical composion and specific mechanical properties wherever important, may also be specified.

14.2 CEEMICAL COMPOSITION-EFFECT ON STRUCTURE AND PROPERTIES

Cast iron is a multicomponent alloy of iron with carbon and other elements and solidifies as heterogeneous alloy with more than one constituent in the microstructure. Because of the presence of high carbon and silicon in cast irons, usually 1—3%, they possess lower melting temperature, better fluidity and lesser reactivity with moulding materials as compared to steel. Owing to the formation of low density graphite during solidification, liquid to solid volume shrinkage is reduced and production of complex castings is facilitated.

Various types of cast irons are generally designated on the basis of mechanical properties rather than chemical composition. Because of similarities between different types, other important factors which are vital for successful production of different types of cast irons are:

(a) Inocculants which act as nucleating agents in the production of grey iron to control the size and type of graphite,

(b) Trace amounts of minor elements, e.g. use of bismuth and tellurium in the production of malleable cast iron and presence of residual magnesium which causes the formation of nodular graphite in S.G. iron,

(c) Adjustment of composition required for the production of same grade of small castings and large casting.

High alloy Iron castings have a very wide range in base composition and also contain major amounts of other alloying elements and are used in applications demanding extreme abrasive wear, heat or corrosion resistance, besides other specific properties. As these property requirements are often difficult to establish and verify, high alloy irons are usually specified by their chemical composition.

Carbon Equivalent

Iron-carbon alloys have a eutectic at 4.3 pct. carbon but owing to the presence of silicon, the carbon content of eutectic is decreased. This is a linear relation and it is convenient to express the combined effect of carbon and silicon as carbon equivalent (C. E.)

$$\text{C.E.} = \%\text{C} + \tfrac{1}{3}\%\ \text{Si}$$

In presence of appreciable amount of phosphorus, the phosphorus content of the iron is also included in carbon equivalent

$$\text{C.E.} = \%\ \text{C} + \frac{\%\ \text{Si} + \%\ \text{P}}{3}$$

The carbon equivalent establishes the solidification temperature of alloys as also their foundry characteristics and properties. However, it is noteworthy that irons of constant carbon equivalents but with appreciably different carbon and silcon contents will have different casting properties, e.g. carbon is more than twice as effective in reducing solidification shrinkage and silicon is more effective in preventing thin section from becoming hard and this is not adequately represented by the carbon equivalent equation.

Silicon

It is invariably present in all cast irons, decreases the solubility of carbon in liquid and solid solutions and encourages graphilization. With an increase in carbon equivalent the amount and size of graphite inclusions increase and the proportion of pearlite and its degree of dispersion decrease resulting in impairing the mechanical properties.

Manganese

It combines with sulphur to form manganese sulphide which occurs as randomly distributed particles in the microstructure. In the absence of Manganese, iron combines to form a low melting eutectic (m.p. 985 °C) at the grain boundaries during solidification and this impairs the mechanical properties. For effective neutralisation of harmful effect of sulphur, the manganese to sulphur ratio should be 4 to 5.

Manganese in excess of this amount aids in the formation of final structure as pearlite and increases strength and hardness.

Sulphur

Sulphur can dissolve in unlimited quantity in liquid iron but has insignificant solubility in solid iron. With an increase in sulphur content of iron to about 0.12 pct. or more, fluidity of liquid iron decreases sharply and hard spots appear in the castings. This is due to higher amount of cementite and pearlite in the structure resulting from reduced solution of carbon and silicon in iron and slower distingration of cementite.

Phosphorus

It reduces the temperature of eutectic transformation, aflects eutectic composition and promotes metal fluidity, therefore high phosphorus content (1 pct and above) is preferred to produce thin sections and highly detailed grey cast iron castings. It reduces the solubility of carbon in iron and has insignificant effect on graphitisation. It is soluble in iron upto 0.3 pct. but in larger amounts forms a separate phosphide eutectic, $Fe_3P - Fe_3C\text{-}Fe$ (steadite) which melts at 950 °C. With phosphorus content of about 0.6 pct. a cellular net work of steadite is formed which appreciably reduces the machinability and impact strength of iron. A phosphorus content of 2 pct. or less does not have any measurable adverse effect on the properties of grey iron because of good dispersion of small phosphide particles in the structure. For some specific applications, high phosphorus grey irons provide satisfactory wear resisting and non glazing surface.

Other Minor and Alloying Elements

Additions of alloying elements over 0.1% and up to 3% are generally made to modify or enhance the properties of base iron, whereas alloy contents from 3 to 30 pct. are used to produce irons with entirely different properties. The latter class are designated as high alloy cast irons and have unusual properties.

The effect of various alloying elements on structures and properties of cast irons are given in Table 14.2.

14.3 TYPES OF GRAPHITE

The shape, size and distribution of graphite in the structure exert a pronounced influence on the properties of cast irons. ASTM (A – 247) has classified seven basic types of graphites (Fig. 14.2) which occur in cast irons. Flake graphite form has been further classified into five types from A through E (Fig. 14.3).

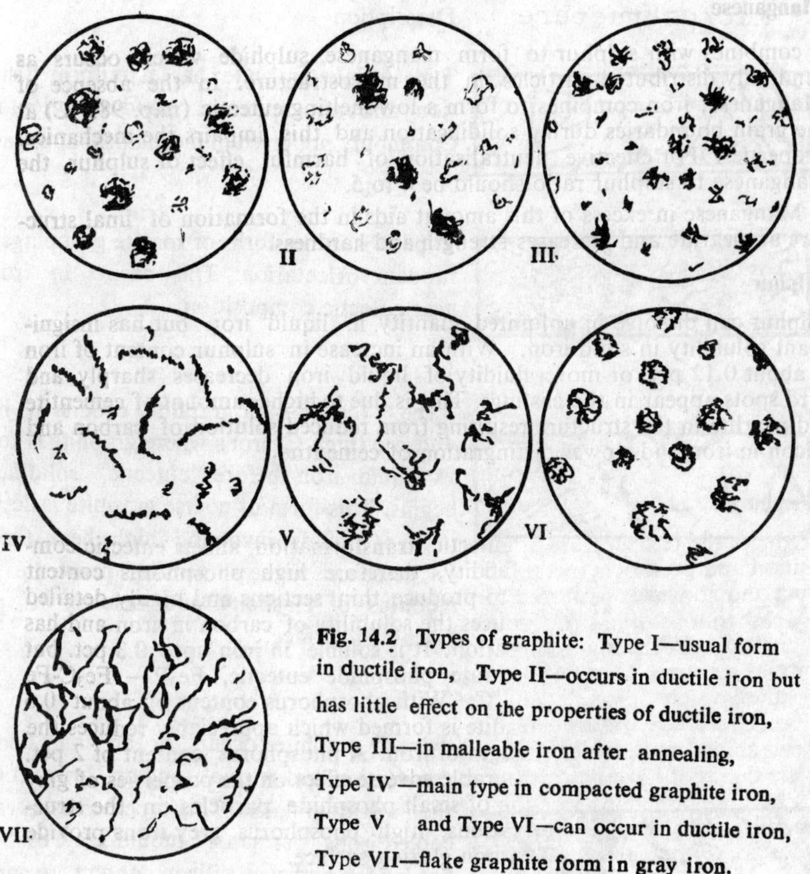

Fig. 14.2 Types of graphite: Type I—usual form in ductile iron, Type II—occurs in ductile iron but has little effect on the properties of ductile iron, Type III—in malleable iron after annealing, Type IV—main type in compacted graphite iron, Type V and Type VI—can occur in ductile iron, Type VII—flake graphite form in gray iron.

14.4 GRAPHITE SIZE

In addition to the form and type a standard for evaluating the size of flakes and nodular forms of graphite has been provided by ASTM specifications (A – 247). These sizes increase geometrically from 1 mm in length or diameter at 100 magnification for size 8 to 128 mm for size 1.

14.5 METALLURGY OF CAST IRONS

The properties of different cast irons are established by the following factors:

(a) The form in which the major portion of carbon occurs, i.e. as carbide or in some shape of graphite.

Microstructure	Description
	Type A flake graphite has a uniform distribution and an apparant randam orientation, Commonly preferred in mechanical applications.
	Type B is in the form of rosette groupings with randam orientation. This occurs in irons of near eutectic composition.
	Type C (called high graphite) occurs in hypereutectic (high C) irons where graphite is formed in liquid iron before eutectic solidification begins. This form of coarse graphite is desirable in ingot moulds involving high heat transfer. Its presence reduces tensile properties and modulus of elasticity and results in pitted machined surface.
	Type D graphite occurs in an inter dendritic segregation and with a randam orientation of the individual flakes. This type occurs in rapidly cooled irons (permanent mould grey cast iron) but having sufficient silicon content to prevent the retention of iron carbide. Irons with this type of grophite have higher tensile properties. However it is difficult to obtain an entirely pearlitic matrix.
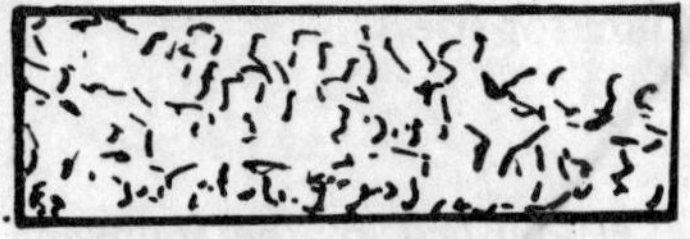	Type E occurs in low total carbon irons (strongly hypereutectic) with a dominant primary structure of austenitic and a normal eutectic solidification. The graphite formation is confined to the interstics between the austenite dendrites so that the graphite is classified as interdendritic with preferred orientation.

fication of flake graphite i.e. Type VII.

TABLE 14.2 Effect of Various Alloying Elements on Structures and Properties

Element	*Effect on structure during solidification*	*Effect onstructure during eutectoid reaction*	*Effect on properties*
1. Aluminium	Strong graphitizer	Promotes ferrite and graphite formation	Improves oxidation resistance. Small amount is used to control chill and annelability in malleable iron. Risk of pinholing increases. Eliminates the effect of nitrogen in producing compacted graphite.
2. Antimony	Little effect in amounts used.	Strong pearlite stabiliser	Lowers tensile strength, transverse strength and impact strength and raises hardness.
3. Arsenic	Effect is insignificant in amounts usually present.	Stabilises pearlite	Increases tensile strength at high temperatures.
4. Bismuth	Carbide promoter but not carbide former	Very mild pearlite stabiliser, promotes widmanstaten graphite	Marked reduction in tensile strength. Small amounts (0.002%) control chill and annealability of malleable iron.
5. Boron upto 0.15 pct.	Strong graphitizer	Promotes graphite formation	
Boron 0.15	Carbide stabilizer	Strong pearlite retainer	Interferes with chill control methods.
6. Chromium	Strong carbide former. Forms complex carbides which are very stable.	Strong pearlite former	Prevents formation of free ferrite in critical sections. Machinability dec-creases with increase of chromium content.

contd.

7. Copper	Mild graphitizer	Promotes pearlite formation	Improves tensile strength, hardness and corrosion resistance by many acids and sulphurus fumes in atmosphere. Increases hardenability.
8. Nickel	Graphitizer	Mild pearlite promoter	Improves tensile strength, hardness and corrosion resistance. Increases hardenability.
9. Manganese	Mild carbide former	Pearlite former	
10. Molybdenum	Mild carbide former	Strong pearlite former	Increases the tendency of a casting to chill and to form free carbides. Increases the strength of casting.
11. Silicon	Strong graphitizer	Promotes ferrite and graphite formation	
12. Tellurium	Very strong carbide promoter but not stabilizer	Very mild pearlite stabilizer	When used below 0.0007 pct, it is most potent chitl producer. Excess tellurium gives widmanstatan or mesh graphite with grave loss of tensile strength.
13. Tin	Little effect with amount used	Strong pearlite retainer	Improves tensile strength, hardness and modulus of elastisity, particularly in hypereutectic irons caused by elimination of free ferrite.
14. Titanium under 0.5 per cent	Graphitizer	Promotes graphite formation	Eliminates the effect of nitrogen in producing fissure defects in sand castings, or compacted graphite in ingot moulds.
15. Vanadium	Strong carbide former	Strong pearlite former	Improves wear resistance, tensile strength, hardness and resistance to growth and creep.

(b) The structure of the matrix (ferrite, pearlite, bainite, martensite or austenite) in which the iron carbide or graphite occurs.

Both the above factors exert a pronounced influence on the properties and determine the grades of cast irons.

14.6 MOULDING PRACTICE FOR GRAY CAST IRON CASTINGS

For the production of gray cast iron castings, several moulding processes such as green sand moulds, core moulds, shell moulds, CO_2 process moulds and permanent moulds are used, but the greatest tonnage of castings is produced by green sand moulding. Owing to the lack of suitable die materials, die castings of gray iron are restricted to a few highly specialised applications. It is possible to cast gray iron in invertment moulds, but such castings are seldom produced owing to economic considerations.

Owing to the precipitation of graphite during solidification, the volume shrinkage in gray iron is largely compensated and the grade of the iron poured governs to some extent the design of the gating and feeding system, e.g. a class – 20 gray iron may actually expand slightly rather than shrink on solidification and a class – 50 iron will show only 4 per cent volumetric shrinkage. In general, gates for gray cast iron castings are 50 to 80 per cent smaller than those required for comparable castings in steel or aluminium.

14.7 MALLEABLE IRONS

In malleable iron the major portion of its carbon appears in the microstructure as nodules of temper carbon. For the purpose of casting into moulds, the composition of molten metal is suitably adjusted particularly in regard to carbon and silicon content, so as to produce white iron castings completely free from primary graphite. The white cast iron castings are then heat treated at a temperature above 950 °C for an appropriate period to dissociate the iron carbide in the matrix as temper carbon, because it is forming in the solid state during heat treatment.

Commercially two types of malleable iron are known as *white heart* and *black heart* basically derived from the colour of the fracture of the annealed material. Alloyed malleable irons or special malleable irons are modifications of the above two types and are known as pearlitic grade. These irons are most important in various engineering applications and are of comparatively recent development. Rapid solidification required to form white iron in the as cast condition, limits the thickness for a malleable casting to about 75 mm. Apart from other factors, the practical section thickness limitation has been a major factor in the growing popularity of ductile iron over malleable iron castings.

As compared to grey iron, malleable iron is stronger, ductile and possesses greater toughness and fatigue strength. Its machnining characteristics are comparable to grey iron. It can be surface hardened to develop good wear resistance. Automotive components, agricultural equipment and rail road industry are major users of malleable iron castings.

For the production of white heart malleable iron castings, the base metal generally has relatively high carbon content (3 per cent and above) and a low silicon content (around 0.5 to 0.6 pct.) which on solidification essentially produces a white cast iron structure. The castings are annealed in a decarburising atmosphere in order to eliminate as much carbon as possible from the surface of the casting by oxidation, thereby producing a soft and ductile skin merging into a high carbon centre consisting of nodules in a pearlite matrix. Owing to non-uniformity of structure throughout the casting, this material is not considered as pearlitic malleable cast iron.

The decarburisation in the heat treatment process is brought about at about 1000 °C by packing the castings in contact with hematite iron ore. The carbon at the surface cf the castings is oxidised and is lost as CO_2 thereby leading to the diffusion of more carbon out wards from the interior and its oxidation. The final structure of thin section may be completely ferritic which on fractering appears steel white and hence the name white iron.

The age old packed decarburisation process has been replaced by gaseous decarburisation process in which air as well as saturated steam are circulated in the heat treatment furnace leading to the formation of the reaction gas containing approximately 0.02 per cent carbon so that a slightly reducing atmosphere is produced to avoid oxidation of iron.

14.8 MOULDING PRACTICE FOR MALLEABLE CAST IRONS

As malleable iron is mainly employed for castings produced on an mass scale, moulds are generally made by machine moulding. In view of greater shrinkage of white iron as compared to grey iron, the gating system is designed to enable progressive solidification of the casting, by running hot white iron first in thick members of the casting (the reverse is true for grey iron). Risers should be set up at the places where the iron enters thick section and about one-third of the height of the riser should lie in that well. Owing to lower fluidity of white iron, a larger cross section of gates is required and the relation between the cross sections of gate, runner, and the spruse are different than those used for grey iron castings. The above ratio may be 1:1:1.5 and could be suitably modified to suit the requirement of individual casting design.

The technology of moulding and the moulding processes employed, i.e. sand moulding, shell moulding, CO_2 moulding, and core making are in general similar to those employed in foundry producing grey iron castings in similar sizes and quantities.

The properties of malleable iron are dependent on the carbon content, a marked reduction in strength and ductility occurs as the carbon content is increased to the range of 3 per cent. At low carbon levels when optimum properties are achieved, problems of high melting point, high solidification shrinkage and more expensive processing are encountered. To have better strength and hardness, pearlitic malleable irons have become more popular instead of malleable irons having a ferritic matrix.

Pearlitic malleable irons are used where higher strength and resistance to mechanical wear are the prime requirement in the service and comparatively less resistance to shock is involved. The composition of malleable irons is closely controlled because the requirements for producing a white iron casting with high carbide stability are the opposite to those for rapid annealing where strong graphitising tendency is desired. In many foundries castings are classified according to section thickness, i.e., 6 mm, 12 mm, 18 mm etc. and the composition is suitably adjusted for the above section thickness to achieve the optimum annealing cycle. For very thin castings which cool faster, higher carbon and silicon content can be used and these can be annealed in appreciably shorter period of time. None of the carbide forming elements which stabilise the carbide structure can be tolerated, e.g. the presence of 0.5 per cent of Cr may prolong the annealing cycle to make it uneconomical. For this reason, majority of malleable castings are made with an average section thickness of 12 mm or less. It has been found that such elements as Te, Bi and B used in small quantities promote carbide formation during solidification, but do not stabilise carbide stability which is deterimental to annealing cycle of a ressonable period. It has thus become possible to increase the thickness of malleable iron castings to about 75 mm by using these special elements. However, such castings constitute a comparatively minor part in the commercial production of malleable iron.

14.9 SPHEROIDAL GRAPHITE CAST IRON (SG IRON)

A new family of engineering materials known as S.G. Iron (also known nodular and ductile iron) was born in 1948 as a result of the pioneering research work of K.D. Mills and his colleagues of International Nickel Company. During the brief history of about 35 years there has been a phenomenal growth in the production and applications of S.G. Iron castings. Owing to the presence of graphite in the form of flakes which interrupt the continuity of the matrix and the notch like shape acts as stress raiser, grey iron is brittle and lacks toughness and ductility. With the invention and development of ductile iron where graphite is present in the form of nodules or spheroids, the great draw back of inadequate toughness of cast irons for several engineering applications has been largely over come. This new engineering material shows appreciable elongation in as cast condition which is substantially improved by a short cycling annealing treatment. It retains the good foundry characteristics, i.e. fluidity and low melting point similar to that of grey cast iron and does not have the limitation of section thickness as in the case of malleable irons. Ductile iron therefore possesses the optimum combination of the product advantages of steel and the process advantages of grey cast iron. In Fig. 14.1 the microstructures of grey cast iron, malleable cast iron and ductile cast iron have been shown to illustrate the shape and distribution of graphite.

Uses of SG Iron Gastings

The automative, agriculture implement industries and pipe and pipe fittings are major uses of ductile iron castings. It has been estimated that automotive applications such as crank shaft, rocker arms, spacer blocks, spindles, cam shafts, connecting rods account for approximately 55 per cent of the total ductile iron castings production.

S.G. iron is finding increasing applications as a substitute of non-ferrous metals and alloys as well as steel and malleable castings due to its superior properties for specific applications.

Mechanical properties of S.G. iron are given in Table 14.3 as per IS 1865 of 1974.

TABLE 14.3 Mechanical Properties

Characteristic	*Requirement for Grade*					
	SG 800/2	*SG 700/2*	*SG 600/3*	*SG 500/7*	*SG 400/12*	*SG 370/17*
(i) Tensile strength N/mm^2 Min	800	700	600	500	400	370
(ii) Proof stress, 0.2 per cent N/mm^2, Min	480	440	370	320	250	230
(iii) Elongation, per cent, Min	2	2	3	7	12	17
(iv) Impact value J, Min	—	—	—	—	—	
(a) V Notch (Charpy)						
Average	—	—	—	—	—	13
Individual	—	—	—	—	—	12
(b) U Notch (Mesnager)						
Average**	—	—	—	—	—	15
Individual	—	—	—	—	—	13

Chemical Composition

Carbon: The iron can be hypoeutectic, eutectic or hyper—eutectic. For heavy section castings above 75 n m a carbon equivalent (total C + 0.31 (Si + P) − 0.07 Mn) value of 4.1 to 4.3 is maintained, whereas for thin castings, it is generally kept greater than 4.3. The total carbon in the base iron is critical for treatment with cerium which requires hypereutectic iron with C.E. above 4.3.

Silicon: In fully treated and inocculated ductile irons, the silicon content is kept within 2.2 to 2.5 per cent and should not exceed 2.75 pct. particularly for impact resisting grades. Considering the increase in silicon content in subsequent treatment, the base metal should contain about 1.5 pct. silicon when Si-Mg alloy is used and 2.0 pct. when Ni-Mg alloy is used.

Manganese: Manganese is a carbide stabiliser and retards the breakdown of pearlite. For the production of ferrite grade of S.G. iron, it is desirable to keep manganese around 0.5 pct. in heat treated grades and around 0.2 to 0.3 pct. in as cast ferritic grade.

Sulphur: Both cerium and magnesium which are used for the production of S.G. iron, are strong desulphurisers and first combine with sulphur and only the residual amount acts as spherodising agent. The sulphur content in the base iron should be kept as low as possible to prevent the wasteful loss of magnesium as also appreciable fall in the temperature of metal. It is necessary to limit the sulphur content to .035 pct. for economic treatment.

Phosphorus: Phosphorus is kept as low as possible preferably below 0.05 pct. It is desirable because of its pronounced effect in decreasing the impact strength as well as lowering the elongation value. For grades which do not call for high impact strength, phosphorus content normally should not exceed 0.08 pct. In special applications, e g., piston rings and rail-road brake-shoes, phosphorus is intentionally kept high (0.3 to 0.6 pct.) for high wear resistance.

Nickel, Copper and Chromium: Nickel improves the ultimate strength and yield strength of ductile irons. However, nickel content beyond 1.0 pct. is considered disadvantageous because it appreciably lengthens the annealing cycle to obtain a full ferritic structure required for optimum ductility and toughness. Copper in excess of 1 pct adversely affccts the formation of spheroids. Chromium as such does not markedly influence the formation of spheroidal graphite, but for obtaining ferritic structure chromium content should be maintained below 0.1 pct owing to its carbide forming tendency.

Subversive elements: More than a dozen chemical elements are known to interfere the formation of graphite nodules and encourage the formation of degenerate graphite shapes. The maximum permissible percentage of deleterious elements is given below.

Elements	*Permissible maximum per cent*
Al	0.10
Sb	0.002
As	0.02
Bi	0.002
Cd	0.01
Pb	0.002
Se	0.03
Te	0.02
Ti	0.10
Zn	0.10
Zr	0.10

The harmful influence of the deliterious elements can be encountered by the addition of small quantity of misch metal (cerium) about 0.02 pct. However, the influence of such subversive trace elements is complex and efforts should be made to control them by careful and judicious selection of raw materials.

Production Technique

Owing to the requirements of composition control of base iron, melting of charge material for the production of S.G. iron is generally carried out in electric induction melting furnances. Air furnaces either reverberatory or rotary type are unsuitable as the oxidising flame leads to high carbon loss. Basic lined direct are furnaces can give base iron of very low sulphur content but are not commonly used. Indirect arc and electric resistance furnances are not commercially viable owing to poor melting economy. Acid lined hot blast cupolas are also unsuitable as the high carbon per centage required cannot be obtained using the high steel scrap charges. Basic lined or lineless hot blast cupolas with basic slag practice may be used for the production of S.G. iron, since the sulphur content can be reduced to the desired level and sufficiently high carbon in the melt can be obtained.

The molten cast iron of suitable chemical composition is then treated with magnesium (some times with cerium), to obtain nodular form after inocculation with multicomponent ferrosilicon. After magnesium treatment, only a small quantity of residual magnesium is left in the metal. For getting the proper spherodisation, it is necessary to control the physical and chemical conditions of the base iron.

Graphite nodules can be obtained by a large number of processes and techniques which can be broadly ciassified into following two categories:

(a) Processes which use magnesium master alloys.

(b) Processes which use pure magnesium.

The proccsses are described in Table 14.4.

Mouldiug and Casting Practice

The moulding practice is similar to that of grey iron with the exception that allowance has to be made for slightly different shrinkage characteristics of nodular iron. Some modification to the gating system is also necessary because of the dross forming tendency of nodular iron. The pattern shrinkage is 13 mm per meter with substantially pearlitic matrix and on annealing to completely ferritic structure, the pattern shrinkage is recommended to about 4 mm per meter.

To counteract the deleterious effect of the dross forming tendency of the metal, the gating system is designed to retain the slag and dirt in the gating system ahead of the mould cavity, the metal is introduced into the mould with minimum turbulence possible and the rate of entry of the metal into the mould cavity is controlled. For reducing the turbulence of metal, the gates are generally positioned at the bottom of the casting and the ingates are taken from the bottom of the runner sections. To trap the dross in the runner and prevent its entry into the casting cavity, the runner cross section should be twice to four times that of the minimum sprue cross section.

The cost of normalising or annealing is usually 10 to 25 pct. of the total cost and therefore, efforts should be made to produce every single type or grade of ductile iron without any heat treatment. As cast delivery presents the most significant saving potential in the overall economy of S.G. iron castings.

14.10 VERMICULAR IRON

This represents a new cast iron alloys containing vermicular graphite structure obtained by treating ductile base iron with controlled addition of

TABLE 14.4 Spherodising Techniques

A. Processes using Mg master alloys	*Mg Recovery*	
(i) Open ladle process	About 25%	(i) Matter alloy added to the metal stream as the metal is tapped from the furnace or transferred from pouring ladle into the treatment ladle.
(ii) Sandwich process	About 50%	(ii) Mg alloy pieces placed in a recess at the bottom of the treat ment ladle and covered by steel punchings 1.5 to 3 mm thick constituting from 1-1.5% by mass of the iron treated. Iron to be treated is poured into this ladle disecting the stream away from the sandwich.
(iii) Puddling process	About 30%	(iii) Uses two melting units. Small unit holds the liquid iron at a low temperature between 1200 to 1300 °C and a small amount of this iron is over treated with Mg-alloy (about 80 per cent recovery). The over treated iron is then diluted with iron held at higher temperature suitable for pouring.
(iv) Trigger process	About 60%	(iv) Similar to sandwich process. Steel punchings replaced by a thin layer of resin bonded sand which gets fused and forms a shell. Reacrion is triggered by breaking the shell with a heated rod.
(v) Plunging process	About 60%	(v) Mg alloy contained in a refractory or graphite plunger bell is lowered into the ladle filled to 2/3 in height with iron to be treated.
(vi) Porous Plug practice	About 30-50%	(vi) Bottom of the ladle is fitted with a refractory plug which is per meable for gases but not for liquid iron, Treatment alloy thrown on top of the liquid and is carried below the surface by the pressurised inert gas (usually nitrogen supplied through the plug)

contd

(vii) In mould process	About 35-65%	(vii)	The reaction chamber holds the treatment alloy and inocculant is moulded into the gating system. Process suited to completely automated pouring system. Requires iron with sulphur below 0.02 pct.
B. Processes which use Pure Mg			
(i) Pressure Ladle and Pressure chamber technique	About 80%	(i)	The ladle is completely sealed or placed into a pressure tight chamber and pressure raised to 15-30 times the atmospheric pressure to prevent boiling of Mg on addition.
(ii) Injection process	About 40-70%	(ii)	Method similar to injection of CaC_2 for desulphurisation. Fine particles of Mg are carried in a stream of inert gas through a graphite tube immersed in a liquid metal.
(iii) George Fishes Converters	About 80%	(iii)	Mg is placed in a chamber filled with a perforated refractory attached to the side. The converter is rotated into the vertical position to permit the metal to enter the chamber and allow the Mg to dissolve in the iron.
(iv) Use of Magnesium inpregnated ferro-coke	About 40-50%	(iv)	Mg impregnated lumps are kept at the bottom which is detachable to facilitate the removal of spent coke after treatment.

magnesium resulting in magnesium residuals between 0.005 and 0.015 pct. and an alloying element like titanium in excess of 0.23 pct. This class of iron approaches mechanical properties intermediate between grey iron and ductile iron and shrinkage and chill characteristics of grey iron. It is thus possible to produce thin walled castings (2.25 to 3 mm thickness) leading to reduction in vehicle weight using interchangeable pattern and casting design for parts specified in grey cast iron and vermicular iron. Rest all other techniques of moulding, melting, etc, are identical to those employed for S.G. iron castings.

15

Production of Steel Castings

Steel castings find very wide uses in different branches of industry and the steel casting industry is one of the great basic industries of the worrld. Steel castings are generally used as constituent parts of manufacturing equipment, machinary and other durable goods and may range in mass from several grams to a few tonnes (some times) as heavy as 100 tonnes or more. The configuration and shape of these castings may be very complex. The steel possesses a combination of several properties such as high strength and ductility, ability to withstand substantial stresses under variable and impact loads which are not usually met by other metals and alloys. With the addition of suitable alloying elements, important service properties of steels such as wear resistance, corrosion resistance, scaling resistance and mechanical properties at high temperature are substantially improved. The range of properties obtained can be modified by varying the carbon and alloy contents of the steels as also by giving suitable heat treatment. thereby constantly enlarging the uses of steel in new fields of industry.

15.1 CLASSES OF STEEL CASTINGS

Iron-carbon alloys containing upto 2 pct. carbon along with small amount of other elements are included under the category plain carbon steels. The impurities present in their usual amounts are as follows.

Mn	0.5 to 1 pct.
Si	0.2 to 0.8 pct.
P	0.05 maximum
S	0.05 maximum

Other impurities may also be present as tramp elements coming chiefly from purchased scrap and the situation is becoming aggrevated in recent years owing to increased production of alloy steels and consequently higher percentage of alloys in practically all purchased scrap.

Carbon steels. These are most extensively used for shaped castings and are divided into three classes i.e. low, medium and high carbon steel cast ings.

Low carbon steels. These contain below 0.20 pct C 0.35 to 0.65 pct Mn, 0.15 to 0.25 pct. Si, and 0.05 pct S and P maximum each. The strength of these castings after annealing and normalizing is 395–410 MPa with an elonga tion of about 24 pct. and are used mainly in electrical engineering and machine building industry. Low carbon steels have lowere fluidity and greater susceptibility to form hot cracks in castings.

Medium carbon steels. The great majority of the output of steel castings is in the medium carbon range (0.2 to 0.5 pct.) and represent what is normally termed as carbon steel castings. As compared to low carbon steels these steels have better casting properties as these are more fluid and less susceptible to hot cracking .These steels are used for small medium siz ed and large castings in diversed fields of industrial applications. The tensile stress varies from 410 to 540 MPa (40 to 50 Kg/mm^2) and elongation from 12 to 23 pct., thereby making them suitable for different applications.

High carbon steels. Contain above 0.5 pct. C and possess high fluidity and low susceptibility to hot cracking. The castings tend to develop high internal stresses owing to low thermal conductivities of these steels. These castings are used for hot rolls, gears, drums of mobile hoist machines, pump impellers and other particular products and uses requiring specialised properties. These castings are generally suitable for wear resisting applications not involving shocks. The effect of carbon content on the mechanical properties of cast carbon steels in the annealed condition is illustrated in Fig. 15.1).

Cast alloy steels Although it is difficult to draw a definite line of demar cation between carbon and alloy steel castings, a differentiation can be made on the basis of alloying elements specifically added to produce a desired improvement or change in properties, the resultant product being designated as cast alloy steel. Alloy steels possess better strength under static and dynamic loading conditions, greater heat and corrosion resistance, enchanced resistance to wear, response to hardenability and other special properties, thereby considerably enlarging the area of their use. Although a large number of alloy steel castings are available, broadly they may be classified as (a) low alloy steels usually with an alloy content less than 8 pct. and (b) high alloy steels which comprise heat resistant, corrosion resistant and wear resistant steels, apart from other compositions.

15.2 SPECIFIC CHARACTERISTICS OF CASTING OF STEELS

The physical and metallurgical characteristic call for special consideration in the moulding and casting of steels are detailed below:

(i) In view of the higher melting point of steels as compared to cast iron, higher pouring temperatures are involved. This requires sands of greater refractoriness such as high silica sand bonded with refractory clays.

(ii) Molten steel on cooling from the pouring temperature to the solidi fication temperature contract about 1.6 pct. per 1000 °C, then it undergoes solidification contraction of 3 pct during freezing and finally a further contraction of 7.2 pct. when the solidified metal cools to room tempera ture. Therefore when casting steels, it is important to provide an ample supply of molten metal from risers to compensate for the decrease in

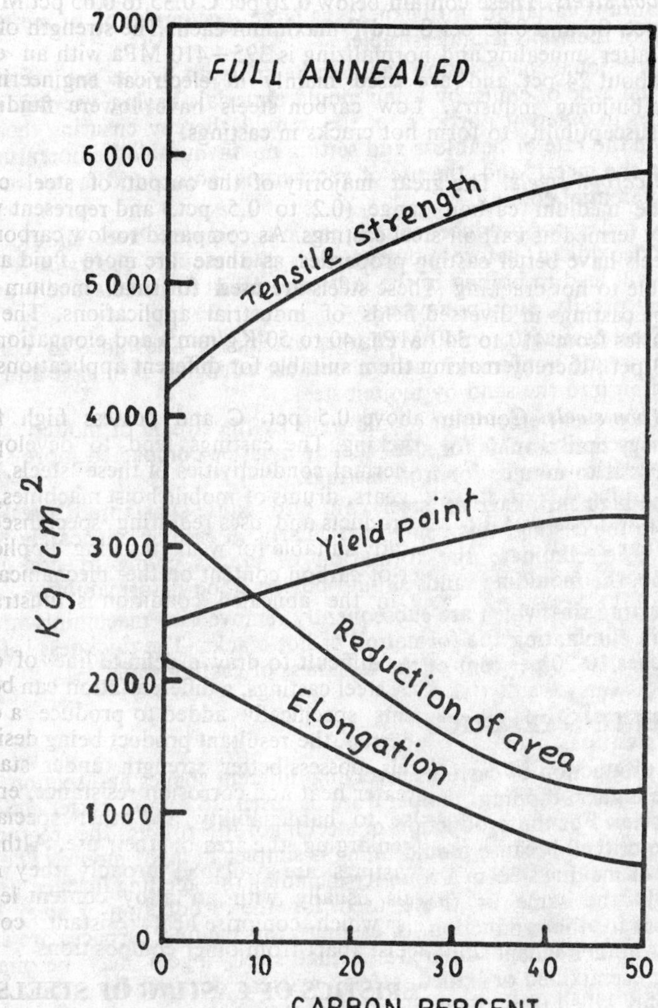

Fig. 15.1 Effect of carbon content on the mechanical properties of cast carbon steels in the annealed condition.

volume to avoid shrinkage cavities. In case of cast iron this is somewhat less critical because of relatively low solidification shrinkage.

(iii) No carbonaceous material or washes can be used on the mould faces because of the tendency of molten steel to dissolve carbon.

(iv) The design of the steel castings should ensure progressive solidification and adequate feeding of the casting. In such directional solidification, the metal starts solidifying at places farthest from the risers and proceeds progressively towards the risers, the latter being last to solidify. The casting walls should be as uniform in thickness as possible and there should not be abrupt variation in thickness as large accumulation of metal at the junction of the walls is undersirable.

(v) Directional solidifcation can also be achieved by the use of external or internal type of chills, by using mould materials having varying thermal properties in different portions of the mould, thereby ensuring the desired change in the rate of heat loss and setting up favourable temperature gradients in the metal and the use of exothermic compound in the mould to have differential cooling.

(vi) In order to produce sound castings, molten steel should be the roughly deoxidised before pouring into the mould otherwise dissolved oxygen will lead to boiling in the ladle and result in defects such as blow holes, gas prosity, and spongyness.

(vii) Steel castings should have larger machining allowance as compared to cast iron and nonferrous castings because of higher shrinkage and deeper penetration into the sand by molten steel.

(viii) To prevent erosion of mould walls by the stream of molten steel, it is necessary that moulds for steel casting must be compacted more strongly as compared to moulds for iron castings.

(ix) The high shrinkage of steel can also cause substantial stresses and crackes in the casting, this being particularly in case of intricately shaped castings. To eliminate the formation of cracks owing to non uniform shrinkage, the moulding sands should possess good deformability.

(x) Casting ribs which are subsequently removed in machining are also helpful in eliminating the formation of hot cracks. The thickness of these ribs reaches 10-20 per cent of the thickness of casting walls.

15.3 MELTING PRACTICE

For the production of castings involving different shapes and section thickness, the higher tapping temperature is used to attain better fluidity of molten steel. For the production of steel ingot for rolling, the requirement is not so critical because mould filling is simpler in ingot mould. The basic features of making steel in a foundry including the melting furnace used are essentially the same as those used for the production of steel iugots, excep that for the production of castings, the melting furnaces are smaller in size. Another salient difference is that steel ingots may be made as rimming, semikilled or killed steels, whereas for foundry products only thoroughly killed (fully deoxidised) steel is used.

Now a days overwhelming majority of steel foundries use either electric arc or electric induction type of furnaces because of better metallurgical control which results in cleaner steel with superior properties. The requirements of stringent specifications in various grades of steel castings particularly for the production of alloy steel castings has encouraged the use of electric furnaces. In the electric arc furnace, it is possible to refine the metal to achieve low sulphur and phosphorus contents even when using inferior grade scrap as melting stock. In electric induction furnaces, generally more refining is attempted because the stirring action of the bath makes it difficult to maintain a slag cover on the metal. The melting stock therefore consists of selected grade *pedigree* scrap. The induction furnace possesses the advantages of flexibility in operation and is particularly suitable for the production of low carbon steels because there is no carbon pick up as may occur in an electric arc furnace from the electrodes. Small lots of alloy steel castings and special steel castings are generally produced using induction melting.

Other melting units such as open hearth furnace and other processes involving the use of oxygen steel making are comparatively not popularly used for the production of steel castings owing to their large production capacity. To ensure low hydrogen content in steel as also to reduce non-metallic inclusions to minimum, vacuum degassing of molten steel may be used for making special purpose castings to meet stringent service requirements.

15.4 MOULDING PRACTICE

15.4.1 Green Sand Moulding

Because of versatility and favourable economics, small steel castings are produced in green sand mouldgs than by other methods. Owing to higher pouring temperature, only synthetic sands are used in mould mixtures for steel and the binder used is generally restricted to bentonite. Mould coatings of a refractory type consisting of silica flour, zircon flour, or chromite flour are used to obtain better finish on the casting including smoother surfaces, decreased scabbing and buckling and less metal penetration.

15.4.2 Dry Sand Moulding

Dry sand moulding is commonly used for producing large steel castings weighing from 1 tonne to well over 100 tonnes because of its superiority over green sand moulds for large castings. Dry sand moulds are stronger, permit closer dimensional tolerances and withstand more handing. This being particularly important when core setting may take 3-4 days. Dry sand moulds are similar to green sand moulds except that silica flour is generally added for enhanced rammed density of the mould and a thermosetting resin may also be incorporated in the mix. Moulding is usually done with a sand slinger and after coating with a mould wash, drying is done at 320 – 360 °C for 10 – 40 hours depending on the size of the mould.

For very large castings which can not be accomodated in flasks or extremely heavy section castings which might breakout of flask because of high ferrostatic pressure are generally produced by pit moulding which is a variant of dry sand moulding.

Shell moulding is used for producing castings in carbon or low alloy steels of the section thickness in the range of 6 mm to 50 mm and weighing upto about 150 kg. Investment casting is used for steels of various compositions to produce intricate shapes, usually range not more than 5 kg. Ceramic moulding has proved specially, useful for casting intricate shapes for making cast to shape tools and dies from tool steels, because of the advantage of good dimensional accuracy and smooth surfaces. Steel can not be cast into plaster moulds because of the destruction of plaster caused by high pouring temperature of steel.

15.4.3 Core Making

Fine grained silica, zircon, chromite or olivine sand is used for core making in steel foundries because of its ability to withstand liquid steel temperature upto 1630 °C. Cores may be produced by oil binding, resin binding, carbon dioxide process, or by using chemically setting core binders. Oil bonded or resin bonded cores require a schedule for production atleast 3 days prior to moulding. For carbon dioxide process cores the lead time is shortest of all core making methods being only 2-4 hours prior to mould production. Shell cores possess best dimensional tolerance ($\pm$ 0.008 cm/cm) compared to 0 020 for CO_2 process and $\pm$ 0 031 for baked oil sand cores. Another advantage of shell cores over cores bonded with other binders is better collapsibility as little resistance is offered to solidification contraction of steel by the shell core. Thus, for the production of thin walled castings where core restriction is a problem, shell cores are generally employed. With shell cores, it is possible to produce casting sections as thin as 4 mm as compared to other types of cores where the usual minimum casting section is 6.5 mm. Shell cores also give considerably better casting surface smoothness than is possible with other sand cores. Shell cores are generally avoided in the production of low carbon steel castings ($<$ 0.15 pct C) because of carbon pick up at the surface from resin unless the core surface is to be removed by machining. Chemically setting core binders are used for the production of medium and large cores as the need for core reinforcement is greately reduced and both these binder systems require no ramming. Although the cores may be removed from core boxes in 1 hour to 8 hours depending upon the cross sectional area of the cores and the binder system used, core finishing operations for medium and large cores require a minimum lead time of 2-3 days before moulding.

15.4.4 Pouring Gating and Risering

Optimum pouring temperature is very important for proper quality of steel castings and molten steel is generally teemed from the ladle into the moulds at a temperature of 1450 to 1550°C. The ladles used for pouring into moulds include the bottom poured ladle, tea pot ladle, and lip pour ladles depending upon the size of casting and other considerations. For the production of large castings, bottom pouring is preferred because it is difficult to control the stream of molten steel by it filting a large ladle. Bottom pouring also greatly reduces the risk of non-metallic particles such as pieces of ladle lining, slag etc. close to the top of the ladle, thereby, delivering cleaner metal to the mould. However, bottom poured ladles generally are unsuitable for pouring into smaller sand moulds. The molten steel is delivered too fast because of the pressure head and the sand mould is not strong enough to withstand the impact of molten stream. Also frequent opening and closing of the stopper leads to leaking of the ladle. Tea pot ladles in various sizes cover the entire range of casting sizes for which the bottom poured ladle is unsuitable. As the metal flows from the bottom of the ladles up the spout and over the lip, it is free of slag and eroded refractory. Lip poured ladles are generally not popular for pouring steel eastings because the hot metal is not taken from the bottom of the ladle, thereby considerably increasing the risk of inhomogenity in chemical compositions as also abundence of nonmetallic inclusions from slag and other extraneous matter which flows unchecked.

Before tapping steel from the furnace it is necessary to preheat ladle linings to 800 – 1050 °C using oil fired or gas fired burners. This reduces the thermal shock to the lining when the molten steel comes in contact, thereby prolonging the life of the ladle lining. Preheating also reduces the formation of skull owing to fast freezing of molten steel specially with casting of low carbon steels, The ladles are lined with firebricks or by monolithic lining depending upon the capacity of the ladle. Monolithic lining is commonly used for small ladles generally below 400 kg capacity for prevention of erosion of the lining which reduces its life as also increase dirt in steel, it is necessary to pay proper attention to the quality of the refractory material used as also its maintenance.

Ideally the optimum pouring time should be established taking into consideration the temperature of molten steel, its solidification characteristics, heat transfer characteristics and thermal stability of mould as also the size, shape and mass of the casting. From practical consideration, however, control on pouring time is difficult because generally it is necessary to pour many different castings from one ladle. An attempt is made to control the speed at which molten steel enters the mould cavity by suitable design of the gating system.

The design of gating system and risers for steel castings calls for special care in view of low fluidity, large volume shrinkage, tendency to form hot cracks and absorption of gases. Rules for gating of steel castings include the following:

(i) The length of gates should be as small as possible to enable better filling of the mould. Usually round gates and runners are preferred to square cross-sections because the former cause less surface friction and allow more metal to flow per unit time.

(ii) During pouring the mould, gate should be filled completely with metal to exclude air and gases from the mould which may dissolve in the metal or form cavities in the casting.

(iii) Whenever possible, the gates should form a part of the main casting pattern as these are less susceptible to erosion as compared to cut gates.

(iv) To promote directional solidification, the metal should enter the mould cavity at the places which solidify last. The metal should enter the mould cavity at as many points as possible to minimise hot spots.

(v) To eliminate the possibility of pulling a piece out of the casting during solidification, it is necessary that the cross sectional area of the gate must be smaller than that of the casting at their junction.

(vi) The gating system for pouring steel unlike that for cast iron calls for setting up large risers to supply feed metal to compensate for shrinkage in the casting. The mass of gating system and risers may account for 25-50 pct. of the casting. The importance of decreasing the amount required for gates and risers by proper designing is quite obvious.

(vii) An effective gating system for pouring steel should provide for filling the mould as rapidly as possible without excessive turbulance. If the metal does not rise rapidly to provide the necessary support, the mould can collapse owing to the destruction of the binder in the moulding sand because of the radiated heat of the molten metal.

(viii) In view of the tendency of large flat surfaces to develop *rat tails*, the gated end of the mould should be low so that the metal runs up a slight inclined. The same also concerns thin walls of the cavity. It is a common practice to use high pouring rates for light sections.

15.4.5 Risering Practice

In view of the high shrinkage, risering of steel castings deserves special attention as detailed below:

(i) Risers are usually placed over thick section of the casting. A thicker section of the casting acts as a reservior for feeding the thinner sections which solidify first and thicker sections are fed by the risers. Because feed ing from the risers depends upon gravity they are usually located at the top of the mould. Such open risers also serve as collectors for nonmetallic inclusions which float up to the surface apart from feeding the casting to compensate for the shrinkage.

(ii) Most risers are cylinderical in shape with their height approximately equal to their diameter, this configuration delays the solidification of steel owing to surface area to volume ratio.

(iii) Exothermic sleeves made of antipiping compound which mainly consists of thermit formulation are used to maintain the metal in the liquid condition in the riser. However, for large castings, the use of such exothermic sleeves is ineffective because sleeves burn out before the riser stops feeding the casting.

(iv) Where the use of open risers may entail a large requirement of metal, blind risers may be used.

15.5 CLEANING AND INSPECTION

The cleaning of steel castings can not be over emphasized because it often represents a large proportion of the total cost of producing the castings Reduction in cleaning cost is frequently feasible by a judicious designing of the casting, e.g., considerable saving in costs would be obtained if the *pads* of metal at specific location could be incorporated in the design of the casting and retained in the finished part. These pads are added to enable proper feeding of metal from risers and constitute a significant cost element by their subsequent removal by power cutting, grinding, or machining.

Carbon steel castings for general purpose which do not experience shrinkage retardation are removed from the mould at about 880 °C. Castings with nonuniformly varying wall thickness are withdrawn at about 600 °C and are held in soaking pits after stripping. Special purpose castings with retarded shrinkage which are susceptible to working and cracking are removed from the moulds at about 200 °C.

The fettling operations for steel castings differ from those employed for cast iron castings. It is more difficult to remove gates and risers from steel castings which require chipping with pneumatic hammers, oxyacetylene flame cutting, abrasive cut off wheels or bend saws, sledge hammers and small power shears. The choice of equipment and methods for cleaming operations are largely governed by the desired level of the castings with respect to dimensional accuracy, surface finish and cost involved. Certain very fragile castings may involve the use of special processing equipment and technique to prevent mechanical damage. A flow chart for cleaning and other associated operations for small castings is shown in Fig. 15.2.

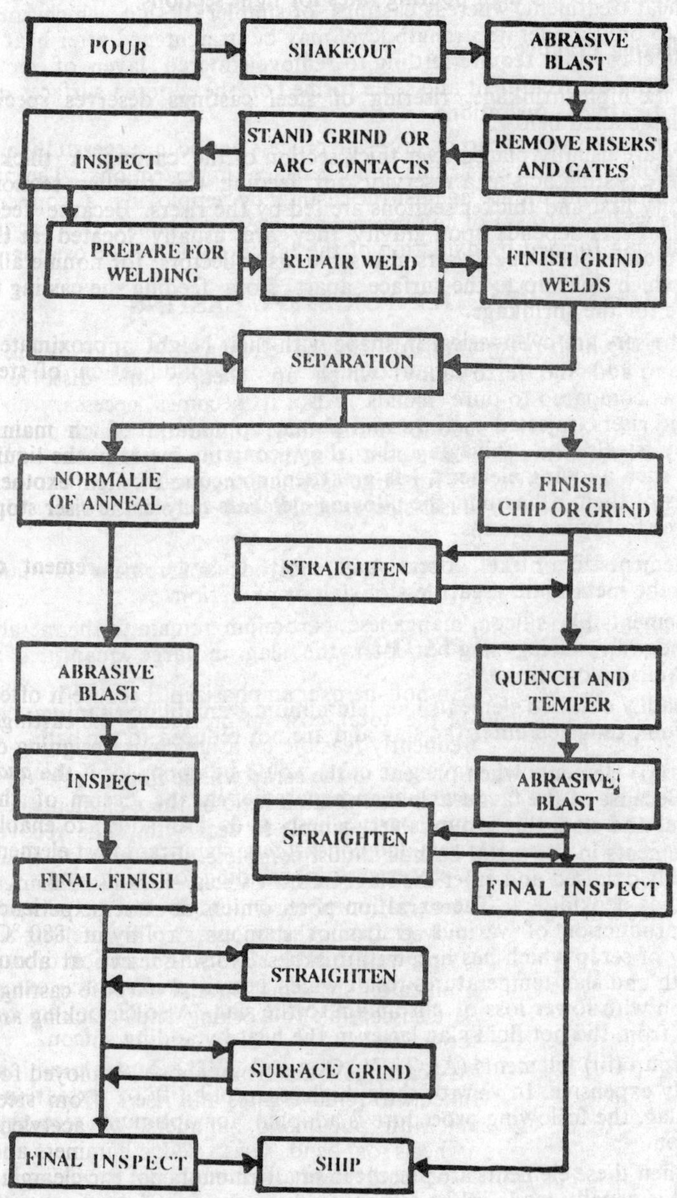

Fig. 15.2 Flow chart for cleaning and other operations for small steel castings.

Some castings require straightening which is usually done by hydraulic press and alternatively sometimes by hand hammer and die operation. Because of better ductility higher strength castings are usually straightened before heat treatment, whereas castings of simpler design which are heat treated to only moderate strength level may be straightened after heat treatment. Steel castings require fettling to remove sintered layer of moulding sand before heat treatment and scale formed on the exposed surfaces during the heat treatment operations.

With in recent years, carbon arc – air process has been successfully applied for removing gates, risers, fins, and surface imperfections. The process consists of striking an arc against the casting by employing a copper clad electrode while a stream of compressed air directed behind the arc flushes away the slag formed by oxidation of the liquid metal of the casting.

15.6 ALLOYING PRACTICE FOR STEEL CASTING

In view of the high premium paid for the elements in pure form, it is formore to add the ferro alloys which are cheaper and dissolve more readily as compared to pure metals. Also it becomes necessary to reuse gates and risers, rejected castings and also scrap material of various grades of alloy steels for salvaging the alloy content by remelting. Proper yield of such alloying elements has great techno economic significance. On the basis of their behaviour, the alloying elements may be broadly divided into three following groups.

(i) Elements like nickel, copper and molybdenum remain substantially in the metal bath regardless of slag or oxidation.

(ii) Elements like silicon, manganese, chromium remain in the metal bath under a reducing slag but enter the slag in large quantities under oxidising condition.

(iii) Readily oxidised elements like aluminium, vanadium, titanium, columbium, tungsten enter the slag and are not reduced to the bath.

Group (i) elements when present in the scrap will be found in the finished metal. Because of the favourable distribution ratio of these elements between the metal and slag, this group is very simple to deal with. The recovery of these elements in the metal bath is almost complete. Elements of group (ii) are readily oxidised and enter the slag. In most steels, silican and manganese areu sed as deoxidisers. The oxidation of chromium is extremely important for the production of various grades of stainless steel by melting a large quantity of scrap which has become available. By using injection of oxygen, high bath and slag temperatures result which permit preferrential oxidation of carbon with lower loss of chromium to the slag. Also chromium can be reduced from the hot fluid slag latter in the heat by adding silicon.

All group (iii) elements (Al, Ti V, Cb, and W) except aluminium are relatively expensive. In view of their high susceptibility to oxidation and loss in slag, the following procedure is adopted for optimum recovery and utilization.

(a) When these elements are present in small amounts not exceeding 2 pct., the steel is usually made in the same way as an unalloyed heat, deoxidised with aluminium or silicon and the ferro alloys of required elements are subsequently added.

(b) when large quantities of these elements are contained in the scrap, e.g., with highly alloyed tool steel, a simple reducing slag is used for dead melting.

(c) Melting and casting under vacuum gives exceptionally good recovery of oxidisable elements besides bringing about effective degassification. The induction furnace is an ideal melting unit when protective atmosphere or vacuum melting is desired. The charge has approximately the same composition as the casting and very little refining is attempted because the slag is colder than the metal and failure of the furnace wall which are thin can be accelerated by slag erosion and stirring action of the induction current. These furnaces are becoming increasingly popular for the melting of high alloy and tool steels.

15.7 HEAT TREATMENT OF STEEL CASTINGS

In the as cast conditions, steel castings may possess coarse grained structure and comparativly poor mechanical properties. To refine the grains and break up the dendrititic structure, relieve internal stresses, and develop optimum physical and mechanical properties, steel castings are subjected to appropriate heat treatment. Chemical composition of steel, design and section size of the casting, grain size, and the desired mechanical properties influence the heat treatment of the castings. Although steel castings respond to heat treatment similarly to wrought steel products, special care is required because of the variations in size, sections and design of the castings as also placing the casting in the furnace properly to minimise the danger of warping, distortion, or cracking during heat treatment. Common heat treatment processes for steel castings are as follows:

15.7.1 Annealing

This type of heat treatment releives casting stresses, eliminates the dendecritic structure and refines the grain size there by raising the tensile and yield strength and improving ductility and machinability. The process consists of heating the casting slowly in a furnace usually 30-50 °C above the Ac_3 point. The time of holding varies from about one hour for thin walled castings to 4-6 hours for large thick walled castings, general rule being one hour per 25 mm thickness of the heaviest section of the casting. The casting should be held at the annealing temperature ensuring uniform heating throughout the section of the casting. The holding period being completed, the casting are cooled in the furnace. It is necessary to exercise proper control over the temperature of the furnace and optimum holding period, or else the fine structure may again change to coarse strueture.

15.7.2 Normalising

This treatment is similar to the annealing process, except that the castngs are removed from the furnace at the end of the soaking period and allowed to cool in still air. This results in higher yield and tensile strengths as compared to the annealed product and with approximately the same ductility values. Such a treatment however, does not relieve the internal stresses to the same extent as in the annealing process. To achieve more uniform grain size and further improving the ductility, double normalising is often employed and to remove internal stresses or reduce the hardness of normalised steels, tempering treatment is freqvently used. This consists of heating the casting to a temperature below the critical range (260-650 °C) and allowing it to cool in the furnace.

15.7.3. Quenching and Tempering

This treatment is applied mainly to high carbon and alloy steel castings where high strength and resistance to impact and or abrasion is required. The treatment consists of annealing or normalising the casting, reheating and quenching. The following precautions are necessary to minimise internal stresses and susceptibility towards cracking.

(a) The tempering treatment should immediately follow quenching. To avoid cracking and wide variations in hardness values, sometimes it is necessary to use a time quench treatment. This consists of immersing in the quenching bath for a predetermined time interval and subsequent tempering treatment immediately.

(b) In case of steels which are susceptible to temper britleness, e.g. straight manganese steels (1-2) pect. Mn) are cooled quickly to tempering temperature by quenching or air cooling as ductility and impact values are lowered significantly when cooled slowly from the tempering temperature.

15.7.4 Flame Hardening

Parts of the casting which are subject to extreme wear are differentially hardened. In such cases the castings are first annealed or normalised and only the surfaces to be hardened are heated to the hardening temperature by a torch or induction heating apparatus followed by quenching, e.g., in a pinion gear, the teeth are made hard and wear resistant with a machinable bore which is left unhardened.

16

Foundry Practice for Non-Ferrous Metals and Alloys

Generally the moulding and casting practices for ferrous and non-ferrous castings are similar, but the following common characteristics are relevant particularly for the production of non ferrous castings:

(i) Most commercial nonferrous alloys possess good casting properties and offer the possibility of producing castings by a variety of casting processes such as sand casting, permanent mould casting, die casting, plaster casting, investment casting etc.

(ii) Generally bench moulding is done for the production of large proportion of small size work in a non-ferrous foundry.

(iii) Owing to relatively faster rate of solidification and tendency for chilling, it is necessary to have larger areas of the metal feeding system.

(iv) Special care is necessary to avoid shrinkage cavities and cracks owing to larger contraction in most of the nonferrous metals and alloys during solidification.

(v) In view of higher fluidity and seeking tendency of nonferrous metal at the pouring temperatures, the sand used for moulding has comparatively finer grain size. The cores used in such moulds are prepared from materials which should permit free contraction of the metal during solidification.

(vi) In view of higher susceptibility to oxidation in case of most non ferrous metals, proper control on pouring is of great importance to minimise turbulance and air aspiration.

(vii) Fluxing and degasging practice including removal of reaction products deserves for greater attention to ensure proper quality of the casting.

Wide use of nonferrous castings, i.e. aluminium, copper, magnesium, zinc, nickel and their alloys is made in machine building industries. Different nonferrous alloys require some specific precautions and techniques to produce sound castings. This chapter is thus devoted to the specific features involved in foundry practice of various non-ferrous metals and alloys.

16.1 PRODUCTION OF ALUMINIUM AND ALLOY CASTINGS

Aluminium is plentiful and comparatively in-expensive and possesses an impressive combination of properties such as lightness, high strength-weight ratio, good corrosion resistance, good formability including high malleability and ductility, superior finishing quality and pleasing appearance combined with high thermal and electrical conductivity. These character istics have already made aluminium a leader in the family of nonferrous metals and the progress of aluminium uses has progressed at an ever accelerated rate in wide ranging industries such as air craft, automotives, electrical installation, chemical industries, dairy industry, building and hardware and several others.

Both in respect of composition and applications, aluminium casting alloys cover a wide field. These include from the pure or nearly pure metal for high thermal or electrical conductivity to several modern high strength alloys of complex composition for different applications.

16.1.1 Aluminium Casting Alloys

Aluminium casting alloys contain many alloying elements which perform the following functions:

(a) Eutectic forming elements, usually silicon which imparts sufficient fluidity to feed the shrinkage occurring in casting.

(b) Strengthening elements to improve the mechanical properties.

The amount of eutectic formers to a certain extent depends upon the casting processes. Sand casting alloys generally contain lower eutectics as compared to those for casting in metal moulds. This is because sand moulds can tolerate greater degree of hot shortness that could lead to pronounced cracking in non yielding metallic moulds. In die casting process, fine dispersion of strengthening and eutectic forming constituents can be obtained because of extremely high cooling rates. Thus it is possible to obtain reasonably good castings inspite of impurity contents that would adversely affect sand or plaster mould castings.

Although a large number of aluminium casting alloys have been developed, most of them belong to six basic types i.e. Al-Cu, Al-Si-Cu, Al-Si, Al-Mg, Al-Z_n, and Al-Sn alloys and 5 different heat treatments are commercially available. A four digit numerical system for the identification of aluminium and its alloys in the form of castings and foundry ingot is given in Table 16.1. The main series of aluminium casting alloys are shown in Fig. 16.1.

Depending upon the alloying additions, aluminium casting alloys range from pure or nearly pure metal (unalloyed aluminium) to the modern high strength alloys of complex composition. These are described below:

(i) *Unalloyed aluminium.* For all practical purposes unalloyed aluminium has no freezing range and the pure metal is not easy to cast as shrinkage becomes a problem and the casting becomes prone to hot-shortness. The castings of pure metal are soft, possess good colour and appearance as well as high corrosion resistance. Pure aluminium castings are used for specific type of applications where high electrical conductivity is required,

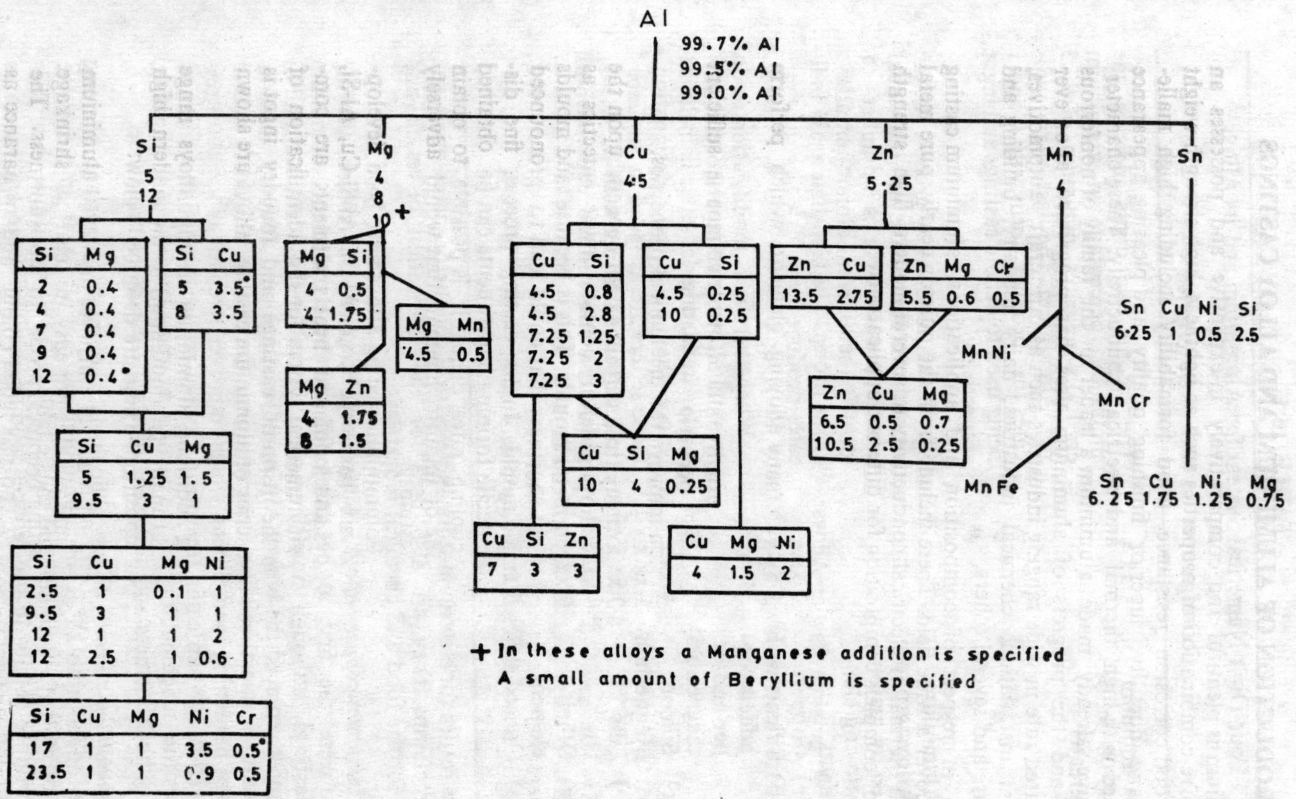

Fig. 16.1 Main Series of Aluminium casting alloys.

TABLE 16.1

Four Digit Numerical System for Aluminium and its Alloys

Alloy	*Designation*	*Description*
1. Aluminium 99.00 pct or greater	1 xx.x	The second two digits indicate the minimum percentage of aluminium eg. 150.x indicates minimum of 99.5 pct. Aluminium. The last digit indicates the product form with 0 and 1 being used to denote castings and ingot respectively
2. Aluminium Alloys groupted by major alloying elements		The 2 xx.x to 9 xx.x groups. In the second two digits have no individual significance but serve as a number to identify the different aluminium alloys in the group. The last digit again indicates product form.
(a) Copper	2 xx.x	
(b) Silicon with added copper and/or magnesium	3 xx-x	When there is a modification to on original alloy, or to the normal impurity limits, a serial letter is included before the numerical designation. These letters are assigned in alphabetical sequence starting with I, Q, Q, and XIX is reserved for experimental alloys.
(c) Silicon	4 xx.x	
(d) Magnesium	5 xx.x	
(e) Zinc	7 xx.x	
(f) Tin	8 xx.x	
(g) Other element	9 xx.x	
(h) Unused series	6 xx.x	

e.g. cast rotors for electric motors and electrical conductor accessories such as cable clamps. They are also used for the fabrication of vessels and fittings for the food and chemical industries.

(ii) *Alnminium-copper alloys.* Alloys with copper as a major alloying addition were the first to be developed and widely used for aluminium alloy castings, although now terrany alloys containing Al, Cu and Si have almost replaced the binary Al-Cu alloys, Additional alloying elements are present in most existing compositions. Copper forms $CuAl_2$ – most important hardening constituent which improves strength as the alloy responds well to age hardcning treatment. Copper reduce internal shrinage and improves the machinability of the casting, however, it reduces the fluidity, promotes hot shortness and adversely affects the corrosion resistance.

(iii) *Aluminium-silicon alloys.* Silicon is the most important alloying element (added upto 25%) in aluminium casting alloys. It improves the fluidity of molten aluminium and thus enables it to flow and fill thin walls and reproduce details. It also reduces solidification shrinkage and improves strength and wledability but does not assist machinability. Alloys with silicon content above 8 pct. need rapid chilling during casting or moidification treatment to prevent the formation of primary silicon crystals which are deterimental to the achievement of optimum mechanical properties. As a class, binary aluminium alloys are used for permanent mould and sand castings for the production of domestic cook wares, certain automobile castings and pump cases.

(iv) *Aluminium-magnesium alloys.* After silicon and copper, magnesium is the most important alloying element and can harden aluminium in non heat treated and heat treated conditions, particularly when magnesium content is 10 pct. These alloys possess high resistance to corrosion, good machinability, but are prone to more casting difficulties as compared to Al—Si alloys. This requires greater care in melting and pouring because the presence of magnesium increases oxidation in the molten state. Moisture intends to produce dross (MgO) and hydrogen which cause roughening and blackening of the casting surface. As a group, aluminium-magnesium alloys require more care in running and gating to produce sound castings because of poor feeding characteristics. Casting characteristics are improved by ternany additions of silicon and zinc.

(v) *Aluminium-zinc alloys.* This system was one of the first developed but binary Al—Zn alloys are now obsolete except for compositions which are used as sacrificial anodes to protect steel structures, pipe lines etc. Alloys for engineering applications now generally contain both aluminium and zinc together with minor addition of one or more of the elements copper, iron, manganese and chromium.

The casting characteristics of these alloys are relatively poor owing to pronounced hot shortness and high shrinkage as well as considerable coring that can occur even to the point at which pure zinc may be present. These alloys are generally sand cast because casting in permanent mould tends to cause hot cracking.

These are natural aging alloys and harden over a period of weeks. As the eutectic melting points are relatively high, castings of these alloys are used where assembly by brazing is required. Other desirable characteristics are good machinability, corrosion resistance, anodising characteristics and pressure tightnees. The alloys are unsuitable for elevated temperature applications because over aging causes rapid softening.

(vi) *Aluminium-copper-silicon alloys.* These are the most widely used aluminium casting alloys and have almost completely replaced original binary alloys of aluminium and copper. The amounts of both these additions are variable over a wide range in view of the specific characteristics and mechanical properties required. Copper contributes to strength and silicon improves castability and reduces hot shortness. Alloys with higher silicon are normally used for more complex castings and also for permanent mould and die casting processes where hot shortness becomes an important consideration.

Aluminium-copper-silicon alloys with more than 3-4 pct. Cu are heat treatable and these alloys usually contain magnesium which increases their response to heat treatment Alloys with more than 10 pct. silicon content possess low thermal expansion which makes them suitable for high temperature applications. With silicon content above 13 pct. (silicon content as high as 22 pct.), primary silicon crystals when properly distributed, impart excellent wear resistance. Automotive engine blocks, pistons, aircraft fittings compressor connecting rods and general purpose castings with pressure requirements are importgnt uses of these alloys.

(vii) *Aluminium—tin Alloys*:—Alloys containing about 6 pct. tin with small additions of copper and nickel for strengthening possess good bearing characteristics imparted by tin. It is fairly easy to add tin after aluminium becomes molten although considerable stirring is required because of high density of tin. For obtaining best bearing characteristics it is necessary to have fine interdendritic distribution of tin which is obtained with rapid cooling in the casting method. Tin addition is also used to improve the machinability of aluminium-copper alloys.

(viii) *Complex aluminium aasting alloys*:—Most aluminium alloys specifications contain a large number of elements because of their manufacture from reclaimed scrap from which this is difficult and uneconomical to entirely remove the tramp elements. Besides, as already been discussed, binary alloys have many limitations which have led to the development of complex alloys in which the specifications of tramp elements are limited to amounts which will not adversely affect their properties. The effects of major alloying elements like copper, silicon, magnesium, and zinc have already been discussed. Iron is generally considered as undesirable impurity as it promotes hot shortness, acts as a stress raiser and reduces the response to age hardening. It is sometimes added to reduce shrinkage in alloys with more than 5 pct. silicon and it acts as a grain coarsener producing brittle structure. Nickel is primarily added to improve high temperature properties and aluminium-silicon-nickal-magnesium-copper alloys are so well suited for pistons of internal combustion engines that these alloys are known as piston alloys.

Addition of titanium (0.05 to 0.2 pct.) or titanium and boron (seldom over 0.1 pct.) achieves grain refinement and promotes good feeding. Zirconium also acts as a grain refiner but larger quantities are needed to impart the same effect as titanium. Beryliium in low concentration (0.0005 to 0.005 pct.) may be used with magnesium containing alloys to minimise oxide formation, thus imparting fluidity and better feeding. Bismuth, lead and cadmium when present below 1 pct. improve macnihability and bearing property. Chromium reduces stress corrosion cracking in Al-Zn-Mg alloys.

16.1.2 Modification of Aluminium Silicon Alloys

Slow solidification of binary Al-Si alloys produces a very coarse microstructure consisting of enutectic with large flakes or needles of silicon. The brittle nature of large silicon flakes reduce the mechanical property, particularly ductility of the alloy. Generally below 8 pct. silicon, the levels of ductility are reasonable because of the presence of relatively large amounts of primary aluminium phase. In alloys containing higher amounts of silicon, the mechanical properties and ductility are seriously impaired due to (a) coarse microstructure of the eutectic and (b) formation of coarse grains of virtually pure silicon as the second phase.

The mechanical properties may be substantially improved by modification which brings about refinement of microstructure. Modification of the structure is accompalished by small amount of sodium or strontium (0.02 pct.).

TABLE 16.2

Factors Affecting Selection of Casting Process for Aluminium Alloys

Factor	*Sand casting*	*Permanent mould casting*	*Die-casting*
Cost of equipment	Lowest cost if only a few items required	Less than die casting	Highest.
Casting rate	Lowest	11 kg/hr common, higher rates possible	4.5 kg/hr common, 45 kg/hr. possible.
Size of casting	Largest of any casting method	Limited by size of machine	Limited by size of machine.
External and Internal shape of casting	Best suited for complex shapes where coring required	Simple sand cores can be used but more difficult to insert than in sand castings	Cores must be able to be pulled because they are metals, under cuts can be formed only by collapsing cores or loose pieces.
Minimum wall thickness	3 to 5 mm required, 4 mm normal	3 to 5 mm required, 3.5 mm normal	1 to 2.5 mm depending upon casting size.
Type of cores	Complex baked sand cores can be used	Reuseable cores can be made of steel or non-reuseable baked cores can be used	Steel cores must be simple and straight so that they can be pulled.
Best linear tolerance obtainable	300 mm/m	10 mm/m	4 mm/m.
Surface finish	6.5-12.5 μm	4.0-10 μm	1.5 μm.

contd.

Gas porosity	Lowest with good technique	Best pressure tightness, low porosity with good technique	May be present
Cooling rate	0.1-0.5 °C/S	0.3-1.0 °C/S	50-500 °C/S
Grain size	Coarse	Fine	Very fine on surface.
Strength	Lowest	Excellent	Highest, usually used in the as cast condition.
Fatigue properties	Good	Good	Excellent
Wear resistance	Good	Good	Excellent
Overall quality	Depends on foundry technique	Highest	Tolerance and repeatability very good.
Remarks	Very versatile as to size, shape, internal configuaration	—	Excellent for fast production rates.

Because of the reactive nature of sodium, it is introduced into the liquid alloy either in the form of salts which are rabbled into the surface or as metallic sodium packed in suitable container. Care is required to prevent accident. Use of sodium presents founding problems such as reduced fluidity and rapid loss of element by evaporation and/or oxidation resulting in fading of the effect. Efforts are being made to use strotium as Al-Sr or Al-Sr-Si master alloy as the loss of strontium, during melting is minimum and the modified microstructure is retained even after remelting of the alloy.

For hypereutectic Al-Si alloys, the modifying agent used is usually phosphorous (0.05 pct.) which only refines the primary silicon phase and not the eutectic mixture, thereby providing moderate improvement in strength and machinability. Addition of magnesium to the alloys which have been treated with phosphorous can bring about some eutectic refinement, although not comparable to that in the alloys treated with sodium.

16.1.3 Metallurgical Factors Affecting Foundry Practice of Aluminium and its Alloys

The following metallurgical factors have great importance in the melting and casting of aluminium and its alloys:

1. Aluminium being very reactive, gets readily oxidised and forms dross which is wetted by molten aluminium and has about the same density. It gets entrained in the melt during melting, handling and casting and is the principal cause of metal loss during melting.

2. The melt readily absorbs hydrogen from moisture and other contaminants which gets evolved during solidification of the casting to form porosity.

3. Owing to large heat capacity (0.282 Cal/°C/gm) and large latent heat of fusion of aluminium (2.55 K Cal/gm atom), almost three times as much heat is required to melt it as compared to melting the same weight of a copper alloy.

4. Owing to large shrinkage during solidification (about 7 pct.) and subsequent cooling to room temperature (about 5 pct.) it is necessary to ensure that the casting is wellfed by fairly large size risers in adequate number.

5. Hot shortness, i.e., tendency to crack at temperature near the melting point shown by some aluminium alloys requires special care in the design and use of moulds and cores to prevent the development of excessive stresses in the casting when it solidifies and cools.

6. The low density of aluminium alloy melts (about 2.38 gm/Cm^3) results in low hydrostatic pressure in the mould. This combined with high contraction during solidification, necessiates careful risering practice to avoid the formation of cavity and shrinkage defects in castings.

7. The oxide film in the melt can provide a high degree of protection against gas absorption and dross formation provided this film is not excessively broken as a result of stirring and turbulence during the melting and casting operations.

8. The surfaee tension of molten metal is high and gets further enhanced by the formation of the oxide film thereby causing difficulty in casting thin sections. The high surface tension of molten aluminium also keeps the fines

in the charge floating on the top of the bath thereby increasing oxidation losses. Alloy additions reduce surface tension but increase the solidification range which may aggrevate shrinkage problems.

9. The high thermal conductivity and low melting point (660°C) of aluminium are responsible for easy melting. However, its good reflectance to radiant energy, throughout the entire spectral range adversely affects heat transfer in reverberatory type furnaces where the maximum rate of melting in limited to about 2.8 Kg/hr/cm^2 of exposed surface.

10. For all for aluminium alloys, the grain size increases as the pouring temperature is increased. For this reason aluminium should be poured at the lowest temperature that will result in sound casting.

11. Molten aluminium dissolves iron from cast iron crucible, and steel tools. Care is therefore necessary to minimise contaminations from these sources by providing protective coating for crucible and tools. It is advisable to use refractory containers when the specification for iron content are stringent.

The melting and casting practice followed should take into consideration, the above characteristics of aluminium to produce sound castings.

16.1.4 Melting of Aluminium Alloys

Furnaces for melting of aluminium and its alloys can be broadly classified into following three groups:

(i) *Indirect fuel fired furnaces* (Crucible or pit furnaces): In these furnaces contamination and absorption of hydrogen is comparatively low as there is no direct contact between the melt and the products of combustion. The thermal efficiency of crucible furnace is low (15-30 pct.) owing to loss of sensible heat to the flue gases. These furnaces are popular for batch melting of smaller quantities of metal owing to the lower capital cost and lesser melting losses.

(ii) *Direct fuel fired furnaces*: Reverberetary type furnaces are more widely used when large melting capacity is required. Tilting circular furnaces and Sklenar furnace are frequently used in large foundries. Judicious care is required during the operation to reduce the melting losses which otherwise could be unduly high.

(iii) *Electrically Heated furnaces*: Elecrrical resistance heating is used mainly in holding furnaces to ensure proper temperature control and is generally not used for meltidg purposes. Because of the speed, cleanliness and the contols they provide, electrical induction furnaces are becoming popular particularly in pressure die casting processes.

The melting loss in crucible furnaces (fuel fired and electrically heated) is lower (0.5 to 1.0 pct) whereas in reverberatory type furnaces the melting loss may range from 1 to 2.5 pct. or more depending on the charge material and the melting practice. The melting loss of silicon varies from 0.5 to 1.0 pct. and seems to depend little on the conditions of the charge and the type of melting unit. Magnesium suffers a loss of 2-3 pct. and may even increase to 5 pct. or more in melting a heavily oxidised charge. The alloying elements lost in melting are present as oxides in dross which should be removed before pouring to obtain a sound casting.

16.1.5 Metal Treatment

During melting aluminium easily gets oxidised and also absorbs gases particularly hyrogen by reacting with moisture and other contaminants. Hydrogen is soluble in solid as well as liquid aluminium, the solubility being approximately 20 times more in molten than in solid aluminium. Owing to a sharp change in the solubility of hydrogen at the freeziing temperature, there is a marked tendency of unsoundness in the casting due to porosity caused by gas evolution. To minimise hydrogen absorption, care should be taken in degreasing oily scrap, adequate preheating of charge and tools, minimum disturbance of molten metal surface, freedom of gases and solid fluxes from moisture and avoiding excessively long holding periods after metal treatment. Metal treatment may consist one or more of the following operations:

(i) *Degassing*: Degassing is carried out to remove dissolved hydrogen in aluminium. This treatment also brings oxides and other nonmetallics to the surface of the melt which are skimmed off. Fluxes also remove the oxides and non-metallics and therefore degassing and fluxing are terms which are used interchangeably in many foundries. However, it is wrong to look at degassing as an alternative to careful melting and pouring practice.

The gas level that can be tolerated in the melt is determined to a certain extent not only by the alloys and their use, but also by the casting process itself. For sand castings, the gas content must be very low. In permanent mould casting a carefully controlled amount of gas in the metal may help to avoid gross shrinkage defect. Tiny globules of gas released during solidification can counter act and distribute solidification shrinkage. In pressure die casting operations, gas content is even less critical because the manufactured part is generally not a stressed component and the lubricants and the process itself tend to cause porosity.

Gases are widely used for the degassing purpose. The common degassing reagents are nitrogen and chlorine gas or solid compounds which evolve chlorine on decomposition at the temperature of molten aluminium. Treatment with nitrogen alone requires prolonged bubbling and is not effective in the removal of oxides and nonmetallic inclusions. It may lead to the formatioh of nitrides of aluminium and magnasium above 700 °C. Removal of dissolved gases is also somewhat unsatisfactory.

Chlorine gas has a powerful degassing and fluxing action. It directly reacts with molten aluminium forming $AlCl_3$ (volatilises at 180 °C) in gaseous form. This leads to the formation of large number of fine bubbles with large surface area. The dissolved hydrogen diffuses into the rising bubbles and escapes to the atmosphere. As chlorine is a toxic and corrosive gas, it can be used only when proper ventilation and handling equipments are available. It also causes losses of aluminium (0.15 to 0.6 pct.) and magnesium. For this reason a mixture of 90 pct. nitrogen and 10 pct. chlorine is generally preferred now. Another modified practice known as trigas fluxing uses a mixture of nominal composition 80 pct. N_2, 10 pct. Cl_2, and 10 pct, CO. It has been reported that the effectiveness of trigas fluxing is equivalent to 100 pct. chlorine. This method avoids air pollution and removal of magnesium is slower due to reduced conecentration of chlorine.

(ii) *Fluxing*: Removal of oxides and other non metallic inclusions in the melt is essential because these serve as nuclei for molecular hydrogen formation during solidification which evolves and gives porosity. Their presence can also impair the mechanical properties of the product and increase tool wear in machining and polishing. The fluxes used serve the following purposes:

(a) They form a protective covering during melting which minimises the formation of oxides and absorption of gases particularly hydrogen.
(b) Recovery of aluminium is improved as a result of separation of entrained metal from the dross.
(c) The molten metal is cleaned by dross removal.
(d) Most fluxes also exert some degassing effect.

Fluxes that accomplish (a) are used when melting alloys containing more than 2 pct. magnesium as these alloys have a greater tendency to form dross and absorb gas more rapidly than magnesium free alloys. For effectively covering the molten metal, fluxes for magnesium bearing alloys are usually composed of low melting mixtures of chlorides and fluorides of sodium, magnesium and potassium. These fluxes are sometimes called liquid fluxes or passive fluxes.

Fluxes used for refining generally contain fluorides and chlorides of sodium and potassium in the ratio of 1 : 1 and form a low melting eutectic. The addition is usually 2 to 3 pct. of charge mass and the flux is added in several steps till the charge melt down period is over. A flux addition of 0.5 to 1 pct. is made after the melt down to cover the bath followed by addition of sadium fluorides on the top for thickening the flux. The molten metal is poured after raking off the flux layer.

Degassing and fluxing may also be carried out with the help of hexachlorethane, C_2Cl_6 which is non-hygroscopic, yields gaseous reaction products and shows a higher gas evolution capacity. Hexaclorethane is added (0.5 to 1 pct.) of the charge mass) in several portions. This ensures more effective removal of hyrogen as they provide greater opportunity for the diffusion of hydrogen into the bubbles of tetrachlorethane C_2Cl_4 formed as a result of the following reaction:

$$3C_2Cl_6 + 9Al \longrightarrow 2C_2Cl_4 + 2AlCl_3$$

(iii) *Vacuum Degassing*: This is based on the principle of a reduction in the solubility of hydrogen in the melt on application of vacuum of 1-10 mm of mercury for 10-15 minutes. Hydrogen dissolved in the alloy in the ionised or atomic state becomes molecular and forms bubbles which rise to the surface. For the production of high quality alloy castings, melting and pouring is carried out under vacuum. Melting is carried out in vacuum induction furnace and pouring under vacuum prevents the oxidation of the alloy and ensures progressive crystallisation of the casting owing to the destruction of the stream of metal that runs into the mould.

(iv) *Ultrasonic Treatment*: When ultrasonic waves are passed through the melt, elastic vibrations of the particles are induced. Temporarily a discontinuity of the liquid mass is caused because separate particles execute a reciprocating motion with different speeds resulting into formation of microwires similar to those formed under devacuum. The gases dissolved in the melt fill the voids and form hydrogen bubbles. The molecular hydrogen does not go into solution again when the particles move in opposite directions. After the growth of bubbles to critical size, they are evolved to the surface and escape to the atmosphere.

16.1.6 Preminim Quality Castings

For aerospace and other critical applications, premium quality castings are made conforming to strict radiographic and mechanical property specifications. This is accompalished by using composite moulds combining several mould materials and using special chills to control the solidification process and obtain high mechanical properties in designated areas. It is necessary to have good understanding of the solidification process to take optimum advantage of the special properties of each casting process. Depending on the mould material, equipment and fixtures, it is possible to have a linear tolerance of $\pm$ 15 mm per meter and minimum wall thickness of 3 to 8 mm is typical.

16.1.7 Casting Processes

Aluminium possesses the advantage of being cast by all of the processes used for the casting of metals namely, pressure die casting, gravity die casting, sand casting, plaster casting, investment casting, in decreasing order of the total amount of aluminium castings produced. Pressure die casting is becoming increasingly popular. Continuous casting of aluminium has been adopted on an industrial scale with higher yields and many other beneficial effects.

With minor exceptions, most aluminium foundry alloys can be cast by all processes, and choice of the casting technique is usually controlled by factors other than alloy composition. Table 16.2 gives details of some of the important factors affecting selection of casting processes for aluminium alloys. However, (a) feasibility and cost factor and (b) quality factors are most important. These may be summarised as follows:

(i) Very large castings are generally produced as sand castings since the weight of metal mould is 10-100 times as much as the casting weight.

(ii) Small castings are generally made with metal moulds for greater dimensional accuracy.

(iii) Complex shapes with many cores are more easily made by sand casting and it may be practically impossible to cast the shapes in metal moulds even by using sand cores.

(iv) As a general rule, the cost of tooling (patterns, moulds, and auxiliary equipments) significantly influences the cost of producing each piece. Therefore, when only a limited number of pieces are to be made, the casting process involving least tooling is preferred even if the cost of each casting piece is high.

(v) On the other hand, the cost share of elaborate tooling per piece decreases when a large number of castings are required. Die casting is a typical example, where toolinge cost is minimised owing to large number of castings produced.

16.1.8 Casting Quality

(i) A sound casting should be free from posity, cracking and surface imperfections.

(ii) For optimum level of mechanical properties, strength and ductility, grain size and their shape. dendrite arm spacing and size and distribution of second phase particles and inclusions are important.

(iii) Fine equiaxed grains generally give the best combination of strength and ductility in castings. By addition of grain refiners, the size and shape of the grains can be controlled to some extent, although low pouring temperatures and high cooling rates are preferred for this objective.

(iv) In all commercial aluminium alloys, multiple phases as well as oxide and gas inclusions are present. It is desirable to have small inter dendritic spacing as during freezing, inclusions and second phase particles segregate in the spaces between the dendrite arms. Coarser distribution of these micro constituents, adversely affects the mechanical properties. Finer dispersion of inclusion and second phase particle obtained by small inter-dendritic spacing gives better properties in the castings.

(v) The shape and size of microconstituents can be controlled to some extent by controlling composition, but is accomplished more efficiently by rapid cooling which minimises the time for growth of micro constituents. In die casting, even though the cooling rates are very high, appreciable amount of porosity at the centre of casting occurs because of air entrapment in the casting. For this reason, pressure die castings are of ten lower in strength than low pressure or gravity fed permanent mould castings which are more sound inspite of slower cooling rate.

16.2 PRODUCTION OF COPPER AND ALLOY CASTINGS

Copper and its alloys constitute a major group of engineering metals. Copper is the only metal other than gold that possesses natural colour, the deeper reddish tint of ten noticed is probably due to a very thin oxide film on the surface. Excellent thermal and electrical conductivity, superior resistance to corrosion good, strength and fatigue resisiance, good bearing surface qualities and wear resistance, ease of joining with different methods like soldering, brazing, welding, etc. are some of the outstanding properties which account for their wide spread use in a large variety of industrial applications. Pure copper has comparatively low mechanical properties and inferior casting characteristics which limit its use for cast parts. It is more commonly used as alloys. Typical applications include components requiring high electrical and thermal conductivity and good corrosion resistance, such as electrical fittings and fixtures, cams, bushings, bearings, valves, pump parts, gears, structural parts, resistance welding electrodes and inserts, moulds for plastic parts, precision cast parts for communication, textile, aerospace, fire arms, instruments, ordnance, etc. In effect the uses are so varied and diverse, that copper and its alloys would be the first choice in most such applications. However, concerted efforts have been made for substitution of copper and its alloys by other metals and alloys owing to economic constraints of high cost and scarcity of supply of copper.

16.2.1 Metallurgical Factors Affecting Foundry Practice

(i) *Susceptibility to oxidation*: As a classs copper base alloys are not particularly susceptibile to oxidation and losses in melting except high zinc containing alloys such as brasses which are prone to high losses to zinc under adverse conditions of melting. Lead and iron may also be lost owing to oxidation and reaction with slag. Phosphorus used for deoxidation and improvement of specific properties is readily lost owing to (a) its high affinity for oxygen and (b) oxidation product is a fluid phosphate which exposes new surfaces to oxidation owing to dewetting on the metal surface. In alloys such as aluminium bronzes and high tensile brasses, preferential oxidation of aluminium occurs forming a protective oxide film on the surface of the melt which prevents further oxidation. If reasonable precaution is taken to avoid agitation during melting, oxidation losses are generally not serious.

(ii) *Constitution and Shrinkage Characteristics*: Based on this consideration, copper base alloys may be classified into two main groups, (a) deoxidised copper, 60—40 brasses, aluminium bronzes and high tensile brasses having short freezing ranges and (b) low zinc brasses (70—80 pct. Cu), cupro nickels, tin bronzes, gun metals, silicon bronzes, nickel silvers, and nickel bronzes having large freezing ranges varying from 30 to 200 °C or more.

Alloys with short freezing range require comparatively large feeders to avoid shrinkage cavity, thereby decreasing the yield of castings. As a consequence, the costing techniques for such alloys are comparatively expensive but it is usually feasible to produce castings of these alloys with remarkably uniform properties and virtual freedom from shrinkage porosity. While producing castings of intricate shapes in such alloys, it becomes necessary to use chills at local heat centres which are not easily fed.

Because of larger freezing range, alloys of second group tend to have widely dispersed shrinkage porosity in the castings. Moderate amount of such dispersed porosity makes it possible to use simple gating and feeding techniques, thereby giving high yield of castings. These characteritsics are particularly notable in tin bronzes and gun metals which do not form tenacious oxide film and this characteristic also helps in using simple and inexpensive methods of gating and feeding. Other alloys such as silicon bronzes, high copper brasses, nickel silvers, cupronickels, and nickel bronzes have comparatively shorter freezing ranges and require somewhat larger size of feeder to avoid shrinkege unsoundness.

Although such alloys with longer freezing range possess superior casting characteristics, caution is necessary because the shrinkage porosity is not completely eliminated, and the apparently sound casting may get rejected after machining. Many of these alloys are commonly used for pneumatic and hydraulic components, and are subjected to pressure test for conforming their suitability for service.

(iii) *Susceptibility to hot tearing*: The freezing range of copper base casting alloys also determines their susceptibility to hot tearing when chill cast into intricate shapes. Alloys with short freezing range, e.g. 60 — 40 type brasses and aluminium bronzes are extensively used for gravity die castings as these are virtually immune from drawing and tearing. Alloys with longer freezing range, however are generally considered unsuitable for gravity die casting process.

(iv) *Susceptibility to gas unsoundness*: Copper-zinc alloys with 20 pct. or more of zinc including high tensile brasses and nickel silvers are generally immune from hydrogen unsoundness owing to high vapour pressure of zinc in such alloys. Other alloys, particularly tin bronzes and gun metal which possess large freezing ranges are very susceptible to hydrogen gas unsoundness because gas rejected from solution during freezing is not able to escape from the casting.

Reaction gases resulting from the combination of oxygen with hydrogen, carbon or sulphur during solidification can produce unsoundness in some of the alloys leading to undesirable porosity. Alloys such as aluminium bronzes which have low oxygen solubility are comparatively free from such unsoundness. To reduce the susceptibility to reaction gas unsoundness, copper base alloys are deoxidised, the commonly used elements for this purpose are prosphorus, zinc and manganese. Silicon in silicon bronzes, magnesium in nickel silvers and other nickel containing alloys and chromium in special bronzes act as deoxidants.

(v) *Oxide entrapments*: It is advisable to ensure quiet pouring into moulds without whirling and splashing to prevent oxidation of copper alloys in pouring and the breaking of oxide coating and its entry into the mould cavity, which may cause decrease in pressure tightness.

(vi) *Melting Point and density* The melting points vary between 900 and 1200 °C for different alloys and the pouring temperatures have to be carefully controlled to suit each alloy to obtain optimum advantage of fluidity and avoid oxidation, extra shrinkage and gas absorption resulting from very high pouring temperature. For copper alloys, natural moulding sands and high clay content can be used to get good surface finish as these sands are sufficiently refractory to prevent any burning on and the high density of alloys (8-9 gms/cm^3) permits lower permeability in moulding sand without undue risk of blowing defects. This is in contrast with the requirement of highly refractory sands for ferrous castings and high permeability sand for light alloy castings.

16.2.2 Copper and Copper Base Alloys

Most of the copper casting alloys belong to the following six main groups:

(i) High conductivity copper and alloys.
(ii) Copper and zinc (brasses).
(iii) Copper and tin (bronzes).
(iv) Copper, tin and zinc (gun metals).
(v) Copper and aluminium (aluminium bronzes).
(vi) Other alloys.

A number of alloying elements like Si, Pb, Fe, Mn and Ni are added in different proportions to modify the properties for specific applications. The effect of various alloying additions on the above groups of alloys is given in Table 16.3.

TABLE 16.3

Cooling Rates and Dendrite arm spacings for various casting Processes

Casting process	*Cooling Rate, °C/S*	*Dendrite-arm spacing, mm*
Plaster, dry sand	0.05 — 0.2	0.01 — 1
Green sand, shell	0.1 — 0.5	0.05 — 0.5
Permanent mould	0.3 — 1	0.03 — 3.07
Die	50 — 500	0.005 — 0.015

The application of copper alloys used for castings is restricted by their high cost. Various copper alloys with respect to the foundry practice, are discussed below:

(i) *High conductivity copper and alloys*: These alloys do not form difficult oxide film, have short freezing range and require simple running and gating technique except large feeder heads which are necessary to eliminate gross shrinkage cavities. The requirement of high electrical conductivity calls for careful melting practice which will yield melts with consistent and low oxygen content so that rigid control can be exercised on the amount of deoxidant to be added. Phosphorous deoxidised copper when melted rapidly in a crucible furnace under the cover of a reducing flux or charcoal has low oxygen content, but unfortunately yields a melt with high hydrogen content and thus susceptibile to gas porosity. Charging manganese ore at the bottom of the crucible ensures a low oxygen content in the bath and the addition of 0.01 pct. phosphorous is sufficient to take care of small quntity of oxygen and the resultant casting has a conductivity of more than 85 pct. IACS. Because of very adverse effect of phosphorus on electrical conductivity, alternative deoxidants including silicon, calcium and boron are used in some foundries.

For some applications, greater hardness, strength and wear properties are required which are achieved by adding cadmium (0.7 to 1 pct), chromium (0.4 to 0.8 pct.), and tellurium 0.3 to 0.7 pct.), conductivity values in all these alloys are generally above 80 pct. IACS. The melting technique for cadmium and tellurium copper is similar to that for high conductivity copper, whereas for chromium copper special fluxing is necessary to suppress the formation of chromium oxide. The alloying additions are made after degassing in the form of master alloys. When electrical and thermal conductivity are not of importance, copper castings may contain about 2 pct. zinc or tin which increase the hardness and improve machinability. In these cases deoxidation by phosphorus alone is generally adequate.

(ii) *Copper-Zinc Alloys* (*Brasses*): In these alloys gas absorption does not occur because of high vapour pressure of zinc which perclude solution of hydrogen. However, alloys containing high copper percentage (70-80 pct.) are susceptible to dispersed shrinkage without elaborate provision for feeding, whereas low copper alloys (60/40 type brasses) have a short freezing range requiring massive feeder to produce sound castings. In the presence of elements like aluminium and managanese which produce tenacious oxide films, the gating techniques deserve special attention. Deoxidation in brasses is generally not required owing to the presence of sufficient zinc in the bath. However, controlled amounts of phosphorised copper may be added to increase the fluidity of molten alloy and thereby assist the separation of dross and nonmetallics into the slag. Addition of suitable fluxes produces a sufficiently fluid slag which can be removed without loss of valuable metal. The slag on the surface of molten alloy should be retained upto the time of pouring to suppress the escape of zinc.

(iii) *High Tensile Brasses*:—For relatively high strength and high corrosion resistance, high tensile brasses are used. A high tensile brass is a 60-40 alloy to which varying amounts of other elements like Al, Mn, Si, Fe, Pb and Sn are added to modify the properties. A typical composition of high tensile brass is Cu 54-62, Al 0-4, Mn 0-5, Fe 0-2, and Sn 0-1. Close control over alloying elements, particularly aluminium is necessary for optimum mechanical properties, as these properties change rapidly with composition. It is necessary to determine the zinc aquivalent values of these additions and the effect on resulting microstructure. The most commonly used alloys have $\alpha+\beta$ structures with atleast 15% α-phase and show very high strength. The zine equivalent values of various alloy additions are given in Table 16.4. It should be noted that the mechanical properties of these alloys are adversely affected if copper content approaches 70 pct.

TABLE 16.4

Zinc Enquivalents of various Alloying Additions in Brass

Element	*Zinc Equivalent*	*Element*	*Zinc Equivalent*
Silicon	10	Iron	0.9
Aluminium	6	Manganese	0.5
Magnesium	2	Cobalt	0.8
Tin	2	Nickel	1.3
Lead	1		

The foundry characteristics of high tensile brasses and aluminium bronzes are similar and mav be summarised as follows:

(a) These alloys are commonly described as high shrinkage alloys and require the provision of generous feeder to avoid coarse shrinkage porosity. The overall shrinkage is actually lower than for many other copper alloys but the provision of liberal feeder head permits the diffusion of gas from the liquid metal to the advancing solid front and then escapes into the massive feeder head provided.

(b) Because of the presence of aluminium in these alloys, a tenacious alumina film is formed on all the exposed surfaces and if special care is not taken to ensure non-turbulent pouring, the oxide film becomes trapped in the casting owing to folding and crumpling constituting harmful discontinuity, similar in effect to cracks.

(c) Owing to the above casting characteristics the overall yield of castings in these alloys is low and the fettling costs are relatively high.

(d) High tensile brasses are not normally susceptible to gas unsoundness owing to presence of zinc, aluminium bronzes are susceptible to unsoundness due to hydrogen, but not to reaction gas unsoundess because aluminium reduces the oxygen solubility to a very low level. Aluminium bronzes therefore require care in melting because oxidation treatment cannot be employed to remove dissolved hydrogen.

(iv) *Bronzes and Gun Metals*: Very few binary alloys containing only copper and tin are used for castings and it is more usual to add zinc, lead, phosphorus, and nickel in suitable proportions. If substantial quantities of these additions are made, the alloy is named accordingly. Owing to the wide freezing range (approximately 250°C) the 10 pct. tin range alloys give a cored structure which contains other phases as well. The embeded high tin content phase in a soft copper matrix gives good bearing properties and the most common bearing alloys are phosphorus bronzes which contain 0.5 pct. P along with lead and zinc in different proportions to get the desired mechanical properties. When the alloys contain 5-20 pct. lead, they are designated as leaded phospher bronze or leaded bronze.

In view of the large freezing range, and the absence of tenacious oxide film, bronzes and gun metals are frequently cast by top running with pencil gates and progressive solidifieation from the bottom of the casting upward is obtained. Owing to the low surface tension of molten metal, sand penetration is specially marked. In case of phosphor-bronzes which form a fluid phosphate at the metal surface, addition of 0.3 to 0.5 pct, silicon forms a tenacious oxide film which reduces gas absorption and sand penetration giving smooth surafce in the casting. Low pouring temperatures and painting of the mould with Mg—Al alloy mould paint has also been found helpful in reducing the gas absorption and penetration by molten metal.

High lead bronzes and gun metals have pronounced tendency for segregation of lead owing to a wide immiscibility gap and in slowly cooled castings the lead rich liquid tends to form unduly large particles. Gun metals are susceptible to interdendritic porosity owing to wide freezing range and lead when present helps to seal the pores and improves pressure tightness of the casting. Addition of 1-2 pct. nickel and 0.5 pct. sulphur are reported to minimise lead segregation, but the most effective method is to promote rapid solidification by pouring at the lowest optimum temperature and if necessary and practicable by the use of chilis. Special care is also necessary to prevent gas absorption particularly when the alloys are sand cast and the residual phosphorus content used for deoxidation exceeds 0.2%.

(v) *Aluminium Bronzes*: These alloys contain approximately 10 pct. aluminium along with varying amounts of other elements (e.g. Fe and Ni) to modify the properties. Owing to their excellent mechanical properties and good corrosion resistance, these alloys are used in certain environment in preference to high tensile brasses.

Normally the microstructure consists of α and β phases in alloys containing more than 8.5 pct. aluminium. However, owing to the eutectoid reaction at 560°C, a very hard and brittle γ-phase is formed which adversely affects the properties of the castings. To prevent its formation, nickel and iron additions are made which slow down the eutectoid reaction. It is also possible to suppress the reaction by rapid cooling and finally the r-phase is eliminated by heat-treatment.

(vi) *Silicon Bronzes*: These alloys contain 1.5 to 5 pct. Si, 0-3 pct. Zn, 0.1 pct. Mn, 0.2 pct. Fe, and balance copper and have the following foundry characteristics:

(a) Owing to freezing ranges between those of copper—tin alloys and copper aluminium alloys, the requirements of feeding are greater than tin bronzes but less massive as compared to aluminium bronzes which have short freezing ranges.

(b) The oxygen solubility is low owing to high silicon content and the oxidation treatment does not have any marked degassing effect. Although the alloys are susceptible to hydrogen unsoundness, care during melting usually precludes degassing treatment.

(c) The oxide film on the surface is fairly strong but is less troublesome than in case of high tensile brasses and aluminium bronzes and special gating techniques are not often required.

(vii) *Cupronickels, Nickel Bronzes, and Nickel Silvers*: These alloys contain 5 to 30 pct. nickel and the following precautions are necessary to produce sound castings:

(a) These alloys are prone to gas unsoundness due to the reaction of carbon and oxygen resulting in evolution of CO during solidification. Even nickel silvers containing about 20 pct. zinc are susceptible to gas unsoundness which is indicative of very small oxygen content involved in the above reaction. It is therefore, not advisable to use charcoal cover during melting of a nickel containing alloy to avoid dissolution of carbon. The melting is generally carried out under oxidising conditions by using manganese ore in the furnace charge and final deoxidation is carried before pouring with small amounts of manganese, phosphorus and magnesium.

(b) The feeding practice is similar to that used for silicon bronzes in view of their comparable freezing ranges.

16.2.3 Melting Furances

Copper base alloys can be produced by melting suitable copper base scrap and virgin copper in crucible, reverberatory or rotary furnace. The modern trend is to use indirect arc and induction furnaces for the melting of copper base alloys to keep the melting losses on lower side. Induction and crucible furnaces are used generally when straight remelting is required. Considering the free energy values, most impurity elements present in copper can be preferentially oxidised to join slag during melting. The main groups of copper alloys for melting considerations may be classified as follows:

(i) Dilute copper alloys which contain only small amounts of alloying elements like Cr, Ag, Cd, Zr, and Tc, and are used where high thermal and electrical conductivities are required. For these alloys, copper scrap of appropriate composition is blended with virgin copper and is melted in oxidising atmosphere free from SO_2, which may introduce sulphur in copper melt leading to gas porosity and considerable reduction in mechanical properties. The oxidation may be carried by lithium, or calcium boride (CaB_6) Phosphorus as Cu – 15 pct. P alloy may also be used in case electrical conductivity requirement is not critical. Subsequently additions of alloying elements are made into the deoxidised molten copper for obtaining highest possible yield.

(ii) Alloys of copper with tin, zinc, aluminium, lead, etc. are produced by melting the virgin copper and appropriate clean copper scrap of appropriate composition uuder oxidising conditions and then dissolving alloying additions in it. Broazes are usually deoxidised with phosphorus whereas in the case of alloys containing aluminium and zinc, no further addition may be necessary to give sufficient deoxidation. Lead addidition made to improve machinability, wearing and pressure tightness properties requires proper strirring before pouring of the molton alloy to avoid segregation at the bottom of the crucible. To compensate for losses of zinc owing to its high vapour pressure and oxidation, it is necessary to make extra additions to compensate for loss just before pouring.

Casting Processes

Low shrinkage alloys like aluminium bronzes, silicon bronzes, manganese bronzes, silicon brasses and some nickel silvers possess better fluidity and can be easily cast to give high grade castings by sand, permanent moulds, centrifugal and die casting processes. However, it is necessary to pay careful attention to the design to promote directional solidification, avoid abrupt changes in cross section and notches, use of liberal fillets and locate gates and risers at proper places, to minimise internal shrinkage and dross. All copper alloys can be sand cast which is an economical casting method permitting greatest flexibility in casting size and shape.

Permanent mould and die casting processes are suited for tin, silicon aluminium and manganese bronzes and also for yellow brasses, the limitation being reduced die life with larger castings. Centrifugal casting is suitable for almost all copper alloys.

Generally pressure gating systems (2:8:1 or 3:9:1) are used for copper alloys. The riscring is similar to that used for aluminium alloys due to large freezing range involved.

16.3. PRODUCTION OF MAGNESIUM ALLOY CASTINGS

Magnesium alloys are good structural materials in view of their superior specific strength as compared to some structural steels and aluminium alloys. Magnesium metal is about 4.5 times lighter than iron and 1.6 times lighter than aluminium, the density of its ailoys ranging from 1.75 to 1.90 gm/cm^3. In view of their excellent damping capacity, good machinability, and fatigue strength, magnesium alloys are of much importance for air craft and other transport industries and dynamic applications. After requisite treatment, the corrosion resistance of magnesium alloys in dry atmosphere is comparable to aluminium alloys. Magnesium alloys can be cast into parts upto 500 kg in weight and be produced by sand and die casting processes Structural applications include textile and printing machines where light weight minimises inertial forces. Use in materials handling equipment such as dock boards, grain shovels, gravity conveyor, etc. and commercial application such as luggage and ladders lead to convenience and economy.

Combination of good strength and stiffness at both room and elevated temperaturc along with light weight make magnesium alloys valuable for aerospace applications, instrument engineering, automotive industry, radio engineering and other industries.

The casting of magnesium alloys involves distinct specialisation as a result of some distinct properties of the metal itself and piesent formiddable difficulties in their foundry technique. The principal factors are as follows:

(i) In the molten state the chemical reactivity of magnesium is so high that it combines readily with oxygen, nitrogen and all forms of water. Magnesium alloy's surfaces burn rapidly above about 800°C if not properly protected.

(ii) The latent heat of solid magnesium is 0.246 cal./°C/gm. The low heat capacity and insufficient fluidity of magnesium alloys does not permit their casting at a temperature just above 650°C (melting point of pure magnesium) and the normal casting temperatures for the magnesium alloy parts range in between 720 and 800°C. For a given volume; the heat content of molten magnesium is only 70 pct. of that aluminium because of the low heat capacity and low specific gravity of magnesium metal. Use of high casting temperatures thus becomes necessary, thereby increasing susceptibility to oxidation if unprotected to burning.

(iii) Magnesium alloys generally solidify on a large temperature range and are prone to shrinkage porosity defects, their incidence being influenced by the necessity for using relatively high temperature.

(iv) Some magaesium alloy castings are pronc to show pronounced grain size variations from point to point within the casting which strougly influence their mechanical properties.

(v) During heat treatment to improve properties, special care has to be exercised because of the tendancy to oxidise at solution treatment temperature.

(vi) As the metal has high transparency to X-rays, radio-graphic technique is difficult in interpretation and specialised methods of inspection are required to detect microporosity in castings.

(vii) Proper coordination between the designer and foundryman is very relevant to attain optimum mechanical properties of the alloys. This is because the mechanical characteristics of magnesium alloys are similar to those of aluminium rather than of cast iron and it is most desirable to ensure that sharp section changes are eliminated, corners and fillets are rounded as far as possible and highly stressed parts are connected with rest of the body of the casting as generously as possible to ensure proper feeding. These design considerations call for more rigorous application for magnesium alloy castings.

16.3.1. Magnesium Alloys

Most widely known magnesium base alloy systemes are those with manganese, aluminium and zinc. On account of casting difficulties (impaired fluidity and susceptibility to hot cracking) as also low mechanical property, magnesium alloys are not popular. Alloys in magnesium-aluminium-zine system with aluminium as the basic alloying element (5 – 11 pct.) are generally used. Addition of aluminium decreases susceptibility of magnesium alloys to self ignition and brings about improvement in the mechanical and casting properties. Zinc addition upto 6 pct. improves the strength but adversely affects the casting properties. Addition of manganese in magnesium aluminium-zinc casting alloys in amounts of 0.1 to 0.5 pct. confer beneficial effect on the corrosion behaviour and the best corrosion resistance in these casting alloys is obtained when iron and nickel contents are kept as low as possible besides the addition of manganese as mentioned above. The corrosion resistance is highest when iron and nickel are present in concentrations of less than 0.005 pct. and 0.001 pct respectively along with flux inclusions less than 0.2 to 0.3 pct, As Zirconium refines the grain size of alloys and improves their mechanical properties, use of zirconlum containing alloys is inereasing rapidly.

Recent developments in casting alloys include high strength magnesium-zinc-zirconium and the creep resistant magnesium-rare earth-zinc zirconium and magnesium-thorium-zinc-zirconium alloys.

Addition of small amount of beryllium of the order to 0.003 pct., greatly reduces the bnrning tendency of molten magnesium, calcium addition though not so powerful in effect, tends to produce a similar result. Additiou of beryllium, however brings about pronounced grain coarsening in magnesium – aluminium – zinc alloys, thereby seriously impairing their mechanical properties. This calls for severe chilling conditions during solidification of the casting to obtain fine grain size inspite of the coarsening tendency of beryllium addition.

16.3.2 Foundry Techniques

Owing to high chemical reactivity of magnesium alloys, special care is required in casting them. These include the use of mould reaction inbibitors in the mould and core sands and use of special fluxes during the melting operations. If the moulding sands do not contain special protective additive, magnesium reacts with moisture and oxygen of the air in the mould parts and the silica resulting in maximum heat evolution and alloy burning. Boric acid, sulphur and ammonium bi-fluoride are added as principal inhibitors, the usual quantities being of the order of 0.3 to 0.5 pct. boric acid, and 3-10 pct. sulphur, present together. Ammonium bi-fluoride addition is of the order of 5 pet. added to the sand in the form of concentrated aquous solution and sulphur additions are also made when producing heavy castings. Sulphur provides an atmosphere of sulphur vapour and sulphur dioxide and this gaseous shell protects the metal from contact with water vapour and gases in the mould. Boric acid reduces the chemical reaction by fusing and providing an insulating layer between the metal and moisture. The protective inhibitors added to sand result in copious evolution of gases and therefore it is imperative that moulding sands for magnesium alloys should possess high permeability (100 to 120 AFA units) to permeat their escape. As the fumes are unpleasant, it is necessary that the foundry must be well ventilated and should have a good sand handling system as the inhibitor being volatile, requires continuous replacement and also for the purpose of maintaining good sand permeability. It is also essential to keep the sand clean and free from contaminants, specially carbonaceous material, as particles of coke or coal mixed with the sand may glve rise to serious surface blemishes on the casting. For this reason coal or coke fired furnaces are considered unsuitable for melting magnesium.

The fluxes used in the melting of magnesium alloys are usually mixtures of alkaline earth metal chlorides and fluorides, sometimes with an addition of magnesium oxide. As the inclusions of the residual fluxes in the finished castings lead to corrosive attack, it is necessary to have a fiuxing procedure which would ensure that no flux was carried over into the casting itself on pouring. The fluxes suitable to give protection from the atmosphere during melting (melting flux) and for refining and cleaning the metal when molten (refining flux) vary in composition but following are the typical figures.

Melting flux $MgCl_2$, 11, $CaCl_2$ 42, NaCl 27, KCl 18

Refining flux $MgCl_2$ 34, $CaCl_2$ 15, NaCl 9.5, KCl 7, MgO 10.5 CaF_2 18

The refining flux is supposed to exert its cleaning action by thickening highly fluid fluoride particles and in causing them to settle down at the bottom of the pot For this reason, the pot is not entirely empetied during pouring. To retard dross transmission, pouring basins are made large to trap the slag and top gating system with slit gate is used. Mechanical dross extraction is also attempted by the use of steel wool or perforated tinned steel sheet. Usually sulpher is also dusted over the metal during pouring.

16.3.3 Melting of Magnasium Alloys

Magnesium alloys can be melted in gas fired, oil fired, resistance, and induction furnaces. For ordinary purposes, the melting of magnesium alloys is done in mild steel pots because iron pick up by the molten charge is small. These pots are aluminised to prevent iron contamination. However, for producing high purity alloys of low iron content, special steel pots may have to be used. Line frequency induction fruriaces having magnesite lining are progressively becoming popular. Reverberatory furnaces may be employed for continuous melting of magnesium alloys and for remelting bulky scrap.

16.3.4 Shrinkage Effect

The use of relatively high pouring temperature to ensure satisfactory filling of the casting results in setting up adverse temperature gradient within the mould cavity which tends to produce unsoundness in the casting. Cross-sections of the mould traversed by the large quantity of metal tend to become strongly heated and in these parts the metal solidifies late or last thereby giving rise to shrinkage porosity. Some alloys of the magnesium-aluminium-zinc system also exhibit microporsity taking the form of dispersed cavities found in parts of the casting which have solidified last. Even though the percentage of voids may be quite small (of the order of 0.75 pct), the presence of microporosity of this extent may rednce the tensile strength to about 50 pct. of a porosity free part of similar composition. It appears that the main cause of micropprosity is simple shrinkage but gas may also play a part and produce porosity which can not be easily distinguished from the other type. From the practical point of view, therefore the metal should be free from gas and the running and pouring conditions as also the use of chills are so adjusted so as to ensure the solidification of the important parts of the casting first and to have the feeders full of hot metal. If proper care is taken, normal magnesium alloys do not pick up seriout quantities of gas in melting. Degassing is accompalished with chlorine as nitrogen reacts with magnesium. Removal of gas from the molten metal is accompalished in a similar way as degassing of aluminium, i.e. either with chlorine or fully chlorinated hydrocarbons (hexa chloro ethane, C_2Cl_6). Use of hexachloo ethane for degassing also gives the advantage of grain refinement.

16.3.5 Basic Safety Rules in Melting Magnesium-Aluminium Alloys

Magnesium alloys have a tendency to ignite and burn with a dazzling white flame in the open air at 400—500°C depending on the composition. Dry dust of magnesium and its alloys undergo spontaneous combustion with explosion at 400 – 480 °C and the moist dust ignites at 360 – 380 °C, the burning being accompanied by a large quantity of heat and fume. Canventional water or foam fire extinguishers can not be used as these will cause explossion. Reaction with water yields hydrogen and this when mixed with oxygen forms an explossive mixture. Dry ground flux applied in melting of magnesium alloys is used for quenching burning magnesium. Precautions are required while machining magnesium alloys against fire hazards and in all cases the flux should be available at the working places in sealed boxes to avoid reaction with moisture. It is imparative that the foundry shop for casting magnesium alloys should be isolated from other rooms by fire resistant walls and metal beams. The melting shop should also be provided with adequate emergency exit to allow escape in the event of fire.

16.4 PRODUCTION OF ZINC AND ALLOYS CASTINGS

Zinc possesses good resistance to atmospheric corrosion and this property is the basis of a large number of zinc applications. Zinc base alloys are very important for die castings. The automotive industries use the largest number of zinc alloy die castings, both for structural and decorative functions. Important uses include carburettor, fuel pump bodies, speedometer frame, grilles, parts of hydraulic brakes, hub for steering wheel, instrument panel, body moulding and a large number of other components. Die castings are also employed in parts of working machine, vacuum cleaners, motor housings, electric clocks, kitchen equipment and utensils, in business machines like type writers, picture projectors, accounting machine, gasoline pumps, hoists, etc. A large range of products for the electrical industry are manufactured by using diversed alloys of zinc as die castings. Building hardware, toys and novelties pad locks represent a substantial proportion of zinc base die castings used. Zinc die castings are also used for the manufacture of tools and machine shop equipment where steel inserts are often cast in zinc castings as parts of the tools and thus saving substantial assembly cost. It has been estimated that about 20 pct. of the total zinc produced in the world is used for die casting of zinc alloys.

16.4.1 Characteristics of Zinc and its Alloys

(i) In the early days of development zinc alloys in general acquired a dubious reputation because they were prone to distortion, cracking, and ultimately complete disintegration in warm and humid environment. This was traced to the presence of impurities in the alloys particularly lead, tin, and cadmium, which if present in excess of maximum limit (0.005 pct. Pb. 0.004 pct. Cd and 0.003 pct. Sn). The advent of high purity zinc (purity exceeding 99.99 pct.) and minor addition of magnesium has practically eliminated inter crystalline corrosion and disintegration.

(ii) Pure zinc is not very suitable for producing die castings as it attacks cast iron or steel dies and melting pots and forms zinc iron studge. A small proportion of aluminium (about 0.5 pct or more) inhibits the sludging tendency and hence aluminium is present in all zinc die casting alloys. It improves the fluidity of alloys and improves their mechanical properties.

(iii) Owong to eutectic and eutectoid transformations in the zinc aluminium alloy system during solidification in pressure die casting, the zinc rich phase and the aluminium rich phase are greately supersaturated. The relief of this condition of supersaturation by slow precipitation from solid solutions leads to small dimensional changes which zinc alloy die castings undergo. Such changes produce a maximum shrinkage of 0.7 mm per meter. It has been found that about two thirds of the shrinkage occurs in 4 to 5 weeks and the balance continues over a period of years. When castings are required to maintain rigid dimensional tolerances, a low temperature stablishing annealing treatment is given which may be as follows:

(a) 3 to 6 hours at 100 °C.
(b) 5 to 10 hours at 85 °C.
(c) 10 to 20 hours at 70 °C.

(iv) Zinc alloys can be cast in sand moulds but are generally not popular as it leads to very coarse structure resulting in brittle and weak casting.

(v) Zinc alloys possess a number of advantages such as low cost, ability to be die cast at low temperature and close dimensional tolerance permit the thinnest sections produced and are easily machinable. They are characterised by strength greater than all other die cast metals ekcept copper alloys.

(vi) Iron can be maintained easily within limits with aluminium addition which minimises the attak of ferrous metals by molten zinc. If present in excess, it may form brittle iron aluminium compound which may interfere with polishing and machining. Silicon is ordinarily present in very small quantity and like iron if present in excess can interfere with machining.

(vii) Magnesium redues intercrystalline corrosion, although the use of high purity zinc has reduced the need for this addition. Magnesium refines the grain size and increases hardness if present in larger quantities, it introduces hot shortness and has an adverse influence on castability probably because of oxide formation. However, the concentration of magnesium now used in various die casting alloys is low so that adverse effect is negligible.

(viii) Copper forms a solid solution with zinc which improves strength and reduces the susceptibility to intercrystalline corrosion. However, it is necessary to control copper content to produce a dimensionally stable casting. Aluminium is invariably present in all the alloys because of its beneficial effect on castability owing to increased fluidity and improved mechanical properties owing to a fine grain size.

(ix) The melting point of zinc alloys is low and the amount of hcat required to bring zinc alloys to casting temperature is only 25 per cent of that required for aluminium.

16.4.2 Zinc Casting Alloys

(i) *Zinc Pressure Die Casting Alloys*: Compositions of two main zinc pressure die casting alloys (B S 100 4) are given below:

	Al	Cu	Mg	Fe	Pb	Cd
Alloy A	3.9/4.3	0.03	0.03-0.06	0.1	0.005	0.005
Alloy B	3.9/4.3	0.75/1.25	0,03-0.06	0.075	0.003	0.003

	Sn	*Zn*
Alloy A	0.002	Remaining
Alloy B	0.001	Remaining

(ii) *Other Casting Alloys*: Recently two zinc base alloys suitable for sand, permanent mould, plaster moulds, steel mould and investment casting, containing 12% Al and 27% Al are finding increasing applications owing to their good strength and hardness properties to substitute for cast iron and copper alloys in many structural and pressure tight applications. Ease of machining and better inherent corrosion resistance are additional advantages with these alloys. Good bearing and wearing characteristics of these alloys permit their use as bearing, bushings, and flanges. Castings ranging from a few grams upto 50 Kg in weight and from 1.5 mm to 5 mm in thickness have been successfully cast by sand casting or in permanent moulds. They are particularly suitable for the production of castings where the quantity is too small for pressure die casting to be economical. Gravity die casting alloys are also being used for prototype manufacturing, for trial marketing prior to undeertaking large scale production by pressure die casting owing to close similarly in strength to zinc die castings.

16.4.3 Melting and Casting Techniques

Owing to low melting point and excellent fluidity, minute details in the die can be reproduced and complex shapes can be cast with accuracy. For obtaining high standards of quality, it is necessary to pay careful attention to the design of the die and conditions during melting and casting. Superheating of zinc alloys is undesirable and for obtaining optimum mechanical properties and to avoid undue attack on the melting pot and casting machine, both metal and die temperatures should be consistent with the production of sound castings of good surface appearance. Generally die temperature in the range of 180 to 220 °C and metal temperature in the rang of 400 to 525 °C are found to give best results under most casting conditions. It is advisable to avoid sudden change of die sections to minimise turbulance in the metal and strains in the ejected casting resulting from local differences in rates of cooling.

For better appearance many zinc alloy die castings are electroplated and this aspect should be considered during die design because complexity of design may create problems during electroplating and with minor modifications these could be minimised considerably. Sharp corners and notches are undesirable from considerations of strength as well as die design. In view of the profound effect of impurities such as lead, cadmium, tin and their alloys, it is of utmost importance to guard against contamination by exercising visilence coupled with clean and systematic working..

16.4.4 Melting Furnaces

The design and melting capacity of the melting furnace depends on the production pattern of the plant. Open topped oil or gas fired cast iron crucible furnace are used frequently and inside of the crucible is coated with the refractory to minimise attack. Channel type induction furnace can be used for melting of zinc alloys, but the problem of oxide build up in the channel is even more trouble some than melting of aluminium for die casting. Fluxing salts consisting of zinc chloride base mixtures are used to avoid oxidation losses and the metal for transfer to the die casting machine is taken from below the melt surface using a metal pump. This avoids the transfer of oxide and sludge.

17

Modernisation, Mechanisation and Lay Out of Foundry

The foundry is a commercial set up for manufacturing castings. The industry requires collection of huge raw materials such as fuels, fluxes, metal scrap, alloys, etc. and a variety of equipments, e.g. furnaces, moulding machines, fettling equipments, pattern making machines, etc. Broadly foundry can be categorised on the basis of production pattern and metals to be cast. On the basis of production pattern foundry may be categorised as (a) captive foundry, (b) jobbing foundry, (c) production foundry, and (d) semi production foundry.

Captive foundry is an integral part of some manufacturing organisation, where all the castings made are consumed mainly in the products being manufactured by that organisation. *Jobbing foundry* normally produces small number of castings of a given type for different customers. Such type of foundries may also have facilities for mass production. *Production foundry* is highly mechanised and can produce castings economically on mass scale. Semi-*Production* foundry is a combination of jobbing foundry and a production foundry as regards its nature of work is concerned.

On the basis of metal casting, foundries may be classified as ferrous foundries and non-ferrous foundries. Ferrous foundries may be further classified as (a) steel foundry, (b) grey cast iron foundry (c) malleable iron foundry, and (d) ductile iron foundry. Non-ferrous foundries may be classified as (a) light metals foundry (b) copper, brass and bronze foundry and (c) lead, tin and zinc base foundry.

The essential sections of a foundry are represented in Fig. 17.1.

17.1 Need of Modernisation and Mechanisation

A common man considers a foundry as a dark, dirty place with mounds of sand, coal, ashes, and metal scrap filled with obnoxious fumes and smokes.

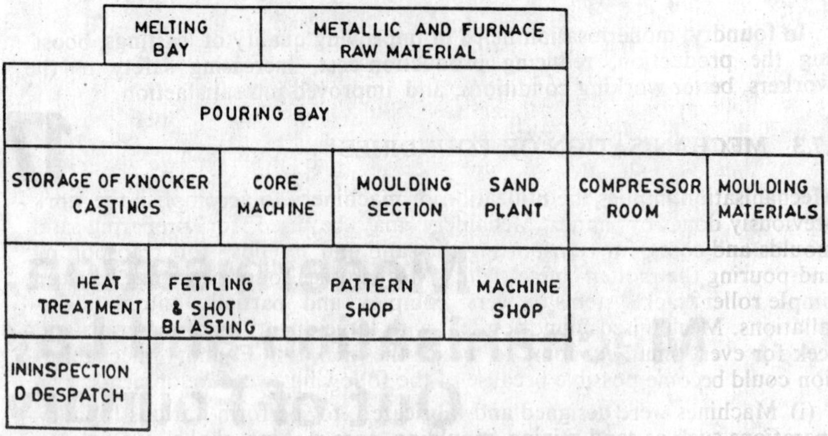

MELTING BAY
METALLIC AND FURNACE RAW MATERIAL
POURING BAY
STORAGE OF KNOCKER CASTINGS
CORE MACHINE
MOULDING SECTION
SAND PLANT
COMPRESSOR ROOM
MOULDING MATERIALS
HEAT TREATMENT
FETTLING & SHOT BLASTING
PATTERN SHOP
MACHINE SHOP
ININSPECTION D DESPATCH

Fig. 17.1 Essential Section of a Foundry.

This image of a foundry was true earlier and is still true toa good extent for many foundries. Therefore, modernisation and mechanisation of foundries is essential which leads to improved working conditions in the foundry ensuring a safe, healthy, and happy life for the workers. The improved working conditions result in increased production, improved quality and reduction in production cost. The mechanisation may prove uneconomical as well as impractical for small orders as well as for the production of large sized castings. On the other hand, mechanisation becomes economical and practical for foundries producing relatively few types of castings. Sometimes these foundries can be mechanised completely, e.g., foundries manufacturing automobile parts, electric motors, portable tools, sewing machine parts, electric fans etc. where bulk orders of small sized parts are undertaken and the jobs are of repetitive nature.

The production of large number of castings does not only depend on the use of material handling equipments, but also depends on the use of new and specialised techniques. Thus for economical and quality production on mass scale in foundries, both modernisation and mechanisation play an important role.

17.2 Modernisation of Foundries

Modernisation of foundry is referred to replacement of old processes and equipments by new ones which result in the production of required quality castings with lower costs. The modernisation of a foundry includes the following:

(i) Changing over to better and newer foundry equipments.

(ii) Employing newer, better and more economical moulding, melting and casting techniques.

(iii) Improving conditions which make a foundry dirty, dusty and smoke filled, i.e. improving working conditions in foundries providing adequate illumination, air circulation, dust extraction, etc.

In foundry, modernisation helps in improving quality of castings, boost ing the production, reducing production cost, increasing safety to the workers, better working conditions, and improved job satisfaction.

17.3 MECHANISATION OF FOUNDRIES

Mechanisation implies the utilization of machinery to accomplish the work previously done by hand. Mechinery may be used for preparing sand moulds and cores, for transporting the sand and moulds, for transporting and pouring the molten metal, etc. Mechanised foundries range between simple roller track systems to very complex and partially automatic installations. Mechanised foundries deal with large quantities of materials and seek for every small economy to reduce the final cost. Foundry mechanisation could become possible because of the following two developments:

(i) Machines were designed and fabricated to perform various foundry operations such as sand mixing, moulding, core making, shake out, cleaning etc.

(ii) These machines were integrated with material handling equipments so that continuous processing could be accompalished in the foundries.

In foundry, mechanisation results in the following advantages:

(i) Increased production for a given foundry floor.
(ii) Closer dimensional tolerance, improved surface finish and higher accuracy in castings.
(iii) Saving in time and labour, since heavy lifting and other laborious operations are carried out mechanically.
(iv) More hygienic and healthy working conditions.
(v) Minimum casting defects.
(vi) Reduced production costs.
(vii) Increased earning for the foundry workers.

However, mechanisation has a considerably impact on employment.

The extent to which a foundry can be mechanised depends upon quantities of various castings to be produced, amount of flexibility desired in producing different castings, methods used to perform various steps in making castings, and to improve working conditions. Various areas for mechanisation include the following:

(i) Materials handling
(ii) Handling of sand
(iii) Moulding and core making
(iv) Melting, pouring and shakeout operations.
(v) Control of dust and fumes.

17.3.1 Materials Handling:

Earlier, the transporting of materials in foundries was usually carried out manually, by truck, or by crane. Later on skips were introduced for the same purpose. Now a days transporting of material is mainly done by

belt conveyer and/or apron conveyor and sometimes overhead belt conveyors are also used. These are very useful for saving in time and labour. A material halding equipment is an invaluable asset in the foundries engaged in mass scale production. Foundry in general is a receiving centre of large quantities of diverse materials such as sand moulds, cores, molten metal, metal scrap, flux, etc. The use of specific materials handling equipments results in following advantages:

(a) Minimum of total handling time.
(b) Easier, safe and cleaner handling.
(c) Idle time of workers and machines is eliminated.
(d) Movement of materials becomes fast.
(e) Fatigue is decreased of the workers.
(f) Safety is increased.
(g) Location and stock of materials better and in less space.
(h) Possibility of automated operations to increase production.

Each material handling equipment possesses certain characteristics with respect to materials, direction of movement, speed, and power required for the operations. The equipments used for handling different materials with their specific use are given below:

(i) *Handling of Sand*: About 4 or 5 tonnes of sand is prepared and handled per tonne of metal poured in a typical ferrous foundry and the amount of sand may still be larger in light non-ferrous foundries. Because of this and the fact that the sand so greatly influences the quality of castings, much attention has been devoted to this part of mechanisation. A simple flow sheet of sand circulation in a mechanised foundry is given in Fig. 17.2.

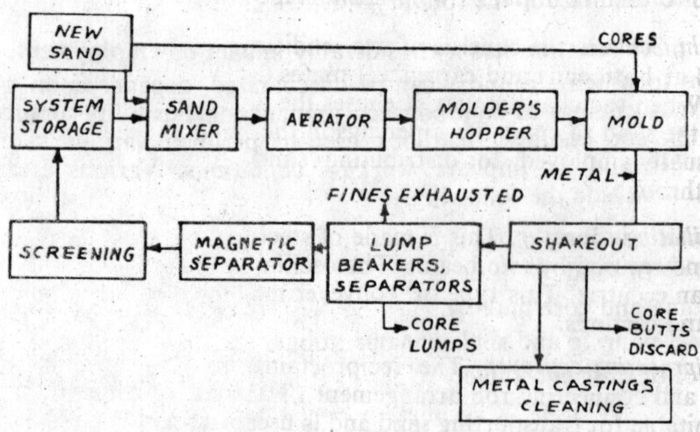

Fig. 17.2 A simple flowsheet of sand circulation in a mechanised foundry,

Sands are to be conveyed from one part of the production foundry to another for various purposes as shown in Fig. 17.2, (i.e. shake out station-Magnetic separator—reconditioning plant). Reconditioned sand is sent to work stations through distribution mains. All this handling of sand is conducted on wide range of sand conveyors covering specified areas of work. The important conveyres employed are belt conveyer, bucket elevater, apron conveyer, flight conveyer, reciprocating conveyer, oscillating conveyor, mono-rail conveyor and crane.

(a) *Belt conveyer*. It consists of an endless belt (made of cotton plies bonded with rubber), two pulleys (called head and tail pulleys) rollers or idlers for carrying the loaded belt and returning the empty belt, belt tightening mechanision and belt cleaners. The belts are available in varying sizes and strengths. The belt conveyor is commonly employed to convey sand from one place to another requiring horizontal or an inclined direction of movement. However, the belt conveyers can be used to transport sand at lower temperature (less than 100 °C) as the conveyer can not wirhstand a higher temperature. For hot sand, metal conveyer is used.

(b) *Bucket Elevator*. A bucket elevator consists of two pulleys (one at the top and the other at the bottom) which carry an endless belt having a number of buckets all around. The whole assembly is enclosed in a steel casing which has two openings, one at the botton for feeding and the other at the top for discharge. The bucket conveyor is an ideal means for conveying the sand vertically upwards. The bucket elevator is usually preferred to inclined belt owing to a saving in space and cost.

(c) *Apron Conveyer*. This consists of over lapping steel plates hinged at the ends and when assembles, it works as a belt. The main advantage of apron conveyer is that it can be used for transporting materials which are too hot for the belt conveyer. The disadvantages of the apron conveyer are (a) lower speed of movement. (b) spillage and leakage of sand through the plates, and (c) higher initial as well maintenance costs.

(d) *Flight Conveyer*. It consists of two endless chains moving on sprockets fixed at both ends and carry steel plates (known as flights) at certain intervals. When the chain moves, it causes the flights to shift in the trough and force the sand to enter the openings in the hopper. This type of conveyer is usually employed for distributing sand to work stations from an overhead through.

(e) *Oscillating conveyer*. This is made of steel plates which are fastened together and secured by two beams. The oscillating action is produced by means of an ecentric. This type of conveyer may be used for transporting hot sand and castings.

(f) *Reciprocating conveyer*. The reciproctaing action is imparted by a crank-pin and connecting rod arrangement. This type of conveyer is occasionally suitable for transporting sand and is used for short travels where the depth of excavation required for continuous conveyance is not important.

(g) *Mono-rail conveyer*. This type of conveyer consists of buckets or containers of the drop bottom type which are run on an overhead mono rail. The conveyer may be operated either manually or electrically. The monorail conveyer is invariably used for carrying sand and other materials.

(h) *Crane*. An overhead traveiling crane is usually more convenient for carrying sand for small foundries engaged in jobbing work. The sand is filled in a bottom discharge type of bucket or container which is transported with the help of a crane.

17.3.2 Production of moulds and their han dling

Hand moulding is the initial point in the consideration of mechanisation of moulding. The first evidence of mechanisation is the use of pneumatic rammer which is a hand tool with compressed air reciprocating a piston through a simple valve arrangement. A sand slunger is used to sling the sand on to the pattern by the moulder as a preliminary compacting operation. The slinger produces a compact mould in a very short time and with less effort than hand moulding.

The mechanical aids employed for stripping are usually of two forms. i.e. *straight draw machine* and the roll-over machine. The former is used to rap or vibrate the pattern plate during the stripping. The latter one may be employed to roll a complete mould either prior to or after stripping. The roll-over machine may be either centre indexing type or swinging arm type.

The moulds may be conveyed from the moulding section to the storage where they remain stationary for pouring or they may be conveyed past a pouring station where molten metal is poured while they are in motion. After cooling, the moulds are conveyed to shake out station and after shake out the emptied moulding boxes, boards, etc. are returned to the moulding section. All these operations of transporting of moulds can be conducted on a suitable conveyer system. The usual conveyers employed for transporting moulds are roller conveyer, pel let (car type) conveyer and overhead conveyer.

(a) *Roller conveyer*. This type of conveyer basically consists of two beams fixed on trestles of suitable height and supporting laterally arranged rollers above. The roller conveyer may be either of the gravity or power driven type. In gravity type of roller conveyer, the ends of the rollers are fitted in sealed bearings and the moulds are pushed by the operator to cause then to move on rollers. The beams may be fixed slightly inclined to facilitate the movement of moulds by the gravitational force. The gravity conveyer is suitable for both jobbing and mass production foundries. In the case of power driven roller conveyers, some of the rollers at fixed intervals are driven by an electric motor having variable speed enabling the coordination of the mould movement with mould production. This type of conveyer is considered suitable only for mass production foundries due to high cost of conveyer.

(b) *Pallet* (*Car type*) *Conveyer*. This consists of pallets or cars made of cast iron or steel plates mounted on wheels which can roll along a narrow gauge track. The moulds are placed on the pailets and pushed manually. The pallets may be connected by a chain and driven from end by power for continuous drives. This type of conveyer is most suitable for transporting moulds in large scale foundries from the moulding section to the pouring section and also to the shake out station after cooling.

(c) *Overhead conveyer*. In small foundries an overhead conveyer of the mono rail type may be used for transporting of moulds. The moulds are placed on the plateform of the conveyer which is carried to the pouring section by an overhead mono rail. The same conveyer can also be used for transporting the moulds after cooling to shake out station and from shake-out station to carry the solidified castings.

17.3.3 Liquid Metal and its Handling

Different furnaces are used for melting of the different metals. In melting section there is not much mechanisation but however, some machines and mechanical operations can be employed to increase the production rate or number of castings. In melting section the mechanisation is done mainly in the shape of the furnace and tapping arrangement of the furnace.

In pouring section, the mechanisation is done mainly in the arrangement of moulds and pouring system in which the losses of molten metal are minimum. Other developments are inert atmospheric or vacuum casting techniques in which the ladle is placed on top of the fuinace. Following two systems can be used to pour the molten in the moulds.

(a) Where the continuous conveyer is not used the completed moulds are carried on a roller conveyer to the pouring section where the moulds are stored. The molten metal is carried in ladle to the pouring section with the help of a travelling crane or hoist and the metal is poured into the stored moulds.

(b) When a continuous conveyer is used and the moulds constantly moving, the molten metal is carried on a mono-rail conveyer and poured into the moving moulds.

17.3.4 Handling of Castings

From the shake out station, removed castings are transported on a suitable conveyer to the cleaning and fettling section. For handling of casting there is a large variety of conveyers which can be used. The important ones include plate band, roller, oscillating, and overhead varieties.

Plate-band conveyer consists of plates which are joined together and mounted on a continuous chain moving on power driven sprockets. It is excellent for both horizontal and inclined travels and for long as well as short distances. A plate band conveyer is employed in carrying the castings to the fettling or inspection section.

The *roller type* can carry the castings in the same way as it carries moulds. The gravity type roller conveyer is particularly economical to install and it also satisfies the requirement of small foundries.

Overhead Conveyers carries the castings by means of hooks suspended from the chain. This is considered suitable for carrying castings from the shake out station to the shot blasting rooms for cleaning.

17.3.5 Dust and Fume control

In a foundry the air is highly polluted which affects the efficiency of workers directly affecting the production rate. Therefore, it is essential to maintain hygienic working conditions by cleaning the air pollution. In general, in a foundry pollution is caused in the following operations.

(a) Fine particles of saw dust in the pattern shop.

(b) Dust and sand particles are exuded when sand is mixed and prepared during moulding, shakeout and fetting operations.

(c) Fumes are produced during melting, metal transfer and pouring operations.

Therefore, it is necessary to provide requisite systems for collection of all the dust and fumes produced and their disposal so that pollution of atmosphere in the foundry is prevented and thereby threat to the health of workers is eliminated. However, provision for control of dust and fume should be made with the layout of foundry as it will be quite difficult to incorporate the necessary equipment latter on.

Broadly the materials causing pollution and unhealthy conditions may be classified into following three categories:

(a) Solid particles, such as dust, fume, smoke and fly ash,

(b) Liquid such as mist and fog.

(c) Gaseous matter where the polluant may be either gas over the entire range of atmospheric and process temperatures and pressures or liquid at lower tempetature and gas at the temperature and pressure at which the gases are released into the atmosphere.

The method of separating the polluant depends on its type. Some operations may remove several polluants where as some may be used only for one of them. The common methods employed in foundries for the control of pollution are discussed below:

(a) *Cylone*: The cyclone is made of conical shape (Fig. 17.3) having a cylindrical top. This works on the principle of centrifugal separation. The feed is introduced tangentially near the top. A spinningg motion is im-

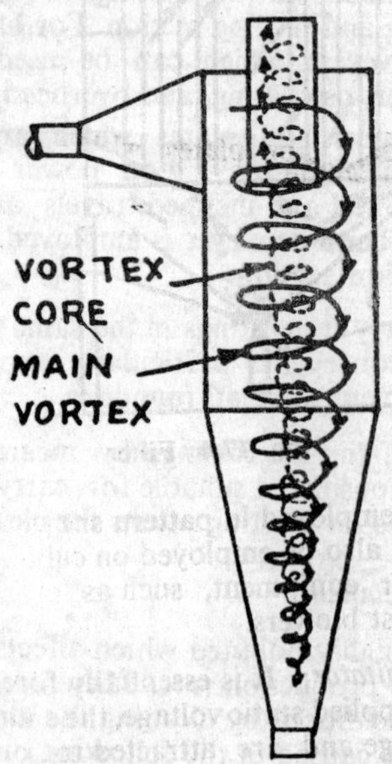

Fig. 17.3 The cyclene

parted to the suspension in the cylindrical portion and a vortex motion of the particulate matter is created along the longitudinal axis. The accompanying centrifugal acceleration increases the settling rate of the particles and tends to throw them radially. The particles spiral down to the cones wall (where a zone of reduced pressure exists) and flow downward to the apex through which they are discharged.

Cyclones may be operated dry or wet. The cyclone is used in sand preparation plants, in moulding shops and on shakeout stations to separate sand dust particles from air. Cyclone may also be used in cleaning the cupola exhaust gas.

(b) *Filter*: The filter (Fig. 17.4) serves the purpose of separating finely divided solid particles from gas or air stream accompalished by driving the solid containing gas through a porous structure (membrane, septum, cloth, or bed of natural or synthetic materical) which permits the passage of gas or air and refines the solids. The filters should be continuously or periodically cleaned or replaced. The filtering action may be obtained by direct interception, impaction or diffusion. In direct interception, the gas having solid particles heads directly towards a part of the solid surface comprising the filter. In impaction, the gas sweeps by the solid material of the filter and allows the particles to touch the filter surface. In diffusion, a blow from the molecule of the gas projects the particles to the filter surface.

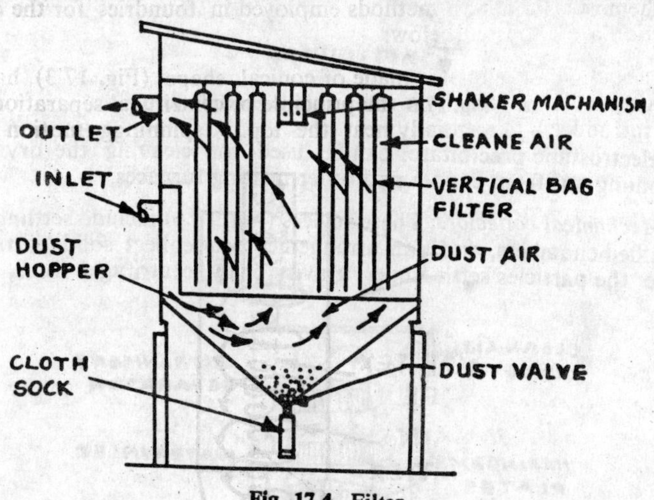

Fig. 17.4 Filter

Filters are commonly employed in pattern shops on various wood working machines. Filters can also be employed on cupola gas collection systems in conjunction with other eqnipment, such as after burners, gas coolers, recuperators, and exhaust blowers.

(c) *Electrostatic Precipitator*: It is essentially based on the principle that under an action of high applied static voltage, the solid particles in gas phase acquire electrostatic charge and are attracted to an electrode of opposite polarity, where they are collected and washed out. This consists of either tubes or plates as one electrode and wires located centrally in each tube or paralled with the plates as the other electrode. Cross sectional view of an electrostatic precipitator is shown in Fig. 17.5. Rectified high voltage is applied across these two electrodes. The gas enters from the bottom and rises upwards. The solid particles are ionised and get attracted to the surfaces of electrodes and thus get precipitated to them.

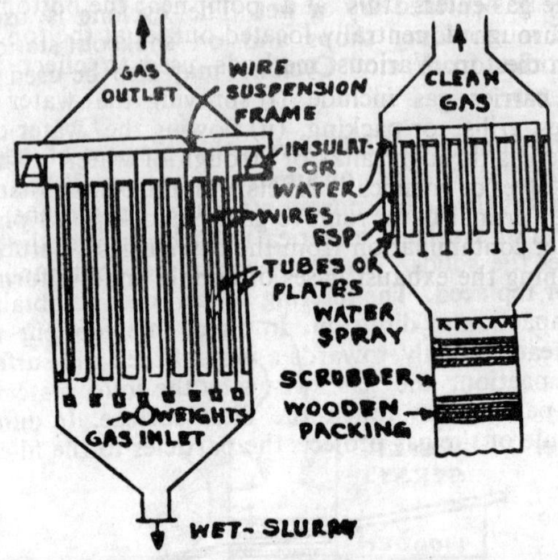

Fig. 17.5 Electrostatic precipitator

The electrostatic precipitator can be used for cleaning the dry effluent gases coming out from cupola and other melting furnaces.

(d) *Mechanical collectors*: These devices (Fig. 17.6) include settling chambers, baffled chambers, and fan arrangement to collect solid particles. In this case the particles settle under gravity or by centrifugal force, but do

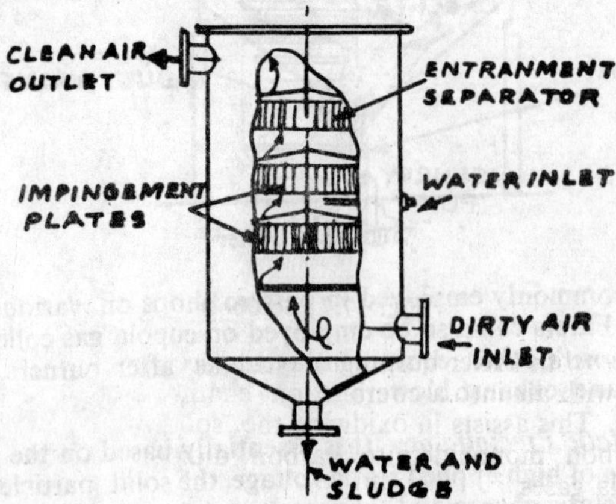

Fig. 17.6 Wet mechanical (centrifugal) duet collector,

not depend on vortex as in the case of cyclones. Since the efficiency of mechanical collectors is low, they are generally used as pre-cleaning devices or in combination with filters or scrubbers. Mechanical collectors are generally employed for cupola exhaust systems.

(e) *Scrubbers*: It consists of a tall cylinder (Fig. 17.7) with conical top and bottom. The gas enters this at a point near the bottom of the cylinder and leaves through a centrally located out let at the top. The water is introduced from the top. Various methods used to effect a contact between water and carrier gas include (i) spraying the water into chamber containing baffles, grills, or packing, (ii) flowing the water over weirs and (iii) bubbling the gas through tanks or trough of water. The water introduced either dissolves or chemically reacts with the contaminants. The scrubber is employed primarily to remove gases and vapour phase as well as particulate matter contamination from the carrier gas. Scrubbers are most suitable for cleaning the exhaust gases of cupola and arc furnaces.

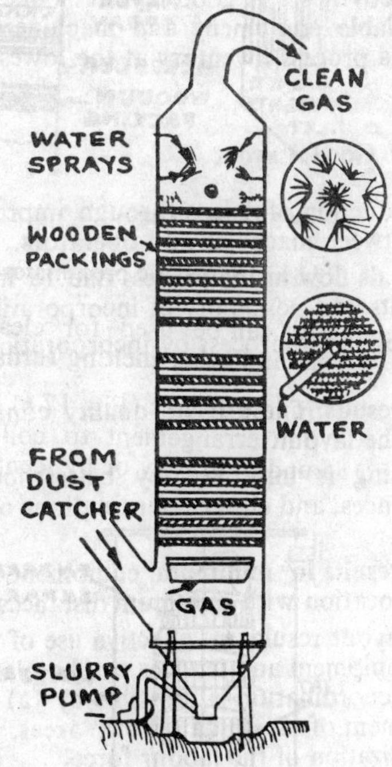

Fig. 17.7 Scrubber

(f) *After burners*: After burning may be accompalished by introdusing fuel oil along with air into a combustion chamber through which the carrier gas passes. This assists in oxidising the solid combustible material and burning of carbon monoxide into carbon dioxide. After burners may be used for cupola gases.

(g) *Combination devices*: Some systems use two or more than two of the above mentioned equipments in combination so as to control dust and fumes most economically and with a minimum pressure drop of gas. For example cyclone and scrubbers can be used in series.

17.4 PLANT LAYOUT FOR FOUNDRIES

Plant lay out may be defined as the arrangement and coordination of equipment work centres, floor area, and facilities specified from flowsheet to achieve maximum efficiency of functioning of plant in combination of man, material and machine.

Plant layout may also be considered as a floor plan for determining and arranging the required equipments and machines in one best place to obtain the quickest flow of material at the lowest cost with least handling of materials in processing.

In order to maintain good production activity with high quality of products and minimum over heads, it is essential to reduce to minimum all non-productive plant activities. A good layout is one which provides for full utilization of available equipment and machinery, material handling devices, manpower, and process inventory at the lowest cost of production and distribution.

17.4.1 Advantages of Good Layout

(i) Elimination or reduction of delays through improved arrangement or better work balance between machines and operators.

(ii) Smoother materials flow in the process due to incorporation of new methods and equipments and possibility of incorporating automation.

(iii) Improved control on the process by incorporating methods to inspect goods in process.

(iv) A good layout results in improved quality control due to inspection locations provided in the layout.

(v) Materials handling is improved by better locationof equipment, reduced handling distances, and closer coordinations of the entire handling activity.

(vi) A good layout results in minimum equipment investment, planned machine balance and location with minimum distances of handling.

(vii) Well planned layout results in effective use of available area. In a well planned layout, equipment and services can be placed in such a manner that the most effective coordination is possible by (a) performing multiple functions, (b) development of up-to-date work areas, and (c) operator job assignment for full utilization of the labour force.

(viii) Good plant layout results in improved utilisation of labour, since it allows the design of individual operations, the process, and the flow and handling of materials in an effective manner.

(ix) A good layout provides convenience and comfort and thus morale of employees is boosted. For examples correct lighting, proper cooling and ventilation, noise and vibration control, lunch and rest facilities, etc. lead to efficient job performance with reduced idle time.

(x) A good layout results in improved efficiency in plant services since problems of maintenance, service equipment, buildings, etc. are considered during the arrangement and layout planning.

(xi) A good layout facilitates future rearrangement or expansion, since new utility distribution system giving maximum flexibility and capacity become an integral part of the good layout.

17.4.2 Principles of Foundry Plant Layout

(i) In order to obtain maximum potential of a layout for a new foundry a systematic procedure must be followed. There are some guiding steps used in making the decision for the plant layout. These are discussed below:

(i) *Analysis of the product to be produced*: This includes (a) annual tonnage, (b) the type of castings and their range of sizes, weights and composition and (c) maximum weight of single casting.

(ii) *Determination of the process and work place required to manufacture the product*: For each manufactured product, operation sheets are developed showing only the sequence of operations.

Castings may be grouped as light, medium and heavy or according to the quality required. Most appropriate locations for moulding, core making, melting inspection, etc. should be suitably determined. Before the details of layout are decided, various processes such as green sand, skin dried, dry sand machine moulding, shell moulding, CO_2 moulding process, etc. should broadly be considered first to suit the casting requirements.

(iii) *Preparation of Layout planning charts*. This is of prime importance as it tabulates and combines the various factors to be provided in the final layout. The various factors to be incorporated are:

(a) The flow process showing all operations, moves, shortages and inspections in sequence.

(b) Standard time required for each operation obtained by time studies or earlier standards.

(c) Selection of machine and equipment, balance etc,

(d) Assessment of man power requirements.

(e) Determination of requirements of equipments and method and load of material handling.

A full review and analysis of each step is required to complete the layout planning chart. Once this is complete, the layout of the manufacturing area involves mainly conversion of the layout planning chart data to the physical plant statistics. This can be achieved by three additional jobs, i.e. (i) determination of work stations (ii) analysis of storage area requirement and (iii) establishing minimum aisle widths.

Work stations are determined by the consideration of requirement of machine, operator, material and service area. This is accompalished by using non-machine and/or operations charts and scaled work station sketches/models.

Before beginning the actual layout, both the size and the location of the storage area should be studied with relation to production activities. In general, three types of storages, i.e., storage of raw material, in process storage and finished good storage are considered.

Before starting the layout, it is essential to establish the clearance around the various machineries, equipment, between departments, etc. Aisle widths should be carefully determined as these are mainly dependent upon materials handling equipment, work station clearance required, and pedestrian traffic.

(iv) *Equipment layout.* Sufficient space should be assigned for individual equipment to have the accessibility to the equipment for its maintenanee. Too close fitting of an equipment into the building may not result into good economy. Though, a slightly larger building than necessary will cost slightly more, the extra cost will in fact be small compared to the penalties due to inconvenience and the need for future expansion. Another essential part of planning a layout is an access for initial construction and maintenance. For example, overhead equipment should have space for lowering it. Space should also be provided for repair and replacement of equipment.

(v) *Safety consideration in layout.* A foundry industry involves handilng and processing of large quantity of material with smooth, dirty and hot conditions and thus special attention has to be given for safety considerations to avoid accidents arising due to physical, environmental or personal factors. Therefore, an acceptable good layout should in corporate safety measures in working areas, provision of safe guarding equipment and power transmitting equipment, proper illumination, ventilation, etc.

(vi) *Plant expansion.* This may include sufficient provisions for addition of new product line or for increased production of the existing product. This should be studied thoroughly while designing a plant layout. An engineering judgement should be taken keeping in view the penalty for bad judgement, scrapping of existing serviceable equipments, and remodeling which may involve much greater losses compared to the losses due to rejected equipment. However, some times due to economics of larger plants, the replacement is essential.

(vii) *Proper Utilization of Floor Space.* The value of land is a considerable factor in making use of floor space. However, a rule of practising economy of floor space should be followed, which should be consistent with good house keeping in the plant with proper consideration to line flow of materials, access to equipment, space to permit working on parts of equipment requiring frequent servicing, and safty and comfort of the operators.

(viii) *Utilities Service.* Tne distribution of various utilities such as water, air, steam, and power generally permits the designing to meet almost any conditions. However, considerations are given for the proper placement of each of these services and practising a good design for an ease of operation and reductions in cost of maintenance. The pipes should not be laid on floor or between the floor and 2.2 m, where the operator passes or works. The unorderly arrangement of piping invites problems in operation of the plant.

(ix) *Building and Office.* After a thorough study of quantitative factors, the selection of building should be considered. Usually, standard factory buildings can be used, but if none is satisfactory to handle the space and process requirements, then a competent architect may be consulted to design a building around the process and not beautiful structure in wnich a process must fit. In foundry industry, much thought should be given for the disposal of hot slag, waste sand, ash, etc. Some part of building may require air conditioning e.g. instrumental analysis laboratory, which may require an elaborate set up.

The requirement of office building will depend on the scope of operational activities. The exact requirements of space should be worked out and provided in the layout.

(x) *Materials Handling*. In foundry industry the consideration of material handling equipment is quite important, since a huge quantity of sand, scrap, finished good, empty boxes, etc. is involved in this process. Whenever possible the advantage of the topography of the site location should be taken. The flow of materials should be smnoth without causing any hindrance to the movement of the persons and running of process equipments. There should be minimum materials handling and transport with adequate supply of materials including scrvices. There should be unrestricted disposal of products, sand, empty boxes. etc. The congestion due to inadequate transport facilities may work out the advantages of improved techniques. Therefore, the planning of sufficient transport system is of great importance for an efficient transport system which is of great importance for an effective layout. The transport system includes the following provisions.

Existing or possible future rail roads and roads adjacent to the plant should be known in order to plan rail sidings and access roads within the plant. The railroads and roadways systems of the correct capacity and at the right location should be provided for traffic. The factors in rail track planning are:

(a) Existing and future off site main rail facilities.
(b) Permissible radius of curvature rail road spurs.
(c) Provision for traffic handling.
(d) Loading and unloading facilities for initial plant construction and subsequent operations.
(e) Storage space for loaded and empty cars.
(f) Space for cleaning and car repairs.

Major provisions in road planning are

(a) A means of interplant movement for road traffic, both pedestrani and vehicular.
(b) Routing of heavy traffic outside the operational areas.
(c) Heavier and wider roads for heavey traffic.
(d) Roadways for access to initial constuction.
(e) Roadways to isolated points such as safety equipments, fuel tanks, gas plant, etc.

17.4.3 Methods of Layout Planning

For a detailed planning study, space requirement should be known for various raw materials, products, by products, and process equipments. A trial plot plan can be made with the help of directional schematic flow pattern. A number of studies and reference will be required before selecting the final plant layout. The various methods of making the layout planning are (a) unit area concept, (b) two dimensional layout and (c) scale models. These are briefly discussed as below:

(a) *Unit Area Concept*. It refers to the use of basic blocks to build an arrangement for layout plans. This method is particularly suitable for large plant layouts. Unit areas are usually delineated by means of distinct process phases and operational procedures, taking into consideration materials flow and safety requirements. Thus, the first task of planning a layout is the delineation of the shape and extent of a unit area and the interrelationship of each area in a master plot plan.

(b) *Two Dimensional Layouts*. In order to visualise the layout problem, two dimensional scaled templates or small cut outs of unit areas and equipment are shifted within each area marked for them on paper. This excer-, cise is to be carried out by a group of engineers. This exercise will provide the basic plot plan, from which two dimensional diagrams can be prepared. The preparation of the scale model can also be started simultaneously in the layout planning procedures.

(c) *Three Dimensional Models* (*Scale Models*). The scale models have the advantage over detailed two dimensional method. A low cost model can be made from blocks of wood and card board on the scaled paper. This low cost model is mainly used to develop plot and elevation plans and can not be used to represent piping and utilities layout. A costly complete model can be used for the detailed layout of process piping, utilities, and control facilities. These isometric layout diagrams to scale provide much more accurate cost estimation than two dimensional models. These expensive scale models are used during the construction and operator training period to great advantage. These models are permanently placed in control room, where an operator can trace lines quickly instead of walking and climbing over an extensive part of the plant. The various advantages of three dimensional models are

(i) Selection of optimum design.
(ii) Effective planning in construction.
(iii) Savings in engineering design construction, operating and maintenance cost.
(iv) Safer and rapid training of personnel.

Index

D

L

M